Springer Complexity

Springer Complexity is an interdisciplinary program publishing the best research and academic-level teaching on both fundamental and applied aspects of complex systems-cutting across all traditional disciplines of the natural and life sciences, engineering, economics, medicine, neuroscience, social and computer science.

Complex Systems are systems that comprise many interacting parts with the ability to generate a new quality of macroscopic collective behavior the manifestations of which are the spontaneous formation of distinctive temporal, spatial or functional structures. Models of such systems can be successfully mapped onto quite diverse "real-life" situations like the climate, the coherent emission of light from lasers, chemical reaction-diffusion systems, biological cellular networks, the dynamics of stock markets and of the internet, earthquake statistics and prediction, freeway traffic, the human brain, or the formation of opinions in social systems, to name just some of the popular applications.

Although their scope and methodologies overlap somewhat, one can distinguish the following main concepts and tools: self-organization, nonlinear dynamics, synergetics, turbulence, dynamical systems, catastrophes, instabilities, stochastic processes, chaos, graphs and networks, cellular automata, adaptive systems, genetic algorithms and computational intelligence.

The three major book publication platforms of the Springer Complexity program are the monograph series "Understanding Complex Systems" focusing on the various applications of complexity, the "Springer Series in Synergetics", which is devoted to the quantitative theoretical and methodological foundations, and the "Springer Briefs in Complexity" which are concise and topical working reports, case-studies, surveys, essays and lecture notes of relevance to the field. In addition to the books in these two core series, the program also incorporates individual titles ranging from textbooks to major reference works.

Understanding Complex Systems

Founding Editor: S. Kelso

Future scientific and technological developments in many fields will necessarily depend upon coming to grips with complex systems. Such systems are complex in both their composition – typically many different kinds of components interacting simultaneously and nonlinearly with each other and their environments on multiple levels – and in the rich diversity of behavior of which they are capable.

The Springer Series in Understanding Complex Systems series (UCS) promotes new strategies and paradigms for understanding and realizing applications of complex systems research in a wide variety of fields and endeavors. UCS is explicitly transdisciplinary. It has three main goals: First, to elaborate the concepts, methods and tools of complex systems at all levels of description and in all scientific fields, especially newly emerging areas within the life, social, behavioral, economic, neuro- and cognitive sciences (and derivatives thereof); second, to encourage novel applications of these ideas in various fields of engineering and computation such as robotics, nano-technology and informatics; third, to provide a single forum within which commonalities and differences in the workings of complex systems may be discerned, hence leading to deeper insight and understanding.

UCS will publish monographs, lecture notes and selected edited contributions aimed at communicating new findings to a large multidisciplinary audience.

For further volumes:
http://www.springer.com/series/5394

Delio Mugnolo

Semigroup Methods for Evolution Equations on Networks

Springer

Delio Mugnolo
Universität Ulm
Institut für Analysis
Ulm, Germany

ISSN 1860-0832 ISSN 1860-0840 (electronic)
ISBN 978-3-319-37474-1 ISBN 978-3-319-04621-1 (eBook)
DOI 10.1007/978-3-319-04621-1
Springer Cham Heidelberg New York Dordrecht London

Printed on acid-free paper

Springer is part of Springer Science+Business Media (www.springer.com)

To Emma, who is as old as this book

Preface

> *In order to become worldly things, that is, deeds and facts and events and patterns of thoughts or ideas, [action, speech, and thought] must first be seen, heard, and remembered and then transformed, reified as it were, into things—into sayings of poetry, the written page or the printed book, into paintings or sculpture, into all sorts of records, documents, and monuments. The whole factual world of human affairs depends for its reality and its continued existence, first, upon the presence of others who have seen and heard and will remember, and, second, on the transformation of the intangible into the tangibility of things.*
>
> *Hannah Arendt,* The Human Condition

In March 2007 Frank Neubrander invited me to hold a lecture series on "Evolution Equations on Networks" at the Mathematical Department of the Louisiana State University at Baton Rouge. At that time my interest in differential equations on networks was quite young and I thus structured my course mostly as an introduction to known results on linear hyperbolic and parabolic equations. Many of the results I mentioned had been obtained over the previous 3 years by a group of a few researchers with a definite operator theoretical background, most of whom were then, or had earlier been, based at the University of Tübingen. A preliminary version of those lecture notes has been circulating over the internet ever since.

Meanwhile, the topic of differential equations on graphs, ramified spaces, and more general network-like objects was gaining momentum and an impressive interdisciplinary discourse began.

Ever since then, more and more often researchers active in different fields—operator theory, mathematical physics, and graph theory, in particular—have been meeting at conferences to discuss and exchange ideas, and many new collaborations have begun. In my opinion, this has been one of the most fruitful examples of interbreeding experienced in mathematics in the last decade. It was out of this cultural climate that I was asked by the Springer-Verlag to rework and update my

Baton Rouge notes. Indeed, this book derives from them, even if hardly a paragraph has been consistently taken over.

While I am writing these lines, a book on the same subject has just been published by Gregory Berkolaiko and Peter Kuchment, and another one by Pavel Kurasov is expected to appear in the near future. There is not (yet?) such a thing as a canonical theory of differential equations on networks, and it is fair to say that both [1, 2] and the present book strongly reflect the mathematical tastes and interests of their respective authors. The present book is specifically devoted to the study of evolution equations—i.e., of time-dependent differential equations, like the heat equation, the wave equation, or the Schrödinger equation—while I have chosen to discuss elliptic equations and further spectral issues only superficially.

The largest part of the literature which has appeared in the last 10 years on the subject of differential equations of graphs is devoted to spectral problems. This certainly has good reasons in view of applications to solid state physics and quantum chemistry, but in this way it covers almost exclusively the theory self-adjoint problems. While the Spectral Theorem is certainly a most efficient tool to deduce features of evolutionary systems from their underlying spectral properties, it is necessary to resort to different, perhaps more sophisticated methods whenever one is interested in more general settings: This is particularly true because when working on networks it is so easy to produce non-self-adjoint realizations of the relevant differential operators simply by playing with the transmission conditions in the nodes. One classical and elegant tool is that of operator semigroups: This book is simultaneously a very concise invitation to this theory, founded 75 years ago to a large extent by Einar Hille and Marshall Stone, as well as a handbook about their applications to differential equations on networks.

Although only a few results in this book are fully proved, I have tried to keep the book as self-contained as possible by recalling all results of abstract semigroup theory the exposition relies upon. For this reason, this is not a classical textbook. While it can be (and has already been) used as a synopsis for a graduate course, in general I have tried to keep the presentation as fluid and self-contained as possible. In particular, most proofs are only sketched, and I have rather referred to the original papers. The audience I had in mind while writing the book consists primarily of graduate students or young researchers with some experience in the classical functional analytical theory of evolution equations. Virtually all abstract methods applied throughout this book rely upon tools coming from operator theory and elementary graph theory.

At this stage it would be probably premature to assert that differential operators on networks—or quantum graphs, to use the name under which they have become famous over the last 15 years—have turned into an independent research field. It seems to me more accurate to say that an increasing number of researchers with different backgrounds have been joining forces to study different, ever broader aspects. Rather than a new field of its own, this research area is probably best described as an exceptionally convenient source of models that are easy to study by merging different, complementary, sometimes only loosely related mathematical theories. Yet the obtained results are often highly non-trivial and difficult to reproduce in more usual, higher dimensional settings.

It is difficult to foresee in which direction differential equations on networks are heading. Nonlinear and geometric evolution equations will probably gain importance in the next 10 years, while numerical analysis may suggest some new tools. Physical developments will surely provide new interesting problems and networks will possibly become an entrenched part of the PDE theorist's toolbox for constructing non-trivial examples or surprising counter-examples.

In any case, I feel it is a good idea to look back now, in 2014, and attempt to write down what has happened in this field in the last few years. *Others who have seen and heard* are certainly not missing; but it is only through transforming *the intangible into the tangibility of things* that they *will remember*, to use Hannah Arendt's expression.

Structure of the Book

Since the early version of these notes was completed in 2007, the theory of differential equations on networks has gone a long way. In my opinion, the single most significant advance has been the ever-growing consciousness that discrete and continuous aspects of networks, and in particular the analysis of graph matrices and differential operators on network-like structures, are two sides of the same coin. This has led to the rediscovery of a number of notions ideas, and methods which have been around for many years already, especially in algebraic graph theory, as well as of some new interesting insights. While elaborating my notes, I have decided to put particular effort in presenting both aspects in a unified framework, following the recent revival of difference operators among analysts and mathematical physicists who work on graphs.

In many of the recent monographs in semigroup theory, differential equations are presented as more or less simple applications of a sophisticated abstract theory. In this book, I try to emphasize the converse aspect: How the theories of semigroups (and other partially related objects, like cosine operator functions and quadratic forms) develop in a natural way from the problem of finding qualitative properties of concrete evolution equations. This is particularly relevant in the case of networks, I find, since many (finite or infinite) matrices originally introduced by graph theorists can be efficiently studied along with usual differential operators by suitable unified methods. Presenting the operators we will encounter in association with all most usual differential equations on networks is the main aim of Chap. 2.

The large part of our investigations of evolution equations will use methods typical of Hilbert spaces. In order to apply them to our purposes, some usual classes of function spaces have to be adapted to our setting of discrete and metric graphs: These objects will be introduced in Chap. 3. In particular the former seem to be not quite frequent in the literature.

The main features of semigroups are summarized in Chap. 4, including some results that perhaps do not belong in the canon of this theory—like the recent characterization of boundedness of semigroups. The first applications of this theory to transport equations on metric graphs are subsequently presented.

While writing this book, I have tried to provide the reader with several examples and applications, in particular to naive neuronal modeling—it is my strong belief that theoretical neuroscience can and should be the ideal testing workbench for many different mathematical results on graphs and networks. For this reason, in Chap. 5 I have felt compelled for the sake of self-containedness to describe at least superficially what is currently known about the structure of animals' brains and their most important region, the cerebral cortex—and how to model the ongoing neural activities. This is meant as a short interlude to mark the half of the book.

Typical diffusion processes are usually governed by special objects that enjoy much better properties than usual semigroups and thus reflect almost all properties of exponentials of self-adjoint operators. One of the main features of these *analytic* semigroups is that their properties can be studied in a much easier manner than in general, using the elegant theory of forms. These both complementary notions as well as their applications to diffusion-type equations on discrete and continuous graphs are the topic of Chap. 6.

Most of the properties of evolution equations that can be deduced from diagonalization of operators have nowadays been extended to equations associated with sectorial operators, but in Chap. 7 we mention a few results that do indeed depend on self-adjointness. Some of them are highly non-trivial, including a characterization of the wave equation with finite speed of propagation (which seems to be little known outside the community of experts on abstract Dirichlet forms) and the trace formula, arguably the deepest and most thoroughly investigated result of the theory of operator on networks.

Chapter 8 is devoted to the interplay between symmetries of discrete graphs and properties of evolution equations on them and their associated metric versions. We use the word "symmetry" in a casual way, to mean a few different notions of structural regularity of the graph connectivity. In particular, we show that a rather weak notion of symmetry (namely, the existence of so-called almost equitable partitions) is the source of a class of non-trivial Lie groups of symmetries that boil down to the usual gauge group $U(1)$ in the case of a graph consisting of a single interval.

Finally, the book is concluded by two appendices devoted to reminders of general graph theory and Sobolev spaces, respectively.

Acknowledgments

I would like to thank Frank Neubrander for his invitation to Baton Rouge 6 years ago. Without it, these notes would probably never have been conceived.

My first steps into this field were suggested and encouraged by Rainer Nagel in the final stage of my doctoral studies in Tübingen.

It was mostly upon my arrival at the University of Ulm that I started focusing on the ubiquitous emergence of networks and on the manifold possibilities of investigating them by functional analytical methods. In doing this I have been massively supported by the Graduate School on "Mathematical Analysis of Evolution, Information and Complexity" and its spokesman, Wolfgang Arendt. I am greatly indebted to both.

During the last 2 years I have greatly profited from a generous grant given to my research project on "Symmetry Methods for Quantum Graphs" by the Land Baden–Württemberg in the framework of its *Juniorprofessorenprogramm.* A large part of this book was written during my research stays at the Mittag-Leffler Institute in Stockholm and at the Center for Interdisciplinary Research (ZiF) in Bielefeld in 2013. I am grateful to these three institutions for their financial support.

Several friends and colleagues have accompanied me in this journey at the border between continuous and discrete mathematics. Above all, I am grateful to my coauthors for many hours of exciting and instructive discussions. I am particularly indebted to Stefano Cardanobile and Robin Nittka—to the former for many flights of fancy and to the latter for bringing me back down to earth.

In many other cases coffee break computations or speculative discussions have not—or not yet—been written down, but have been nevertheless inspiring: I much enjoyed talking, among others, to Felix Ali Mehmeti, Wolfgang Arendt, Fatihcan Atay, Rami Band, Jonathan Breuer, Ralph Chill, Dominik Dier, Sebastian Egger, Pavel Exner, Fritz Gesztesy, Sebastian Haeseler, Vu Hoang, Bobo Hua, Matthias Keller, Stefan Keppeler, Evgeny Korotyaev, Vadim Kostrykin, Hynek Kovařík, Pavel Kurasov, Daniel Lenz, Annemarie Luger, Bojan Mohar, Konstantin Pankrashkin, Gabor Pete, Stefan Rotter, Stefan Teufel, Lutz Weis, and Wolfgang Woess. Several discussions on mathoverflow.net have also helped in the development of this book.

Finally, I warmly thank Jacopo Bertolotti, Stefano Cardanobile, Amru Hussein, James B. Kennedy, Matthias Keller, Stefan Keppeler, Yaroslav Kurylev, Gabriela Malenová, Rainer Nagel, and Silvia Romanelli for their valuable comments on the preliminary versions of this book. Of course, I am the one to blame for any remaining mistakes and imprecisions.

Ulm, Germany
January 2014

Delio Mugnolo

References

1. G. Berkolaiko, P. Kuchment, *Introduction to Quantum Graphs*. Mathematical Surveys and Monographs, vol. 186 (American Mathematical Society, Providence, 2013)
2. P. Kurasov, *Quantum Graphs: Spectral Theory and Inverse Problems* (in preparation)

This picture was taken in Berlin during *Der Berg* (The Mountain), an art installation organized in the summer of 2005 inside the Palast der Republik (Palace of the Republic), the former House of Parliament of the German Democratic Republic. Unfortunately, *Der Berg* was only meat a swan song: In 2006 the Bundestag, the German Parliament, decided that the Palast der Republik was to be torn down. Its demolition was completed in 2008.

Contents

Chapter 1
Introduction

Electric phenomena have been known for centuries. When natural scientists began to investigate them more closely in the eighteenth century, it soon became clear that electric properties can be investigated by mathematical methods.

In 1827 G. Ohm observed a linear dependence between voltage along a wire and current flowing through it—his famous law "$V = IR$". He discovered experimentally that the proportionality factor R, the *resistance*, increases with distance only linearly: This paved the way for the development of electric telegraph networks on a large scale.

In 1845 G. Kirchhoff, then aged 23, derived in [19, p. 513] (later translated in English in [21]) two equations that relate current and voltage in electrical circuits under constant magnetic field:

I. wenn die Drähte $1, 2, \ldots \mu$ in einem Punkte zusammenstoßen,

$$I_1 + I_2 + \ldots + I_\mu = 0,$$

wo $I_1, I_2, \ldots$ die Intensitäten der Ströme bezeichnen, die jene Drähte durchfließen, alle nach dem Berührungspunkte zu als positiv gerechnet;

II. wenn die Drähte $1, 2, \ldots \nu$ eine geschlossene Figur bilden,

$$I_1 \cdot \omega_1 + I_2 \cdot \omega_2 + \ldots + I_\nu \cdot \omega_\nu$$

= der Summe aller elektromotorischen Kräfte, die sich auf dem Wege: $1, 2..\nu$ befinden; wo $\omega_1, \omega_2, \ldots$ die Widerstände der Drähte, $I_1, I_2, \ldots$ die Intensitäten der Ströme bezeichnen, von denen diese durchflossen werden, alle nach *einer* Richtung als positiv gerechnet.[1]

1

I. If the wires $1, 2, \ldots \mu$ *touch in a point, then*

$$I_1 + I_2 + \ldots + I_\mu = 0,$$

D. Mugnolo, *Semigroup Methods for Evolution Equations on Networks*, Understanding Complex Systems, DOI 10.1007/978-3-319-04621-1_1,

These principles, which are nowadays universally known as *Kirchhoff's circuit laws*, can be concisely stated as follows in modern terminology:

I. At any node of a circuit, the sum of currents flowing into that node is equal to the sum of currents flowing out of it.
II. Along any cycle inside a circuit, the sum of voltages inside the conductors that form that cycle is zero.

They are referred to as *Kirchoff's current* and *voltage law*, respectively, and at least the former will be ubiquitous in this book.

The decisive step in Kirchhoff's analysis was the intuition that an electric network is fully described by an intrinsic weight attached to each conductor—nothing but Ohm's resistance—and by a binary relation that describes whether or not two nodes are connected by a conductor, regardless of their three-dimensional geometry. One hundred and ten years earlier, exactly the latter simplification had been crucial in L. Euler's famous solution of the Königsberg bridge problem! But while Euler's solution was truly combinatorial in nature, Kirchhoff exploited his formalism to translate the problem of determining currents (and hence, by Ohm's law, voltages) along a circuit's wires into the language of linear systems.

Two years later, in [20], Kirchhoff refined his own ideas by proposing a linear algebraic algorithm to find the voltages of each conductor in an electric circuit in which the (externally applied) electromotive forces are known. This algorithm led to what is now known as the *Matrix–Tree Theorem*. His first step in this proof is the proof of the following assertion:

> Es sey μ die Zahl, welche angiebt, wie viele Drähte man bei einem beliebigen Systeme *wenigstens* entfernen muß, damit alle geschlossenen Figuren zerstört werden; dann ist μ auch die Anzahl der voneinander unabhängigen Gleichungen, welche man darch Anwendung des Satzes [II.] herleiten kann.[2]

In a graph theoretical jargon, cf. Lemma 2.3 below, this can be rephrased as follows: The null space of a finite oriented graph's incidence matrix has dimension equal to the graph's cyclomatic number. This is arguably the oldest result in the history of algebraic graph theory. In the same article, Kirchhoff went on to

---

where $I_1, I_2, \ldots$ denote the intensity of the currents that flow through those wires—all of them considered as positive in the direction of the touching point;

II. if the wires $1, 2, \ldots \nu$ form a closed figure, then

$$I_1 \cdot \omega_1 + I_2 \cdot \omega_2 + \ldots + I_\nu \cdot \omega_\nu$$

= the sum of all electromotive forces that are met along the path $1, 2..\nu$; where $\omega_1, \omega_2, \ldots$ denote the resistances of the wires and $I_1, I_2, \ldots$ the intensities of the currents by which these wires are traversed—all of them considered as positive in one *direction.*

[2] *Let μ denote the minimal number of wires that have to be removed in order to break all closed figures; then μ is also th number of mutually independent equations that can be derived by applying the Theorem [II.].*

introduce the quadratic form associated with what is nowadays known as the *discrete Laplacian*, cf. Sect. 2.1.4.

Kirchhoff was ahead of his time as the contemporary mathematics could hardly elaborate on his intuitions—in particular, matrices had not yet been formally introduced and algebraically investigated.

While calculus was of course already very well developed in the middle of the nineteenth century, discrete mathematics was still in its crib. One of the earliest attempts to connect these both concepts was pursued by G. Boole in the last years of his life. Before the notion of limit was well-understood, the derivative of a function u at x had been defined by G. Leibniz by means of the difference quotient

$$\frac{u(x + \Delta x) - u(x)}{\Delta x},$$

where Δx was an infinitesimal. But even if infinitesimals were *passé* in his day, Boole saw that analogous definitions may also be relevant if $x + \Delta x$ is interpreted as a point at any finite distance from x. In [4] he thus studied in depth the properties of a new kind of analysis in which limits of difference quotients are replaced by mere differences of values of functions—he thus coined the notion of "finite differences" that is still common in numerical analysis. His new calculus carried over some—if not all—properties of usual, differential calculus; it proves particularly well-behaved whenever performed over somewhat regular networks or grids; and it approximates differential calculus efficiently if mutual distances between the network's nodes tend to 0.

Boole denoted by Δu_x the terms of order zero in Taylor's expansion of a function u at $x + \Delta$ with respect to x, i.e.,

$$\Delta u_x := u(x + \Delta x) - u(x),$$

and thus wrote:

> In the Differential Calculus $\frac{du}{dx}$ is not a true fraction, nor have *du* and *dx* any distinct meaning as symbols of quantity. The fractional form is adopted to express the limit to which a true fraction approaches. Hence $\frac{d}{dx}$, and not d, there represents a real operation. But in the Calculus of Finite Differences $\frac{\Delta u_x}{\Delta x}$ is a true fraction. Its numerator Δu_x stands for an actual magnitude. Hence Δ might itself be taken as the fundamental operation of this Calculus, always supposing the actual value of Δx to be given [...].

If "the actual value of Δx" is taken to be the resistance of a conductor, then $\frac{\Delta u_x}{\Delta x}$ is the current flowing through a conductor in accordance with Ohm's law. Seemingly unaware of Kirchhoff's algebraic investigations, Boole replaced in the setting of standard calculus $\mathbb{R}$ by $\mathbb{N}$ and the first order differential operator $\frac{d}{dx}$ on $\mathbb{R}$ by a difference operator—in fact, by the incidence matrix of the oriented graph associated with $\mathbb{N}$. Armed only with Taylor's formula and an enviable courage to elaborate on what were—at best—purely formal identities, Boole ventured to discretize most of the usual differential and integral calculus and eventually envisaged to study partial *difference* equations. Without even a precise notion of "operator" at his disposal, he could for instance prove in [4, Chap. 14] that the discrete (both in time and space) wave equation is solved by the overlapping of two travelling waves.

Because distances are linearly proportional to resistances in electric networks, the objects introduced by Kirchhoff could now be understood as special instances of discrete differential operators. This, however, did not happen for decades due to the lack of advanced graph theory, linear algebra and operator theory. It took some time for the mathematical community to observe that graphs—as discrete objects—are most naturally described by means of matrices; and it took even longer to realize that certain properties of graphs can be efficiently studied applying linear algebraic methods to said matrices. While D. Kőnig devoted in [18]—the first book on graph theory ever—several sections to the study of matrices (to the "matrix theory of Poincaré–Veblen", as he calls it), his interest mostly lies in the attempt to deduce results in linear algebra *from* results in graph theory. A few years later, however, the converse approach—the one that is usual nowadays—was initiated, see e.g. [5, 31]. Though, as late as 1971 C.St.J.A. Nash-Williams could insert algebraic graph theory in his famous list of "unexplored and semi-explored territories in graph theory" [27].

In the first half of the twentieth century, many classes of dynamical systems on discrete structures had already come to be studied. The easiest such systems are arguably random walks: An early, beautiful instance of these investigations is G. Polya's celebrated result on recurrence/transience of random walks on lattice graphs, which was first obtained in [30] and then neatly re-proved in [26] by methods of algebraic graph theory. In the 1930s a few chemists, including E. Hückel and L. Pauling in [16, 29], respectively, tried to circumvent the mathematical difficulties of the newly invented Schrödinger equations by studying quantum mechanical properties of complicated molecules in a simplified form, where differential operators are replaced by suitable matrices. In 1944 L. Onsager extended in [28] the naive model of ferro-magnetism proposed 20 years earlier by E. Ising—essentially, a nonlinear dynamical system on a one-dimensional lattice graph—to two-dimensional lattice graphs, proving that phase transition may occur. A few years later, D.O. Hebb proposed in [14] a pioneering psychological model in which a discrete dynamical system is deputed to govern the time evolution of the weights of edges of so-called *cell assemblies*—graphs whose nodes are synapses of a small ensemble of neurons. Likewise, discrete models began to be proposed, accepted, and studied in sociology [33], ecology [25], neurobiology [32], etc. Some of these models were mathematically refined and even subtle, albeit their justification was seldom other than heuristic. Boole's dream of a parallel development of two theories—*differential* and *difference* calculus—was coming true, and graph matrices were beginning to be looked at and studied as operators on finite dimensional vector spaces.

While we will occasionally emphasize the interplay of network theory with both differential and finite difference calculi, the present book is *not* devoted to these both concurrent and complementary theories. Rather, its central topic is the interplay of differential and difference operators with the functional analytic theory of evolution equations—arguably one of the most lucid, elegant and successful creations of the mathematics of the twentieth century.

The earliest purely mathematical study of evolution equations associated with difference operators—which was likely performed in [3], although comparable ideas had already been roughly sketched in [10]—was however most probably not influenced by any of the above-mentioned applied models. Rather, A. Beurling and J. Deny introduced the setting that we are going to present in detail in Sect. 6.4.1 with the explicit goal to provide a model that was both transparent and non-trivial for their new theory of *Dirichlet forms*: It is unclear whether Beurling and Deny were aware of the graph-theoretical interpretation of their setting, but in [2, Sect. 8] they explicitly remarked on its connection with Kirchhoff's circuit formalism. They explain as follows their motivation for introducing a discrete setting:

> Nous avons jugé utile de mettre en évidence [...] quelques propriétés remarquables des espaces de Dirichlet en traitant un cas simple: celui où l'espace de base X n'a qu'un nombre fini de points, la mesure ξ étant constituée par la masse $+1$ en chacun de ces points.
> Des exemples montreront que même ce cas élémentaire n'est peut-être pas sans intérêt, mais notre but est surtout de donner un aperçu des méthodes de démonstration que nous utiliserons dans le cas général.[3]

As it is, both their abstract theory and their "elementary case" have proved most relevant for the later development of the theory of evolution equations on network-like objects. On the one hand, they have suggested an efficient unified framework for the treatment of both difference and differential equations; on the other hand, they have shown that the interplay of discrete mathematics and functional analysis is mighty enough that the network paradigm can be used for actual mathematical analysis of some relevant systems, and not only for merely descriptive purposes. In particular, their elegant abstract formalism characterizes elliptic problems whose resolvent operators are sub-Markovian.

At about the same time a relevant number of quantum chemists started elaborating on Hückel's early theory using the less rough simplification that electrons need not be confined in the point-like atoms of a molecule, but are allowed to move along ideal lines corresponding to atomic bindings. These scientists—including G. Kron, C.A. Coulson, and K. Ruedenberg in [8, 22, 34], respectively—were probably the first to consider differential operators on linear network-like structures, and in some of these papers some classes of self-adjoint realizations were also correctly determined. However, it took over 50 years before also this class of operators on discrete structures was recognized as a further source of Dirichlet forms. A few years earlier, a sketchy investigation of differential equations on networks had already been performed by R.J. Duffin in [9].

---

[3] *We have deemed useful to underline [...] a few remarkable properties of the Dirichlet spaces by considering a simple case: the one in which the underlying space X has only finitely many points—the measure ξ being given by a mass $+1$ in each point.*

Some examples will show that even this elementary case is perhaps not devoid of interest, but our goal is over all to give an overview of the proof methods that we will use in the general case.

The attention of Beurling and Deny was focused on symmetric forms, and hence on self-adjoint problems, but 20 years earlier E. Hille had introduced in [15] a general theory of *analytic* semigroups that convincingly generalized the main properties of self-adjoint semigroups. It took only a few more years for M. Itô to combine both these theories, thus initiating in [17] the modern theory of non-symmetric Dirichlet forms, and hence of non-self-adjoint sub-Markovian semigroups.

Seemingly unaware of the case study in [3], G. Lumer was among the first mathematicians, after Beurling and Deny, who applied pure functional analytical method to operators on network-like structures. But unlike Beurling and Deny, Lumer's main interest in [24] was explicitly aimed at evolution equations—more precisely, at discussing relevant properties of the operator semigroup generated by a differential operator:

> We deal with the classical one-dimensional-ramified-space situation where we have a finite or countably infinite "network" in $\mathbb{R}^n$, with different differential (diffusion type) operators on each arc. [...] The topological connecting conditions are given via "nodal spaces" [...] and the analytical connecting conditions are imposed via local "connecting operators" [...]

If one thinks of one-dimensional differential operators, one is of course drifting away from the coarse discretization of Boole, Beurling, and Deny. Indeed, there is a point in preferring to allow the space to come back into the picture, even if this contradicts Euler's and Kirchhoff's brilliant simplifications. Many systems—be they physical, biological, architectural, ecological, etc.—cannot be properly investigated without embedding them in space, although maybe not the natural, three-dimensional Euclidean space. (This turns out to be a much debated topic in contemporary epistemology, see e.g. the multidisciplinary contributions in [13].)

The revival of the theory of partial differential equations on networks was not only due to Lumer, though. It seems that similar ideas have been developed almost simultaneously but independently in different communities: Let us mention in particular the investigations motivated by quantum mechanical considerations in [1, 12, 36], or by modeling of elastic systems in [6, 7, 23]. All these investigations have eventually begun to converge and to interbreed.

Even if Lumer's axiomatic approach has subsequently been little exploited, his work has nevertheless proved critical in paving the way for the investigations of a new generation of analysts, mathematical physicists, probabilists, and graph theorists, whose results are quoted throughout this book. These mathematicians have greatly revived this field since the beginning of the 1980s, opening it to the fruitful influence of harmonic analysis, spectral theory, cohomology, and other mathematical theories. The early 1990s saw the blooming of potential theory on graphs after many results obtained in the 1970s by M. Yamasaki were first discovered and appreciated outside Japan, while a large community of theoretical physicists became fully aware of networks when in 1997 T. Kottos and U. Smilansky used them to shed new light on the Bohigas–Giannoni–Schmidt conjecture. Ten years ago, graphs had finally become broadly popular outside graph theory, too. What kind of graphs will be the object of our investigations?

Generally speaking, graphs are point structures consisting of isolated nodes that may, or may not, be linked to other nodes by edges. They will be presented in the Appendix A in more detail. The word *graph* was first used in this context by J.J. Sylvester in [35], whose investigations were motivated by chemical applications; but the program of simplifying a problem by introducing an abstract version based on binary relations was already explicitly presented and carried out in [11], which is therefore commonly considered to mark the birth of graph theory.

We have already stressed that our main purpose is to use graph-theoretical formalism with the aim of analyzing concrete models, rather than to merely provide a compact description of relevant systems. This difference may look subtle and the borders are often fuzzy. Let us illustrate our point of view by considering some elementary and admittedly simplistic, yet motivating problems.

- **Model 1.** Some people gather at a party. At the beginning, each of them only knows a few other guests. Compute their degrees of separations.
- **Model 2.** Many cars are driving on a highway at constant (but possibly different) speed. Analyze their flow.
- **Model 3.** A passenger wants to fly from an airport to another. Suggest the most efficient way of doing so.
- **Model 4.** A spider has woven its web, which gets stirred by a breath of wind. Study its vibrations.
- **Model 5.** A sexually transmittable disease is spreading. Determine the lowest number of patients to be cured in order to stop the infection.
- **Model 6.** A black-out has occurred. Investigate its propagation through an interconnected electric distribution network.
- **Model 7.** An electron is confined by a strong potential to move approximately only along the atomic bonds of a conjugated system. Describe its motion.

All of these models can be represented by means of a network formalism, but all underlying problems are essentially different: On the one hand, in models 1 and 5 (and to some extent in 3) the spatial issue can be neglected. Being in touch with an acquaintance, enjoying the possibility of flying to another town, being infected by somebody: all these are *spatially discrete* phenomena. The system might be evolving in time according to a differential equation, but the state space is usually finite-dimensional in applications. They are typical problems of *applied graph theory*: Some relevant matrices for this kind of problem will be introduced and preliminarily investigated in Sect. 2.1.

On the other hand, in models 2, 4, 6, 7 the relevant processes are occurring on the links that connect the network's nodes. Such models are best described by partial differential equations: by a conservation law (2), a wave equation (4), a telegraph equation (6), or a Schrödinger equation (7). They are good examples of *evolution equations on networks*. Graphs whose canonical discrete metric is enhanced by a richer structure based on the representation of edges as one-dimensional intervals are often referred to as *metric graphs* (as opposed to *discrete graphs*, an awkward but occasionally useful expression whenever we want to stress that we are going to make use only of the combinatorial properties of a graph).

Graph-theoretical abstractions of electric circuits have been called *networks*, too, at least since [5]. By a *network*, a (discrete) graph is nowadays usually meant, whose edges are equipped with two additional pieces of information: a (binary) direction and a (numerical) value, sometimes called *capacity*—which is however usually interpreted as a resistance, or a conductance, and which we will usually interpret as a length. Indeed, the notions of (oriented) metric graphs and networks tend to melt together. We will see that for many purposes the network formalism is indeed both necessary and sufficient to capture the complexity of both difference and differential equations on graphs. For this reason in this book we have chosen to use the word "network" as a generic term to refer to both discrete and metric graphs, whenever it is not relevant to distinguish between them.

As we will see, there are relevant differences but also tight relations between the theory of operators on discrete and metric graphs. Indeed, one aim of this book is to affirm and emphasize that discrete and differential operators on networks are often two sides of the same coin.

References

1. S. Alexander, Superconductivity of networks. A percolation approach to the effects of disorder. Phys. Rev. B **27**, 1541–1557 (1983)
2. A. Beurling, J. Deny, Dirichlet spaces. Proc. Natl. Acad. Sci. USA **45**, 208–215 (1959)
3. A. Beurling, J. Deny, Espaces de Dirichlet. I: Le cas élémentaire. Acta Math. **99**, 203–224 (1959)
4. G. Boole, *A Treatise on the Calculus of Finite Differences* (Macmillan, Cambridge, 1860)
5. R.L. Brooks, C.A.B. Smith, A.H. Stone, W.T. Tutte, The dissection of rectangles into squares. Duke Math. J. **7**, 312–340 (1940)
6. G. Chen, M.C. Delfour, A.M. Krall, G. Payre, Modeling, stabilization and control of serially connected beams. SIAM J. Control Opt. **25**, 526–546 (1987)
7. F. Conrad, Stabilization of beams by pointwise feedback control. SIAM J. Control Opt. **28**, 423–437 (1990)
8. C.A. Coulson, Note on the applicability of the free-electron network model to metals. Proc. Phys. Soc. Sect. A **67**, 608 (1954)
9. R.J. Duffin, Nonlinear networks. iia. Bull. Am. Math. Soc. **53**, 963–971 (1947)
10. R.J. Duffin, Discrete potential theory. Duke Math. J. **20**, 233–251 (1953)
11. L. Euler, Solutio problematis ad geometriam situs pertinentis. Comm. Acad. Scient. Petropol. **8**, 128–140 (1741)
12. P. Exner, P. Šeba, Free quantum motion on a branching graph. Rep. Math. Phys. **28**, 7–26 (1989)
13. S. Günzel (ed.), *Raumwissenschaften*. (Suhrkamp, Frankfurt am Main, 2008)
14. D.O. Hebb, *The Organization of Behaviour*. (Wiley, New York, 1949)
15. E. Hille, Notes on linear transformations. II: Analyticity of semi-groups. Ann. Math. **40**, 1–47 (1939)
16. E. Hückel, Quantentheoretische Beiträge zum Benzolproblem. Z. Physik A **70**, 204–286 (1931)
17. M. Itô, A note on extended regular functional spaces. Proc. Japan Acad. **43**, 435–440 (1967)
18. D. Kőnig, *Theorie der endlichen und unendlichen Graphen* (Teubner, Leipzig, 1936)
19. G. Kirchhoff, Ueber den Durchgang eines elektrischen Stromes durch eine Ebene, insbesondere durch eine kreisförmige. Ann. Physik **140**, 497–514 (1845)

20. G. Kirchhoff, Ueber die Auflösung der Gleichungen, auf welche man bei der Untersuchung der linearen Vertheilung galvanischer Ströme geführt wird. Ann. Physik **148**, 497–508 (1847)
21. G. Kirchhoff, On the solution of the equations obtained from the investigation of the linear distribution of Galvanic currents. IRE Trans. Circuit Theory **5**, 4–7 (1958)
22. G. Kron, Electric circuit models of the Schrödinger equation. Phys. Rev. **67**, 39–43 (1945)
23. G. Leugering, E.J.P.G. Schmidt, On the control of networks of vibrating strings and beams, in *IEEE Conference on Decision and Control (Proceedings 1989)*, pp. 2287–2290, IEEE, Providence, 1989
24. G. Lumer, Connecting of local operators and evolution equations on networks, in *Potential Theory (Proc. Copenhagen 1979)*, ed. by F. Hirsch (Springer, Berlin, 1980), pp. 230–243
25. R. MacArthur, Fluctuations of animal populations and a measure of community stability. Ecology **36**, 533–536 (1955)
26. C.St.J.A. Nash-Williams, Random walk and electric currents in networks. Math. Proc. Cambridge Phil. Soc. **55**, 181–194 (1959)
27. C.St.J.A. Nash-Williams, Unexplored and semi-explored territories in graph theory, in *New Directions in the Theory of Graphs (Proc. Ann Arbor 1971)* (Academic Press, New York, 1973), pp. 149–186
28. L. Onsager, Crystal statistics, I. A two-dimensional model with an order-disorder transition. Phys. Rev. **65**, 117–149 (1944)
29. L. Pauling, The diamagnetic anisotropy of aromatic molecules. J. Chem. Phys. **4**, 673–677 (1936)
30. G. Polya, Über eine Aufgabe betreffend die Irrfahrt im Strassennetz. Math. Annalen **84**, 149–160 (1921)
31. W. Quade, Matrizenrechnung und elektrische Netze. Arch. Elektrotechnik **34**, 545–567 (1940)
32. W. Rall, Branching dendritic trees and motoneurone membrane resistivity. Exp. Neurol. **1**, 491–527 (1959)
33. N. Rashevsky, Outline of a mathematical theory of human relations. Philos. Sci. **2**, 413–430 (1935)
34. K. Ruedenberg, C.W. Scherr, Free-electron network model for conjugated systems. I. Theory. J. Chem. Phys. **21**, 1565–1581 (1953)
35. J.J. Sylvester, Chemistry and algebra. Nature **17**, 284 (1878)
36. H. Watanabe, Spectral dimension of a wire network. J. Phys. A **18**, 2807–2823 (1985)

Chapter 2
Operators on Networks

In this chapter we are going to review a manifold of operators defined on networks. We will see later on that most of these operators arise in connection with some relevant evolution equation on networks. However, here we are not yet going to discuss any dynamical system or partial differential equation. Rather, our aim is to explain the interplay between the differential operators that are relevant for partial differential equations on domains, their discrete pendants that are more usual in graph theory, and finally their version on metric graphs that are at the basis of the theory of quantum graphs. The analysis of the properties of some associated evolution equations will be postponed to Chap. 4.

Graph theory turns out to be an essential component of our investigations. As one can expect from a field which is almost 300 years old, it is neither possible nor necessary for our purposes to mention all main results in the theory of graphs. We have preferred to simply introduce the basic formalism of graphs in the Appendix A, whereas in the first part of this chapter we follow in the footsteps of a special branch—the so-called *algebraic graph theory*, which often links combinatorics and functional analysis.

We are going to describe two main classes of objects: operators that act on sequences (matrices) in Sect. 2.1 and on functions defined on collections of intervals (differential operators) in Sect. 2.2, respectively.

Throughout Sect. 2.1 we formulate our results for a graph whose weight is γ, but of course they prevail if γ is consistently replaced by μ. The rationale for considering both weight functions is explained in Sect. 2.1.4.1 below.

2.1 Difference Operators on Graphs

We adopt throughout the notations and notions summarized in the Appendix A and assume that

$$\boxed{\mathsf{G} = (\mathsf{V}, \mathsf{E}, \mu) \text{ is a weighted oriented graph.}}$$

D. Mugnolo, *Semigroup Methods for Evolution Equations on Networks*,
Understanding Complex Systems, DOI 10.1007/978-3-319-04621-1_2,

We use the shorthand

$$\gamma(\mathsf{e}) := \frac{1}{\mu(\mathsf{e})} \quad \mathsf{e} \in \mathsf{E},$$

which is justified by the fact that a weight function takes values in $(0, \infty)$, cf. Definition A.14. We define two $\mathsf{E} \times \mathsf{E}$ matrices by

$$\mathcal{M} := \operatorname{diag}(\mu(\mathsf{e}))_{\mathsf{e}\in\mathsf{E}} \quad \text{and} \quad \mathcal{C} := \operatorname{diag}(\gamma(\mathsf{e}))_{\mathsf{e}\in\mathsf{E}}. \tag{2.1}$$

Unless otherwise stated we do generally not assume G to be finite. The need for orientation is mostly due to technical reasons, as almost all physical and computational phenomena described by equations considered in the next chapters do not depend on directions.

2.1.1 The Incidence Matrix

All relevant information about the connectivity of an oriented graph is encoded in a matrix that turns out to be the most fundamental object for functional analysis on graphs. Here and in the following, due to historical reasons we have chosen to keep the traditional representation of operators on graphs as matrices.

Definition 2.1. The *incidence matrix* of G is the $\mathsf{V} \times \mathsf{E}$ matrix[1]

$$\mathcal{I} := \mathcal{I}^+ - \mathcal{I}^-, \tag{2.2}$$

where $\mathcal{I}^+ := (\iota^+_{\mathsf{ve}})$ and $\mathcal{I}^- := (\iota^-_{\mathsf{ve}})$ are defined by

$$\iota^+_{\mathsf{ve}} := \begin{cases} 1 & \text{if } \mathsf{v} \text{ is terminal endpoint of } \mathsf{e}, \\ 0 & \text{otherwise,} \end{cases} \qquad \iota^-_{\mathsf{ve}} := \begin{cases} 1 & \text{if } \mathsf{v} \text{ is initial endpoint of } \mathsf{e}, \\ 0 & \text{otherwise.} \end{cases}$$

We adopt the notations

$$\mathsf{E}^+_{\mathsf{v}} := \{\mathsf{e} \in \mathsf{E} : \iota^+_{\mathsf{ve}} = 1\}, \quad \mathsf{E}^-_{\mathsf{v}} := \{\mathsf{e} \in \mathsf{E} : \iota^-_{\mathsf{ve}} = 1\}, \quad \mathsf{E}_{\mathsf{v}} := \mathsf{E}^+_{\mathsf{v}} \cup \mathsf{E}^-_{\mathsf{v}}, \quad \mathsf{v} \in \mathsf{V}.$$

Of course, matrix multiplication of $\mathcal{I}$ or $\mathcal{I}^T$ with vectors in $\mathbb{C}^{\mathsf{E}}$ or $\mathbb{C}^{\mathsf{V}}$ can be seen as evaluation of a (possibly unbounded) operator at a function over E or V, respectively. This is actually the point of view we will adopt throughout the book, and we therefore prefer to use the notations

$$(u(\mathsf{e}))_{\mathsf{e}\in\mathsf{E}} \qquad \text{and} \qquad (f(\mathsf{v}))_{\mathsf{v}\in\mathsf{V}}$$

[1] We use this casual notation throughout this book to mean a matrix with complex entries that defines a linear operator from E to V.

for vectors in $\mathbb{C}^{\mathsf{E}}$ or $\mathbb{C}^{\mathsf{V}}$, respectively. In particular,

$$(\mathcal{I}^T f)(\mathsf{e}) = f(\mathsf{e}_{\text{term}}) - f(\mathsf{e}_{\text{init}}), \qquad f \in \mathbb{C}^{\mathsf{V}},\ \mathsf{e} \in \mathsf{E},$$

and

$$(\mathcal{I}u)(\mathsf{v}) = \sum_{\mathsf{v} \in \mathsf{E}_{\mathsf{v}}^+} u(\mathsf{v}) - \sum_{\mathsf{v} \in \mathsf{E}_{\mathsf{v}}^-} u(\mathsf{v}), \qquad u \in \mathbb{C}^{\mathsf{E}},\ \mathsf{v} \in \mathsf{V}.$$

We stress that the incidence matrix is independent of the weight function μ. As each edge has exactly two distinct endpoints, each column of $\mathcal{I}^+$ and each column of $\mathcal{I}^-$ has exactly one entry, while the entries of each column of the incidence matrix $\mathcal{I}$ sum up to 0.

Remark 2.2. If one thinks of G as a *geometric graph* (i.e., a graph whose nodes are identified with points of $\mathbb{R}^3$, and whose edges e are identified with simple arcs of length $\mu(\mathsf{e})$), then $\frac{1}{\mu(\mathsf{e})}\mathcal{I}^T f(\mathsf{e})$ can be seen as a difference quotient that converges to $f'(\mathsf{e}_{\text{init}})$ as $\mu(\mathsf{e})$ goes to 0 uniformly for each e. Thus, $\mathcal{I}^T f$ can be looked at as a discretized version of the first derivative of a function f defined in all points of G. This intuition goes back to G. Boole [27], cf. the Notes of Chap. 3, and has played a relevant role in the development of functional analysis on graphs.

The sets of all functions from V to $\mathbb{C}$ and from E to $\mathbb{C}$ define vector spaces $\mathbb{C}^{\mathsf{V}}$ and $\mathbb{C}^{\mathsf{E}}$ called *node* and *edge space* of G, respectively: Hence, the incidence matrix defines a linear operator from $\mathbb{C}^{\mathsf{V}}$ to $\mathbb{C}^{\mathsf{E}}$. In a less modern but equivalent formulation, the following result appeared already in [86, § 1].

Lemma 2.3. *Let* $\mathsf{G} = (\mathsf{V}, \mathsf{E})$ *be a finite oriented graph with* κ *connected components. Then its incidence matrix* $\mathcal{I}$ *has rank* $|\mathsf{V}| - \kappa$. *Hence,* $\mathcal{I}$ *is never surjective (and* $\mathcal{I}^T$ *is never injective), while* $\mathcal{I}$ *is injective (and* $\mathcal{I}^T$ *is surjective) if and only the cyclomatic number* $|\mathsf{E}| - |\mathsf{V}| + \kappa$ *of* G *vanishes, i.e., if and only if* G *is a forest.*

Remark 2.4. If we discard the orientation of an oriented graph's edges, the connectivity of the underlying simple graph is completely described by the *signless incidence matrix* $\mathcal{J}$ defined by

$$\mathcal{J} := \mathcal{I}^+ + \mathcal{I}^-. \tag{2.3}$$

In analogy with Lemma 2.3, the most relevant property of $\mathcal{J}$ is that for a finite non-oriented graph the rank of $\mathcal{J}$ is $|\mathsf{V}| - \kappa^+$, where κ^+ is the number of bipartite connected components, cf. [71, Thm. 8.2.1]. Hence, for a bipartite graph 0 is an eigenvalue not only of $\mathcal{L}$, but also of $\mathcal{Q}$. Indeed, by [54, Prop. 2.3] the spectra of $\mathcal{L}$ and $\mathcal{Q}$ agree if G is bipartite.

Besides the incidence matrices $\mathcal{I}^+, \mathcal{I}^-, \mathcal{I}, \mathcal{J}$, there are several further relevant graph matrices. Some are more rooted in the classical algebraic graph theory, others arise in applications. In the following sections we are going to present a few of them, before turning to study their operator-theoretic properties in connections with difference and differential equations in the next sections.

2.1.2 *The Degree Matrix*

In the unweighted case, in graph theory it is usual to define the outdegree/indegree/degree of a node v as the cardinality of $\mathsf{E}_{\mathsf{v}}^{+}, \mathsf{E}_{\mathsf{v}}^{-}, \mathsf{E}_{\mathsf{v}}$, respectively. In view of our interest for weighted oriented graphs we introduce the following.

Definition 2.5. The *outdegree* and *indegree* of a node v are defined by

$$\deg_{\gamma}^{\text{out}}(\mathsf{v}) := \sum_{\mathsf{e}\in\mathsf{E}} \iota_{\mathsf{ve}}^{-}\gamma(\mathsf{e}) \qquad \text{and} \qquad \deg_{\gamma}^{\text{in}}(\mathsf{v}) := \sum_{\mathsf{e}\in\mathsf{E}} \iota_{\mathsf{ve}}^{+}\gamma(\mathsf{e}),$$

respectively. The *degree* of v is then simply $\deg_{\gamma}(\mathsf{v}) := \deg_{\gamma}^{\text{out}}(\mathsf{v}) + \deg_{\gamma}^{\text{in}}(\mathsf{v})$, i.e.,

$$\deg_{\gamma}(\mathsf{v}) := \sum_{\mathsf{e}\in\mathsf{E}} |\iota_{\mathsf{ve}}|\gamma(\mathsf{e}).$$

Nodes with vanishing indegree or outdegree are called *sources* and *sinks*, respectively.

The *degree matrix* $\mathcal{D}$ of G is defined as $\mathcal{D} := \operatorname{diag}(\deg_{\gamma}(\mathsf{v}))$. We define likewise the *indegree matrix* $\mathcal{D}^{\text{in}}$ and the *outdegree matrix* $\mathcal{D}^{\text{out}}$.

Finally, G is called *regular* if there exists $k \geq 0$ such that $\mathcal{D} = k\,\text{Id}$, i.e., if $\deg_{\gamma}(\mathsf{v}) = k$ for all $\mathsf{v} \in \mathsf{V}$. If G is bipartite, then it is called *semiregular* if there exist $k_1, k_2 \geq 0$ such that $\deg_{\gamma}(\mathsf{v}) = k_i$ for all $\mathsf{v} \in \mathsf{V}_i$, $i = 1, 2$.

We adopt the notation deg to refer to the degree of the underlying unweighted graph, i.e.,

$$\deg(\mathsf{v}) := \sum_{\mathsf{e}\in\mathsf{E}} |\iota_{\mathsf{ve}}|. \tag{2.4}$$

We stress that the conditions imposed on the connectivity of G by regularity and semiregularity are of combinatorial nature only in the unweighted case.

Definition 2.6. The graph G is called *outward* or *inward locally finite* if for all $\mathsf{v} \in \mathsf{V}$ there is $M_{\mathsf{v}} > 0$ such that

$$\deg_{\gamma}^{\text{out}}(\mathsf{v}) \leq M_{\mathsf{v}} \quad \text{or} \quad \deg_{\gamma}^{\text{in}}(\mathsf{v}) \leq M_{\mathsf{v}}, \qquad \text{respectively,}$$

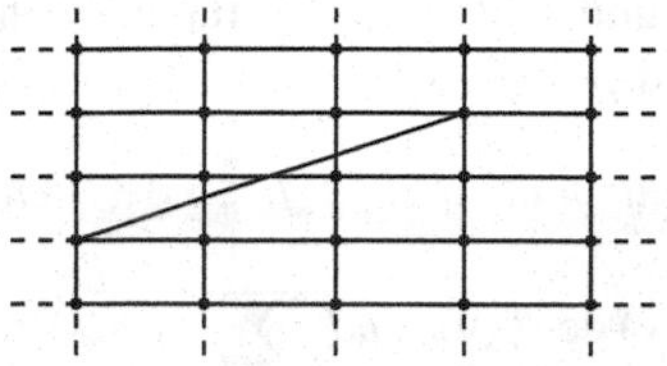

Fig. 2.1 An infinite, uniformly locally finite graph (provided all edges have weight 1, i.e., the graph is effectively unweighted)

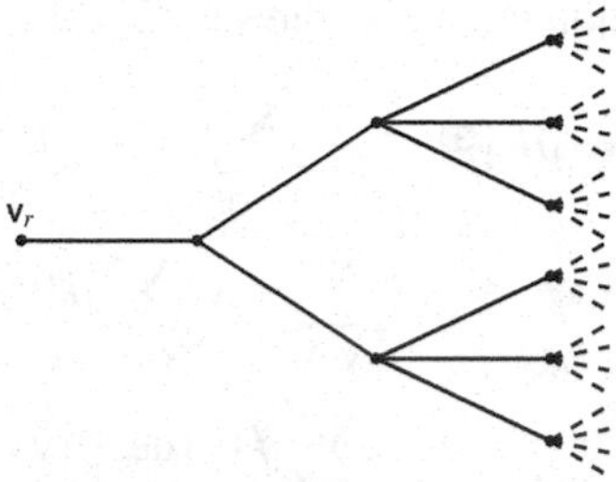

Fig. 2.2 An infinite, locally finite but not *uniformly* locally finite graph (provided all edges have weight 1, i.e., the graph is effectively unweighted). If however the graph is oriented towards the root v_r, then the graph is *outward uniformly* locally finite

Fig. 2.3 The lattice graph $\mathbb{Z}$

and *locally finite* if it is both outward and inward locally finite. If $\left(\deg_\gamma^{\text{out}}(v)\right)_{v\in V}$ or $\left(\deg_\gamma^{\text{in}}(v)\right)_{v\in V}$ are bounded sequences, i.e., if there exists $M > 0$ such that

$$\deg_\gamma^{\text{out}}(v) \le M \quad \text{or} \quad \deg_\gamma^{\text{in}}(v) \le M \qquad \text{for all } v \in V,$$

then G is called *outward* or *inward uniformly locally finite*, respectively. It is called *uniformly locally finite* if it is both outward and inward uniformly locally finite.

Example 2.7. If G is unweighted, then it is locally finite if and only if each node has only finitely incident edges; and uniformly locally finite if and only if there is a uniform upper bound in the number of edges that can be incident to each node (Figs. 2.1 and 2.2).

In the weighted case, uniform local finiteness of a function depends in an essential way on its weight function. For example, in the case of the lattice graph $\mathbb{Z}$ in Fig. 2.3, cf. Example A.11, the weight function defined by $\mu \equiv 1$ induces a uniformly locally finite (in fact, even regular) graph, whereas letting $\mu\big((n, n+1)\big) := |n|$, $n \in \mathbb{Z}$, clearly yields a locally finite but not *uniformly* locally finite graph.

Observe that if G is uniformly locally finite, then $\ell^p(\mathsf{V})$ is continuously embedded in $\ell^p_{\deg_\gamma}(\mathsf{V})$ for all $p \in [1, \infty]$.

Lemma 2.1. *Let* G *be locally finite and let* $f \in \mathbb{C}^{\mathsf{V}}$. *Then*

$$\sum_{\mathsf{e}\in\mathsf{E}} f(\mathsf{e}_{\mathrm{term}})\gamma(\mathsf{e}) = \sum_{\mathsf{v}\in\mathsf{V}} f(\mathsf{v}) \deg^{\mathrm{in}}_\gamma(\mathsf{v}) \quad \textit{and} \quad \sum_{\mathsf{e}\in\mathsf{E}} f(\mathsf{e}_{\mathrm{init}})\gamma(\mathsf{e}) = \sum_{\mathsf{v}\in\mathsf{V}} f(\mathsf{v}) \deg^{\mathrm{out}}_\gamma(\mathsf{v}),$$

provided these series converge.

Proof. Because E is the disjoint union of the sets $\mathsf{E}^+_{\mathsf{v}}$, $\mathsf{v} \in \mathsf{V}$, one has

$$\begin{aligned}\sum_{\mathsf{e}\in\mathsf{E}} f(\mathsf{e}_{\mathrm{term}})\mu(\mathsf{e}) &= \sum_{\mathsf{v}\in\mathsf{V}}\sum_{\mathsf{e}\in\mathsf{E}} f(\mathsf{e}_{\mathrm{term}})\mu(\mathsf{e})\iota^+_{\mathsf{ve}} \\ &= \sum_{\mathsf{v}\in\mathsf{V}} f(\mathsf{v}) \sum_{\mathsf{e}\in\mathsf{E}} \mu(\mathsf{e})\iota^+_{\mathsf{ve}} \\ &= \sum_{\mathsf{v}\in\mathsf{V}} f(\mathsf{v}) \deg^{\mathrm{in}}_\gamma(\mathsf{v}),\end{aligned}$$

and the other identity is proved likewise. □

In particular, taking $f \equiv 1$ one obtains

$$\mathrm{vol}_\gamma(\mathsf{G}) := \sum_{\mathsf{e}\in\mathsf{E}} \gamma(\mathsf{e}) = \sum_{\mathsf{v}\in\mathsf{V}} \deg^{\mathrm{in}}_\gamma(\mathsf{v}) = \sum_{\mathsf{v}\in\mathsf{V}} \deg^{\mathrm{out}}_\gamma(\mathsf{v}). \tag{2.5}$$

In the unweighted case one recovers the usual Handshaking Lemma in (A.2).

2.1.3 The Adjacency Matrix

If one is not interested in the orientation of edges but merely in the connectivity structure, then the following offers a useful alternative to the usage of the incidence matrix.

Definition 2.8. The $\mathsf{V}\times\mathsf{V}$ *incoming* and *outgoing adjacency matrices* $\mathcal{A}^{\mathrm{in}} := (\alpha^{\mathrm{in}}_{\mathsf{vw}})$ and $\mathcal{A}^{\mathrm{out}} := (\alpha^{\mathrm{out}}_{\mathsf{vw}})$ of G are defined by

$$\alpha^{\mathrm{in}}_{\mathsf{vw}} := \begin{cases} \gamma(\mathsf{e}) \text{ if } \mathsf{e} = (\mathsf{w},\mathsf{v}) \in \mathsf{E}, \\ 0 \quad \text{otherwise,} \end{cases} \quad \text{and} \quad \alpha^{\mathrm{out}}_{\mathsf{vw}} := \begin{cases} \gamma(\mathsf{e}) \text{ if } \mathsf{e} = (\mathsf{v},\mathsf{w}) \in \mathsf{E}, \\ 0 \quad \text{otherwise,} \end{cases} \tag{2.6}$$

respectively. The *(weighted) adjacency matrix* $\mathcal{A} := (\alpha_{\mathsf{vw}})$ of G is defined by $\mathcal{A} := \mathcal{A}^{\mathrm{in}} + \mathcal{A}^{\mathrm{out}}$.

Observe that $\mathcal{A}^{\mathrm{in}}$ is the transpose of $\mathcal{A}^{\mathrm{out}}$, so that $\mathcal{A}$ is symmetric.

Remark 2.9. For consistency we have introduced the adjacency matrix $\mathcal{A}$ for oriented graphs. However, given a weighted simple graph the adjacency matrix does clearly not depend on the chosen orientation (although $\mathcal{A}^{\text{in}}$ and $\mathcal{A}^{\text{out}}$ do).

The following is an easy but fundamental combinatorial property of the k-th power of $\mathcal{A}$. Observe that by definition for each n-path from v to w there is also an n-path from w to v. In the following $\text{cap}_\gamma(P)$ we denote by the capacity of a path P, cf. Definition A.17.

Proposition 2.10. *For all $k \in \mathbb{N}$ and $\mathsf{v}, \mathsf{w} \in \mathsf{V}$, the $\mathsf{v} - \mathsf{w}$-entry of $\mathcal{A}^k$ agrees with*

$$\sum_{\mathsf{P} \in \mathfrak{P}^k_{\mathsf{v},\mathsf{w}}} \text{cap}_\gamma(\mathsf{P}),$$

where $\mathfrak{P}^k_{\mathsf{v},\mathsf{w}}$ denotes the set of all paths from v to w of length k. In particular, in the unweighted case of $\mu \equiv 1$, the $\mathsf{v} - \mathsf{w}$-entry of $\mathcal{A}^k$ is simply the number of paths from v to w of length k.

Let us compute the product $\mathcal{I}^+\mathcal{C}(\mathcal{I}^-)^T$, for the incoming and outgoing incidence matrices $\mathcal{I}^+, \mathcal{I}^-$ introduced in Definition 2.1 and the weight matrix $\mathcal{C}$: The $\mathsf{v} - \mathsf{w}$-entry of $\mathcal{I}^+\mathcal{C}(\mathcal{I}^-)^T$ is given by

$$\sum_{\mathsf{e} \in \mathsf{E}} \iota^+_{\mathsf{ve}} \gamma(\mathsf{e}) \iota^-_{\mathsf{we}} = \begin{cases} \gamma(\mathsf{e}) & \text{if } (\mathsf{w}, \mathsf{v}) \in \mathsf{E}, \\ 0 & \text{otherwise.} \end{cases}$$

This and three further analogous computations yield the identities

$$\begin{aligned} \mathcal{A}^{\text{in}} &= \mathcal{I}^+\mathcal{C}(\mathcal{I}^-)^T, \quad \mathcal{A}^{\text{out}} = \mathcal{I}^-\mathcal{C}(\mathcal{I}^+)^T, \\ \mathcal{D}^{\text{in}} &= \mathcal{I}^+\mathcal{C}(\mathcal{I}^+)^T, \quad \mathcal{D}^{\text{out}} = \mathcal{I}^-\mathcal{C}(\mathcal{I}^-)^T, \end{aligned} \tag{2.7}$$

where $\mathcal{D}^{\text{in}}, \mathcal{D}^{\text{out}}$ are the indegree and outdegree matrices introduced above.

These simple formulae are at the basis of the definition of a manifold of operators that we introduce in the next sections.

2.1.4 The Discrete Laplacian

In view of certain applications, in particular to physics, it would be desirable to associate with any graph a semidefinite matrix. The adjacency matrix $\mathcal{A}$ is not a good choice: Because its trace is always zero, one of its eigenvalues is necessarily strictly negative (unless the graph is trivial). One may try of course shift the spectrum of $-\mathcal{A}$ or $\mathcal{A}$ by adding a suitable diagonal matrix with real entries. How large should this perturbation be in order to ensure positive semidefiniteness? A rough estimate is given by Gershgorin's theorem: The new matrix $\tilde{\mathcal{A}} = (\tilde{\alpha}_{\mathsf{vw}})$ is surely positive semidefinite if it is diagonally dominant, i.e., if

$$\tilde{\alpha}_{\mathsf{vv}} \geq \sum_{\mathsf{w}\neq\mathsf{v}} |\alpha_{\mathsf{vw}}| \tag{2.8}$$

(and likewise in the negative semidefinite case). Choosing $\tilde{\alpha}_{\mathsf{vw}}$ to be exactly

$$-\sum_{\mathsf{w}\neq\mathsf{v}} \alpha_{\mathsf{vw}} \quad \text{or} \quad \sum_{\mathsf{w}\neq\mathsf{v}} \alpha_{\mathsf{vw}}$$

yields two matrices with very nice properties: the discrete Laplacian and the signless Laplacian.

Definition 2.11. The $\mathsf{V}\times\mathsf{V}$-matrix

$$\mathcal{L} := \mathcal{I}\mathcal{C}\mathcal{I}^T \tag{2.9}$$

is called the *Laplace–Beltrami matrix* of G, and simply the *discrete* (or sometimes *combinatorial*) *Laplacian* in the unweighted case, i.e., if $\gamma \equiv 1$. A function $f \in \mathbb{C}^{\mathsf{V}}$ is called *harmonic* if $\mathcal{L}f = 0$.

Using

$$\mathcal{D} = \mathcal{D}^{\text{in}} + \mathcal{D}^{\text{out}}, \qquad \mathcal{A} = \mathcal{A}^{\text{in}} + \mathcal{A}^{\text{out}}, \qquad \mathcal{I} = \mathcal{I}^{+} - \mathcal{I}^{-}$$

and (2.7) we promptly obtain the following.

Proposition 2.12. *The Laplace–Beltrami matrix $\mathcal{L}$ of a weighted simple graph and the incidence matrix $\mathcal{I}$ of an arbitrary orientation of the same graph are related by*

$$\mathcal{L} = \mathcal{D} - \mathcal{A}. \tag{2.10}$$

Observe that the diagonal matrix with entries $\sum_{\mathsf{w}\neq\mathsf{v}} \alpha_{\mathsf{vw}}$, $\mathsf{v}\in\mathsf{V}$, is nothing but the degree matrix introduced in Sect. 2.1.2

We can therefore also introduce $\mathcal{L}$ for (simple) non-oriented graphs because neither $\mathcal{D}$ nor $\mathcal{A}$ depend on the orientation of the edges. Actually, $\mathcal{L}$ mirrors intrinsic properties of the underlying (non-oriented) simple graph, cf. Remark 2.9. In other words, all the $2^{|\mathsf{E}|}$ orientations of G have the same Laplace–Beltrami matrix.

A direct computation shows that $\mathcal{L}$ is equivalently given by

$$\mathcal{L}f(\mathsf{v}) = \sum_{\mathsf{v}\sim\mathsf{w}} \gamma\big((\mathsf{v},\mathsf{w})\big)\left(f(\mathsf{v}) - f(\mathsf{w})\right), \qquad \mathsf{v}\in\mathsf{V}. \tag{2.11}$$

Taking e.g. the graph with node set $\mathsf{V} := \mathbb{Z}$, with $(n,m)\in\mathsf{E}$ if and only if $m = n+1$, and

$$\gamma\big((n, n+1)\big) \equiv \frac{1}{a} > 0, \tag{2.12}$$

then one sees that $\mathcal{L}f$ converges to f'' as $a \to 0$.

The following can be deduced from Lemma 2.3.

Lemma 2.13. *Let* G *be finite. Then the multiplicity of* 0 *as an eigenvalue of the Laplace–Beltrami matrix* $\mathcal{L}$ *agrees with the number of connected components of* G.

We have restricted to the finite case in order to avoid to deal with an unbounded version of $\mathcal{L}$, and in particular not to deal with issues related to its domain. All this will be thoroughly discussed in Sect. 6.4.1.

Definition 2.14. The *signless Laplace–Beltrami matrix* of G is defined by

$$\mathcal{Q} := \mathcal{D} + \mathcal{A}. \tag{2.13}$$

Again, it is simply called *signless Laplacian* in the unweighted case of $\gamma \equiv 1$.

Computing like in the proof of Proposition 2.12 shows that for the signless incidence matrix $\mathcal{J}$

$$\mathcal{Q} = \mathcal{J}\mathcal{C}\mathcal{J}^T. \tag{2.14}$$

Thus, $\mathcal{Q}$ does not depend on the orientation of G. We stress that the scaling limit of $\mathcal{Q}$ is not a differential operator—indeed,

$$\mathcal{Q}f(\mathsf{v}) = \sum_{(\mathsf{v},\mathsf{w})\in\mathsf{E}} \gamma\big((\mathsf{v},\mathsf{w})\big)\,(f(\mathsf{v}) + f(\mathsf{w})) \tag{2.15}$$

shows that $\mathcal{Q}f(\mathsf{v})$ diverges to $+\infty$ as $a \to 0$, with a defined as in (2.12).

2.1.4.1 Graphs and Electric Networks

The connection between the theories of graphs and of electric networks is old. There is large consensus that it goes back to [86]. To explain this interplay, represent an electric network as a finite, connected, weighted oriented graph $\mathsf{G} = (\mathsf{V}, \mathsf{E}, \mu)$: Each node v is a junction point, each edge e is a conductor whose resistance is of $\mu(\mathsf{e})$ ohm—and hence its conductance is of $\gamma(\mathsf{e}) := \mu(\mathsf{e})^{-1}$ siemens—and an orientation is taken arbitrarily. Because the resistance of a conductor is proportional to its length, it is natural to think of $\mu(\mathsf{e})$ as a metric parameter—i.e., the length of an interval: This is exactly the idea that leads to the introduction of *metric graphs*, cf. Definition 3.12.

It will be sometimes convenient to regard $(\mathsf{V}, \mathsf{E}, \mu)$ and $(\mathsf{V}, \mathsf{E}, \gamma)$ as two—different, but in some sense mutually dual—weighted oriented graphs: a *resistance network* and a *conductance network*, to put it shortly. In this Sect. 2.1 we are going to focus on operators naturally associated with $(\mathsf{V},\mathsf{E},\gamma)$. Instead, in Sect. 2.2 we will discuss operators defined on $\mathfrak{G}$, the metric graph over $(\mathsf{V},\mathsf{E},\mu)$.

A reference potential can be arbitrarily fixed—in terms of some node v_0 which is set e.g. at 0—and we can thus define a (relative) electric potential in each node of the electric network. Denote by $f(\mathsf{v})$ the potential in the junction point v and by $u(\mathsf{e})$ the *oriented* currents along conductor e: i.e., $u(\mathsf{e}) = c > 0$ (resp., $c < 0$) if a current of c ampere flows from $\mathsf{e}_{\rm init}$ to $\mathsf{e}_{\rm term}$ (resp., from $\mathsf{e}_{\rm term}$ to $\mathsf{e}_{\rm init}$). Ohm's law states that the current vector u and the voltage vector $\mathcal{I}^T f$, which is given by $\mathcal{I}^T f(\mathsf{e}) = f(\mathsf{e}_{\rm term}) - f(\mathsf{e}_{\rm init})$, are related by

$$\mathcal{M}u = \mathcal{I}^T f \quad \text{or equivalently} \quad u = \mathcal{C}\mathcal{I}^T f \tag{2.16}$$

where the resistance matrix $\mathcal{M}$ and the conductance matrix $\mathcal{C}$ are defined as in (2.1).

By Lemma 2.3 $\mathcal{C}\mathcal{I}^T$ is not injective from $\mathbb{C}^{\mathsf{V}}$ to $\mathbb{C}^{\mathsf{E}}$, but it is indeed injective and hence invertible from $\mathbb{C}^{\mathsf{V}}/\mathbb{C} \simeq \mathbb{C}^{\mathsf{V}\setminus\{\mathsf{v}_0\}}$ to $\mathbb{C}^{\tilde{\mathsf{E}}}$ for any arbitrarily chosen $\mathsf{v}_0 \in \mathsf{V}$ and any subgraph $(\mathsf{V},\tilde{\mathsf{E}})$ that is a spanning tree of G rooted in v_0 (and hence, such that $|\tilde{\mathsf{E}}| = |\mathsf{V}| - 1$). Knowing u one can thus reconstruct all values of f, recursively determining the values of potentials of all nodes along the branches of the tree, starting with the neighbors of v_0, until each node is reached. Also the converse is clearly possible.

In the classical mathematical theory of electricity, *to solve an electric network* for given $U \in \mathbb{C}^{\mathsf{E}}$ and $F \in \mathbb{C}^{\mathsf{V}}$ means finding $u \in \mathbb{C}^{\mathsf{E}}$ and hence $f \in \mathbb{C}^{\mathsf{V}}$ that satisfy both Kirchhoff's current and voltage law, i.e.,

$$\mathcal{I}u = F \tag{2.17}$$

and

$$\sum_{\mathsf{e}\in\mathsf{E}} z(\mathsf{e})\left(\mathcal{I}^T f(e) - U\right) = 0 \quad \text{for all cycles } z \text{ in } \mathsf{G}, \tag{2.18}$$

respectively, possibly under further constraints. (Here we have used the notation introduced in Remark A.5.) We refer to [60,135],[22, §§ II.1–3 and § IX.2], and [26, Chapter 20] for details on the interplay between graph theory and electric networks. In particular, combining (2.16) and (2.17) one sees that the potential f in the network satisfies

$$\mathcal{I}\mathcal{M}^{-1}\mathcal{I}^T f = F \qquad \text{on } \mathsf{V}. \tag{2.19}$$

Recall that the distribution of electric potential on a surface Σ satisfies the Poisson equation

$$\Delta_\Sigma u = \phi \qquad \text{on } \Sigma \tag{2.20}$$

for the Laplace–Beltrami operator Δ_Σ and the free charge density ϕ. This suggests that $\mathcal{I}\mathcal{C}\mathcal{I}^T$ can be looked at as a discrete version of Δ_Σ.

(If F is not orthogonal to the constant vector $\mathbf{1}$, then (2.19) can be solved in $\mathbb{C}^{\mathsf{V}}/\mathbb{C}$, i.e., up to a constant: in a more common but equivalent terminology, this means that the solution is given by the pseudo-inverse of $\mathcal{I}\mathcal{M}^{-1}\mathcal{I}^T$. It is well-known that analogous results hold for (2.20).)

2.1.5 The Transition Matrix and the Normalized Laplacian

In this book we will devote most of our attention to time-continuous evolution equations and we will see in Sect. 4.2 that such equations display dissipation or at least conservation of some relevant quantity whenever their numerical range is sufficiently well-behaved. However, one can also consider discrete dynamical systems associated with the powers of a matrix. If e.g. one chooses the adjacency matrix $\mathcal{A}$ of G, usual physical quantities are typically not conserved. Indeed, by Gelfand's formula $\|\mathcal{A}^k\|$ grows as $\rho(\mathcal{A})^k$ for $k \to \infty$, where $\rho(\mathcal{A})$ is the spectral radius of $\mathcal{A}$.

Proposition 2.10 explains why a discrete dynamical system driven by $(\mathcal{A}^k)_{k\in\mathbb{N}}$ cannot converge to an equilibrium unless G is trivial. But a multiplicative perturbation turns it into a more treatable random walk.

Definition 2.15. Let G have no isolated nodes. The *transition matrix* or *random walk Laplacian* is the $\mathsf{V} \times \mathsf{V}$ matrix defined by

$$\mathcal{T} := \mathcal{D}^{-1}\mathcal{A}.$$

The action of $\mathcal{T}$ on a function $f \in \mathbb{C}^{\mathsf{V}}$ can be easily explained: It replaces the value of f in $\mathsf{v} \in \mathsf{V}$ by the average of all values in the neighboring nodes, since the $\mathsf{v}-\mathsf{w}$-entry of $\mathcal{T}$ is given by

$$\begin{cases} \frac{\gamma(\mathsf{e})}{\deg_\gamma(\mathsf{v})} & \text{if } \mathsf{e} = (\mathsf{w},\mathsf{v}) \in \mathsf{E} \text{ or } \mathsf{e} = (\mathsf{v},\mathsf{w}) \in \mathsf{E}, \\ 0 & \text{otherwise.} \end{cases} \tag{2.21}$$

We will see in Chap. 4 that $\mathcal{L}$ is only bounded on relevant sequence spaces if (3.2) holds. A way to avoid this dependence is to replace $\mathcal{L}$ by another matrix that is always bounded—regardless of the connectivity of G—but also encodes most information carried by $\mathcal{L}$.

Definition 2.16. Let G have no isolated nodes. The *normalized Laplace–Beltrami matrix* and *normalized signless Laplace–Beltrami matrix* are the $\mathsf{V} \times \mathsf{V}$ matrices

$$\mathcal{L}_{\text{norm}} := \mathcal{D}^{-\frac{1}{2}}\mathcal{L}\mathcal{D}^{-\frac{1}{2}} = \mathrm{Id} - \mathcal{D}^{-\frac{1}{2}}\mathcal{A}\mathcal{D}^{-\frac{1}{2}}$$

and

$$\mathcal{Q}_{\mathrm{norm}} := \mathcal{D}^{-\frac{1}{2}}\mathcal{Q}\mathcal{D}^{-\frac{1}{2}} = \mathrm{Id} + \mathcal{D}^{-\frac{1}{2}}\mathcal{A}\mathcal{D}^{-\frac{1}{2}},$$

respectively. We call them the *normalized Laplacian* and *normalized signless Laplacian* if $\mu \equiv 1$.

Remark 2.17. If G is unweighted and regular (say, of degree k), then the sets $\sigma(\mathcal{T}), \sigma(\mathcal{A})$ of eigenvalues of $\mathcal{T}, \mathcal{A}$, respectively, are clearly related by $\sigma(\mathcal{T}) = k^{-1}\sigma(\mathcal{A})$, and likewise $\sigma(\mathcal{L}_{\mathrm{norm}}) = k^{-1}\sigma(\mathcal{L})$. Obviously, one also has $\sigma(\mathcal{L}) = k - \sigma(\mathcal{A})$.

Interestingly, the spectrum of $\mathcal{T}$ on a finite, connected G is strongly influenced by several further graph-theoretic properties: For instance, -1 is an eigenvalue of $\mathcal{T}$ if and only if G is bipartite—and in this case the eigenvalues are symmetric about 0, cf. [43, § 1.3].

Again, a direct matrix multiplication shows that the $\mathsf{v}-\mathsf{w}$-entries of $\mathcal{L}_{\mathrm{norm}}$ and of $\mathcal{Q}_{\mathrm{norm}}$ are given by

$$\begin{cases} \frac{-\gamma\big((\mathsf{v},\mathsf{w})\big)}{\sqrt{\deg_\gamma(\mathsf{v})}\sqrt{\deg_\gamma(\mathsf{w})}} & \text{if } \mathsf{v} \sim \mathsf{w}, \\ 1 & \text{if } \mathsf{v} = \mathsf{w}, \\ 0 & \text{otherwise}, \end{cases} \quad \text{and} \quad \begin{cases} \frac{\gamma\big((\mathsf{v},\mathsf{w})\big)}{\sqrt{\deg_\gamma(\mathsf{v})}\sqrt{\deg_\gamma(\mathsf{w})}} & \text{if } \mathsf{v} \sim \mathsf{w}, \\ 1 & \text{if } \mathsf{v} = \mathsf{w}, \\ 0 & \text{otherwise}, \end{cases}$$

respectively. (Recall the notation introduced in Definition A.1: we write $\mathsf{v} \sim \mathsf{w}$ if v, w are adjacent, i.e., if there exists an edge whose endpoints are v, w.)

The difference between the Laplace–Beltrami matrix $\mathcal{L}$ and its normalized version becomes clear writing down explicitly

$$\mathcal{L}_{\mathrm{norm}} f(\mathsf{v}) = \frac{1}{\sqrt{\deg_\gamma(\mathsf{v})}} \sum_{(\mathsf{v},\mathsf{w})\in\mathsf{E}} \gamma((\mathsf{v},\mathsf{w})) \left(\frac{f(\mathsf{v})}{\sqrt{\deg_\gamma(\mathsf{v})}} - \frac{f(\mathsf{w})}{\sqrt{\deg_\gamma(\mathsf{w})}} \right).$$

Like $\mathcal{L}, \mathcal{Q}$ also $\mathcal{L}_{\mathrm{norm}}, \mathcal{Q}_{\mathrm{norm}}$ have by their definition a variational structure given by

$$\mathcal{L}_{\mathrm{norm}} = (\mathcal{D}^{-\frac{1}{2}}\mathcal{I})\mathcal{C}(\mathcal{D}^{-\frac{1}{2}}\mathcal{I})^T, \qquad \mathcal{Q}_{\mathrm{norm}} = (\mathcal{D}^{-\frac{1}{2}}\mathcal{J})\mathcal{C}(\mathcal{D}^{-\frac{1}{2}}\mathcal{J})^T. \tag{2.22}$$

The main reason why we are presenting $\mathcal{T}$ and $\mathcal{L}_{\mathrm{norm}}, \mathcal{Q}_{\mathrm{norm}}$ together is that

$$\mathcal{T} = \mathcal{D}^{-\frac{1}{2}}\left(\mathrm{Id} - \mathcal{L}_{\mathrm{norm}}\right)\mathcal{D}^{\frac{1}{2}} = \mathcal{D}^{-\frac{1}{2}}\left(\mathcal{Q}_{\mathrm{norm}} - \mathrm{Id}\right)\mathcal{D}^{\frac{1}{2}}. \tag{2.23}$$

Thus $\mathcal{T}$ has same eigenvalues as the Hermitian matrix $\mathrm{Id} - \mathcal{L}_{\mathrm{norm}}$—in particular, its eigenvalues are real.

Because $\mathcal{L}_{\mathrm{norm}} + \mathcal{Q}_{\mathrm{norm}}$ agrees with $2\,\mathrm{Id}$ and is hence positive definite, one concludes that the largest eigenvalue of $\mathcal{L}_{\mathrm{norm}}$ is dominated by the smallest eigenvalue of $\mathcal{Q}_{\mathrm{norm}}$.

2.1.6 The Kirchhoff and Advection Matrices

The following was studied by Tutte in [133] as an oriented version of the discrete Laplacian.

Definition 2.18. The *incoming/outgoing Kirchhoff matrices* of G are the $\mathsf{V} \times \mathsf{V}$ matrices given by

$$\mathcal{K}^{\mathrm{in}} := \mathcal{I}^{+}\mathcal{C}\mathcal{I}^{T} \qquad \text{and} \qquad \mathcal{K}^{\mathrm{out}} := -\mathcal{I}^{-}\mathcal{C}\mathcal{I}^{T}. \tag{2.24}$$

By (2.7) one checks that

$$\mathcal{K}^{\mathrm{in}} = \mathcal{D}^{\mathrm{in}} - \mathcal{A}^{\mathrm{in}} \qquad \text{and} \qquad \mathcal{K}^{\mathrm{out}} = \mathcal{D}^{\mathrm{out}} - \mathcal{A}^{\mathrm{out}}.$$

While most of the matrices we have considered so far represent standard ways of defining discrete analogs of second order differential operators, there seems to be no natural, all-round pendant of first order differential operators. A possible version will be discussed in Sect. 2.1.8 below. The following, different one has been proposed in [73, § 2.5.6].

Definition 2.19. The $\mathsf{V} \times \mathsf{V}$ matrix defined by

$$\overrightarrow{\mathcal{N}} := -\mathcal{I}\mathcal{C}(\mathcal{I}^{-})^{T}, \qquad \overleftarrow{\mathcal{N}} := \mathcal{I}\mathcal{C}(\mathcal{I}^{+})^{T}$$

are called *advection matrices*.

Example 2.20. By (2.25), $\overrightarrow{\mathcal{N}}$ and $\overleftarrow{\mathcal{N}}$ are circulant if G is an unweighted oriented cycle of finite length, cf. [73, Fig. 2.27]. More generally: If G is an oriented cycle of length n whose nodes are $\mathsf{v}_1, \dots \mathsf{v}_n$, then $\mathcal{A}^{\mathrm{in}} = (\alpha_{ij})$ satisfies

$$\alpha_{ij} = \begin{cases} \gamma\big((\mathsf{v}_i, \mathsf{v}_{i+1})\big), & \text{if } j = i+1, \\ 0 & \text{otherwise.} \end{cases}$$

If in particular all weights are equal to 1, then this matrix is a projection and same holds for $\mathcal{A}^{\mathrm{out}}$. It follows that $\overrightarrow{\mathcal{N}}$ and $\overleftarrow{\mathcal{N}}$ act as rotations along or against the direction of the orientation, respectively.

Using (2.7) we can check that in fact

$$\overrightarrow{\mathcal{N}} = \mathcal{D}^{\mathrm{out}} - \mathcal{A}^{\mathrm{in}}, \qquad \overleftarrow{\mathcal{N}} = \mathcal{D}^{\mathrm{in}} - \mathcal{A}^{\mathrm{out}}. \tag{2.25}$$

The Kirchhoff matrices and the advection matrices are tightly related. To begin with, they satisfy the symmetry relations

$$\mathcal{L} - \mathcal{K}^{\text{in}} = \mathcal{K}^{\text{out}}, \qquad \mathcal{L} - \overrightarrow{\mathcal{N}} = \overleftarrow{\mathcal{N}}. \tag{2.26}$$

Furthermore, by definition

$$\overleftarrow{\mathcal{N}}^T = \mathcal{K}^{\text{in}}, \qquad \overrightarrow{\mathcal{N}}^T = \mathcal{K}^{\text{out}}. \tag{2.27}$$

These relations will prove useful in the following.

Since the matrices introduced in this section are not Hermitian, their eigenvalues are in general not easy to compute. However, these matrices satisfy an important property. Let us first recall the following fundamental estimate on eigenvalues—*Gershgorin's Theorem.*

Proposition 2.21. *Let $\mathcal{W} = (\omega_{ij})$ be a square matrix of finite size. Then every eigenvalue of $\mathcal{W}$ lies within the balls $B_{R_i}(a_{ii})$ and $B_{C_\ell}(a_{\ell\ell})$ for at least one i and one ℓ, where*

$$R_i := \sum_{j \neq i} |\omega_{ij}| \quad \textit{and} \quad C_\ell := \sum_{j \neq \ell} |\omega_{j\ell}|.$$

In particular, all eigenvalues of a finite square matrix have negative real part if the matrix' diagonal entries are negative and either all columns or all rows sum up to 0.

Now, the diagonal entries of $\mathcal{D}^{\text{in}}, \mathcal{D}^{\text{out}}$ are positive, and therefore so are the diagonal entries of $\mathcal{K}^{\text{in}}, \mathcal{K}^{\text{out}}, \overrightarrow{\mathcal{N}}, \overleftarrow{\mathcal{N}}$. Furthermore, the rows of $\mathcal{K}^{\text{in}}, \mathcal{K}^{\text{out}}$ clearly sum up to 0 and so do the columns of $\overrightarrow{\mathcal{N}}, \overleftarrow{\mathcal{N}}$. Thus, we immediately get the following.

Corollary 2.22. *All eigenvalues of $-\mathcal{K}^{\text{in}}, -\mathcal{K}^{\text{out}}, -\overrightarrow{\mathcal{N}}, -\overleftarrow{\mathcal{N}}$ have negative real part.*

For our purposes, knowing the eigenvalues of an $n \times n$ matrix $\mathcal{W}$ is as relevant as knowing its *numerical range*, i.e., the set

$$W(\mathcal{W}) := \{(\mathcal{W}x|x) \in \mathbb{C} : x \in \mathbb{C}^n \text{ and } (x|x) = 1\}.$$

Let us summarize some basic properties of the numerical range, cf. [97].

Proposition 2.23. *Let $\mathcal{W} = (\omega_{ij})$ be a square matrix of finite size n. Then the numerical range $W(\mathcal{W})$ of $\mathcal{W}$ satisfies the following properties.*

(1) $W(\mathcal{W})$ is a compact convex set of $\mathbb{C}$.
(2) $W(\lambda\mathcal{W} + \mu \,\text{Id}) = \lambda W(\mathcal{W}) + \mu$ for all $\lambda, \mu \in \mathbb{C}$.
(3) $W(\mathcal{W} + \mathcal{Z}) \subset W(\mathcal{W}) + W(\mathcal{Z})$ for any further $n \times n$ matrix $\mathcal{Z}$.

(4) $W(\mathcal{W}) \subset \{z \in \mathbb{C} : \operatorname{Re} z \geq 0\}$ *if and only if* $\mathcal{W} + \mathcal{W}^*$ *is positive semidefinite.*
(5) *All eigenvalues of* $\mathcal{W}$ *are contained in* $W(\mathcal{W})$.
(6) *If* $\mathcal{W}$ *is normal, then* $W(\mathcal{W})$ *is the convex hull of the set of all eigenvalues of* $\mathcal{W}$.
(7) $W(\mathcal{W})$ *lies within the closed convex hull of the union of all balls* $B_{N_i}(a_{ii})$, $1 \leq i \leq n$ *where*

$$N_i := \sum_{j \neq i} \frac{|\omega_{ij}| + |\omega_{ji}|}{2}.$$

(8) *Assume* $\omega_{ij} \neq 0$ *if and only if* $(i,j) \notin \{(1,2),\ldots,(n-1,n),(n,1)\}$. *Then* $W(\mathcal{W})$ *agrees with the ball* $\overline{B_r(0)}$, *where* r *is the largest eigenvalue of the Hermitian matrix* $\frac{1}{2}(\mathcal{W} + \mathcal{W}^*)$.

Example 2.24. The numerical range of the matrices $\mathcal{K}^{\text{in}}, \mathcal{K}^{\text{out}}, \overrightarrow{\mathcal{N}}, \overleftarrow{\mathcal{N}}$ need not be contained in $\{z \in \mathbb{C} : \operatorname{Re} z \geq 0\}$, although all its eigenvalues are positive. This is not really surprising, since these matrices are not normal. For example, consider the oriented cycle on three edges, with weight function $\gamma = (2,1,1)$. Then, its advection matrix is given by

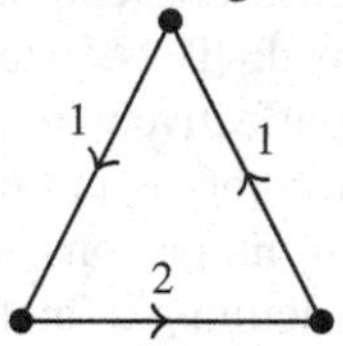

$$\overrightarrow{\mathcal{N}} = \begin{pmatrix} 2 & 0 & -1 \\ -2 & 1 & 0 \\ 0 & -1 & 1 \end{pmatrix},$$

whose eigenvalues are $0, 2 \pm i$ but which satisfies $\left(\overrightarrow{\mathcal{N}}x|x\right) = -5$ for the vector

$$x := \begin{pmatrix} 4 \\ 6 \\ 5 \end{pmatrix}.$$

2.1.7 Generalized Laplacians

Definition 2.25. Let $\mathsf{G} = (\mathsf{V}, \mathsf{E})$ be an oriented graph. A *generalized Laplacian* is any $\mathsf{V} \times \mathsf{V}$ matrix whose off-diagonal v-w-entry satisfies

- $= 0$ if and only if neither $(\mathsf{v}, \mathsf{w}) \in \mathsf{E}$ nor $(\mathsf{w}, \mathsf{v}) \in \mathsf{E}$, and
- < 0 otherwise.

The Laplace–Beltrami matrix, (minus) the adjacency matrix, the normalized Laplacian, and (minus) the signless Laplacian are examples of (symmetric) generalized Laplacians, and so are the examples we consider below.

Fig. 2.4 The graph of possible point mutations among nucleotides

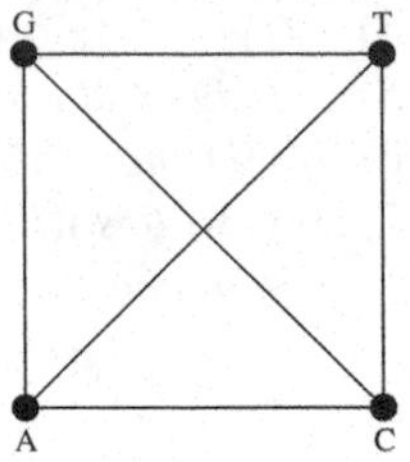

2.1.7.1 The Rate Matrix in Genetics

An application of the discrete Laplacian arose in early models of DNA evolution in the 1960s. The idea, first suggested in [83, 85], is to discuss the possible mutations at a given site of the DNA by means of a Markov chain (Fig. 2.4).

One assigns certain probabilities to each possible change. In the early models the transition matrix $\mathcal{T}$ was assumed to be symmetric (e.g., the probability of a mutation from guanine to cytosine should be the same as that of a mutation from cytosine to guanine), which suggests to actually interpret it as the transition matrix of a (complete) weighted oriented graph, in the sense of Definition 2.15. In later refinements of this model symmetry was dropped: Thus, one simply assigns certain probabilities to each possible change, thus effectively considering a weighted (complete) graph. The construction of these weights is performed in different fashions and each of them has different biological motivations, cf. [142] or [80, § 5.4], but mathematically all of them boil down to defining a (biologically appropriate) generalized Laplacian $\mathcal{A}$ of the above graph—the so-called *rate matrix*.

The evolution of the DNA site is then represented as a dynamical system driven by the rate matrix—a continuous one, due to the frequency of modifications, whose unknown is a probability distribution. Most of these models, but not all of them, require all rows (and sometimes also all columns) of the rate matrix to sum up to 0. As we will see in Example 4.44 below, this algebraic condition is related to the property that the solution defines for all time a probability distribution on $\mathbb{C}^{\mathsf{V}} \equiv \mathbb{C}^4$.

These and similar models have become a relevant part of Darwin medalist Kimura's *neutral theory of molecular evolution*—one of the most important breakthroughs in modern genetics. To better appreciate the revolutionary aspect of these ideas it suffices to recall that both discrete and continuous Laplacians are associated with dynamical systems that can be interpreted as stochastic processes. What Kimura was therefore positing is that at a local level mutations is solely random, so that it is primarily due to statistical fluctuations if genetic change spreads across a population.

2.1.7.2 The Hückel matrix

Sylvester suggested in [130] a formal identification of molecules with graphs whose nodes regard atoms. Furthermore, each of his nodes were labeled (much like in Definition 3.2) by a number that represented their chemical valence.

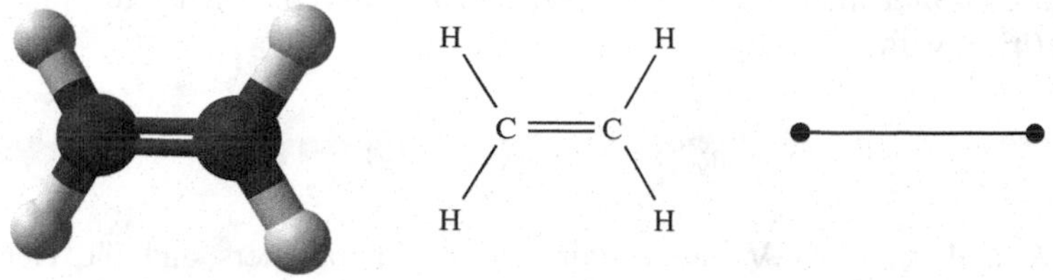

Fig. 2.5 An ethylene molecule (from wikipedia.org), its skeleton and Hückel's graph

In the early 1930s the mathematical foundations of quantum mechanics had already made relevant advances, but quantum chemists had great difficulties keeping pace. In particular, it was a great challenge to compute the spectrum of even the simplest molecules by a formal study of the Schrödinger equation. Hückel proposed in [78] a brilliant *Ansatz* for a semiempirical solution of the spectral problem for a class of aromatic compounds. In its simplest version (much generalized ever since), Hückel's approach to the description of the π orbitals of a conjugated, planar hydrocarbon that is either cyclic or linear is as follows.

Given a molecule, consider its skeletal formula, which can be regarded as a graph (typically, with multiple edges); neglect all the hydrogen atoms and their bonds to the carbon atoms; and finally discard all double bonds. Thus one finds the simple subgraph induced by only those nodes corresponding to carbon atoms: Let us call it the molecule's *Hückel graph* (Fig. 2.5).

Definition 2.26. Let $\alpha, \beta > 0$. Given a Hückel graph $\mathsf{G} = (\mathsf{V}, \mathsf{E})$ with associated degree and adjacency matrix $\mathcal{D}$ and $\mathcal{A}$, respectively, its *Hückel matrix* with parameters α, β is the matrix

$$\mathcal{H}_{\alpha,\beta} := \alpha\mathcal{D} - \beta\mathcal{A}.$$

In the simple case of a molecule of ethylene, for instance, the eigenvalue problem becomes

$$\begin{pmatrix} \alpha & \beta \\ \beta & \alpha \end{pmatrix} \psi = \lambda\psi,$$

where ψ is a general linear combination of atomic orbitals, and thus Hückel's method yields the approximated values of $\alpha \pm \beta$, and in particular the energy gap of 2β between the highest occupied and lowest unoccupied π orbitals.

Given its utter simplicity, *Hückel's Molecular Orbital* theory is surprisingly effective and has some heuristic justification: In a planar molecule the σ and π orbitals are mutually orthogonal: One can thus assume that σ-orbitals exert little influence on the spectral gap. In a first approximation, Hückel opted for neglecting the overlap integrals of distant orbitals, but a more precise description that considered all valence electrons was given in 1963 by the *extended Hückel's*

Molecular Orbital theory. There, Hückel's matrix is replaced by the full matrix $\mathcal{H}^{\text{ext}} = (\mathfrak{h}^{\text{ext}}_{\mathsf{vw}})$, with

$$\mathfrak{h}^{\text{ext}}_{\mathsf{vw}} := \frac{K}{2} o_{\mathsf{vw}} (\epsilon_{\mathsf{v}} + \epsilon_{\mathsf{w}}),$$

where K and ϵ_{v}, $\mathsf{v} \in \mathsf{V}$, are certain physical parameters and the Hermitian matrix $\mathcal{O} = (o_{\mathsf{vw}})$ is the full *overlap matrix* for the given molecule.

2.1.7.3 Pageranks

The mathematics behind internet search engines is largely a well-kept trade secret, but some early rough ideas were disclosed in [29, 114]. Let us begin with the following.

Definition 2.27. The *head* and *tail transition matrices* of G are the $\mathsf{V} \times \mathsf{V}$ matrices defined by

$$\overrightarrow{\mathcal{T}} := \mathcal{A}^{\text{in}} \left(\mathcal{D}^{\text{out}}\right)^{-1} \qquad \text{and} \qquad \overleftarrow{\mathcal{T}} := \left(\mathcal{D}^{\text{in}}\right)^{-1} \mathcal{A}^{\text{in}},$$

provided G has no sinks or sources, respectively.

By (2.25) one sees that

$$\overrightarrow{\mathcal{T}} = \text{Id} - \overrightarrow{\mathcal{N}} \left(\mathcal{D}^{\text{out}}\right)^{-1} \qquad \text{and} \qquad \overleftarrow{\mathcal{T}} = \text{Id} - \left(\mathcal{D}^{\text{in}}\right)^{-1} \mathcal{K}^{\text{in}}.$$

Indeed, applying (2.7) one finds that the $\mathsf{v}-\mathsf{w}$-entries of $\overrightarrow{\mathcal{T}}$ and $\overleftarrow{\mathcal{T}}$ are

$$\overrightarrow{\tau}_{\mathsf{vw}} := \begin{cases} \frac{\gamma(\mathsf{e})}{\deg_\gamma^{\text{out}}(\mathsf{w})} & \text{if } \mathsf{e} = (\mathsf{w}, \mathsf{v}) \in \mathsf{E}, \\ 0 & \text{otherwise}, \end{cases} \qquad \text{and} \qquad \overleftarrow{\tau}_{\mathsf{vw}} := \begin{cases} \frac{\gamma(\mathsf{e})}{\deg_\gamma^{\text{in}}(\mathsf{v})} & \text{if } \mathsf{e} = (\mathsf{w}, \mathsf{v}) \in \mathsf{E}, \\ 0 & \text{otherwise}, \end{cases} \tag{2.28}$$

respectively. Thus, $\overrightarrow{\mathcal{T}}$ is column stochastic and $\overleftarrow{\mathcal{T}}$ is row stochastic. In particular, 1 is an eigenvalue of $\overrightarrow{\mathcal{T}}^T$ with an associated positive eigenvector, but the associated eigenspace need not be one-dimensional.

Definition 2.28. Let G be finite and with no sinks. Let J denote the $\mathsf{V} \times \mathsf{V}$ matrix all of whose entries are $|\mathsf{V}|^{-1}$. A *Google-matrix* is any $\mathsf{V} \times \mathsf{V}$-matrix of the form

$$\mathcal{G}_d := (1-d)\,\mathrm{J} + d\overrightarrow{\mathcal{T}}, \qquad \text{for some } d \in (0,1).$$

Both J and $\overrightarrow{\mathcal{T}}$ are positive matrices, thus each Google-matrix is positive. By properly fitting the value of d one can furthermore enforce irreducibility of $\mathcal{G}_d$ and conclude by the Perron–Frobenius theorem that $\mathcal{G}_d$ has a dominant eigenvalue and that exactly one of the associated eigenvectors, denoted by pr_d, is both strictly positive and normalized. This is called *Perron eigenvector*.

Definition 2.29. Let all nodes in G satisfy $\deg^{\text{out}}(\mathsf{v}) > 0$. Let $d \in (0,1)$ be such that $\mathcal{G}_d$ is irreducible. Then the Perron eigenvector pr_d of $\mathcal{G}_d$ is called *Google-PageRank with parameter* d.

The original Google algorithms also require to deal with a huge digraph, corresponding to the World Wide Web, that is not strongly connected. In particular, the WWW does contain nodes of outdegree 0—*dangling links*, in Google's jargon: think of most files on arXiv.org—but also several further mathematical issues had to be solved: we refer to [95] for a detailed overview.

Google's success relies upon a fine tuning of this basic idea—the first possible tuning consisting, of course, in choosing a suitable value of d (accordingly to [29] "[w]e usually set d to 0.85"). Shifting a matrix to have it satisfy the assumptions of the Perron–Frobenius theorem may appear very artificial. In [29] Brin and Page explain the model behind their idea as follows:

> *PageRank can be thought of as a model of user behavior. We assume there is a "random surfer" who is given a web page at random and keeps clicking on links, never hitting "back" but eventually gets bored and starts on another random page. The probability that the random surfer visits a page is its PageRank. And, the d damping factor is the probability at each page the "random surfer" will get bored and request another random page.*

What the metaphor of the random surfer really says is that the Google-matrices are not quite (oriented versions of) transition matrices of the graph underlying the WWW, but rather generalized Laplacians of the complete graph $\mathcal{K}$ having as many nodes as the WWW. Its actual connectivity is described by the weights of this $\mathcal{K}$.

Google's procedure essentially consists in solving a stationary problem. In fact, by an application of the Neumann series the Google-PageRank can be computed as

$$pr_d = d \sum_{k=0}^{\infty} (1-d)^k (\mathcal{G}_d)^k \mathbf{1},$$

so that efficiency of Google's search algorithm is related to speed of convergence to equilibrium of the discrete dynamical systems governed by the powers of $\mathcal{G}_d$. Alternative pageranks have been proposed that make use indeed of partial differential equations—i.e., of continuous dynamical system. In particular, the following has been proposed by Chung in [44].

Definition 2.30. Let $t \geq 0$ and $f \in \mathbb{R}^{\mathsf{V}}$. The *heat kernel pagerank with parameters* t, f is the vector

$$e^{t(\vec{\mathcal{T}}-\mathrm{Id})} f := e^{-t} \sum_{k=0}^{\infty} \frac{(-t)^k}{k!} \left(\vec{\mathcal{T}}\right)^k f.$$

Let us neglect the damping term $(1-d)\,\mathrm{J}$, which appears mostly for technical reasons. What does the model behind the usage of $\overrightarrow{\mathcal{T}}$ consider as particularly worthy? Apparently, taking $\overrightarrow{\mathcal{T}}\mathbf{1}$ the outcome of the node v is larger (more authoritative WWW-page) if it receives many links (i.e., large $\gamma(\mathsf{e})$) from preceding pages w, and if these are otherwise not linking much (i.e., low $\deg_\gamma^{\mathrm{out}}(\mathsf{w})$). Thus, what this model describes is a transfer of credibility: Each time a page links another, it loses a bit of its opinion-making power and gives it away to the linked page.

Remark 2.31. (1) The advective (rather than diffusive) nature of the pagerank paradigm is even clearer if one considers the heat kernel pagerank instead. Indeed,

$$\overrightarrow{\mathcal{T}} - \mathrm{Id} = -\left(\overrightarrow{\mathcal{N}}(\mathcal{D}^{\mathrm{out}})^{-1}\right)^T,$$

hence $\overrightarrow{\mathcal{T}} - \mathrm{Id}$ can be interpreted as a normalized version of the advection matrix $\overrightarrow{\mathcal{N}}$.

(2) The tail transition matrix $\overleftarrow{\mathcal{T}}$ yields yet other pageranks that measure how much a WWW-page is likely to receive only a few endorsements, but rather strong ones: e.g., this could be used to filter mainstream pages if one is interested in searching only a rather sectorial word usage.

2.1.8 The Dirac Matrix

The Dirac equation is the fundamental law of relativistic quantum mechanics. It is an evolution equation associated with the differential operator we will introduce in Sect. 2.2.3 below. Similarly to the more usual Schrödinger equation, the Dirac equation cannot usually be solved analytically: A favorite approach is rather to discretize it and thus to study the spectrum of the emerging difference operator. Several ways of discretizing the Dirac equation exist, each with its advantages and drawbacks. The one we are going to present is one of the most common.

One discretizes the space-time $\mathbb{R}^4$ replacing $\mathbb{R}^4$ by the *d-dimensional lattice* $\mathbb{Z}^4$. We turn $\mathbb{Z}^4$ into a weighted oriented graph introducing an edgewise constant weight $\mu \equiv a$, i.e., a lattice constant a that—unlike in all previous sections of this chapter—is going to interpreted as the length of an interval, cf. Examples A.11 and A.16. The four-dimensional lattice is bipartite with respect to the partition $\mathbb{Z}^4 = \mathbb{Z}^4_+ \cup \mathbb{Z}^4_-$, where $\mathbb{Z}^4_\pm$ correspond to all nodes x with $x_1+x_2+x_3+x_4$ even or odd, respectively. This naturally defines two sublattices.

Observing that by definition for each $x \in \mathbb{Z}^4$ and $k \in \{1,2,3,4\}$ both $\mathsf{e}^+_{k,x} := (x, x+e_k)$ and $\mathsf{e}^-_{k,x} := (x-e_k, x)$ belong to E, where e_k is the k-th canonical basis vector of $\mathbb{R}^4$, one sees that

$$
\begin{aligned}
f(x+e_k) - f(x-e_k) &= (f(x+e_k) - f(x)) + (f(x) - f(x-e_k)) \\
&= \sum_{y\in\mathbb{Z}^4} \iota_{y e^+_{k,x}} f(y) - \sum_{y\in\mathbb{Z}^4} \iota_{y e^-_{k,x}} f(y) \\
&= (\mathcal{I}^T f)(e^+_{k,x}) + (\mathcal{I}^T f)(e^-_{k,x}).
\end{aligned}
\tag{2.29}
$$

Therefore, one is led to introduce the following.

Definition 2.32. Let $a > 0$. The operator $\tilde{\mathcal{D}}_S$ on $\mathbb{C}^{\mathbb{Z}^4}$ defined by

$$
\begin{aligned}
\left(\tilde{\mathcal{D}}_S f\right)(x) &:= \frac{1}{2a}\sum_{k=1}^{4} \epsilon_k(x)\left(f(x+e_k) - f(x-e_k)\right) \\
&= \frac{1}{2a}\sum_{k=1}^{4} \epsilon_k(x)(\mathcal{I}^T f)(e^+_{k,x}) + (\mathcal{I}^T f)(e^-_{k,x})
\end{aligned}
$$

is called *staggered* or *Kogut–Susskind Dirac matrix* on the four-dimensional lattice $\mathbb{Z}^4$, where

$$
\epsilon_k(x) := \begin{cases} 1 & \text{for } k = 1, \\ (-1)^{x_1} & \text{for } k = 2, \\ (-1)^{x_1+x_2} & \text{for } k = 3, \\ (-1)^{x_1+x_2+x_3} & \text{for } k = 4, \end{cases} \qquad \text{for all } x \in \mathbb{Z}^4.
$$

Of interest for the physicists is the associated Schrödinger-type equation

$$
\tilde{\mathcal{D}}_s f(x) = 0, \qquad x \in \mathbb{Z}^4,
$$

and in the limit $a \to 0$ one expects to recover the "continuous" Dirac operator D acting on functions defined on $\mathbb{R}^4$, cf. Sect. 2.2.3 below. The most relevant property of $\tilde{\mathcal{D}}_s$ is that by (2.29)

$$
\begin{aligned}
\tilde{\mathcal{D}}_s^2 f(x) &= \frac{1}{4a^2}\sum_{k=1}^{4}\left(f(x+2e_k) - 2f(x) + f(x-2e_k)\right) \\
&= \frac{1}{4a^2}\left(-8f(x) + \sum_{k=1}^{4}\left(f(x+2e_k) + f(x-2e_k)\right)\right).
\end{aligned}
$$

Apart from the scaling factor $\frac{1}{4a^2}$, $-\tilde{\mathcal{D}}_s^2$ may thus be interpreted as a discrete Laplacian—or, more precisely, as a pair of discrete Laplacians, each acting on either of the sublattices:

$$
-4a^2\tilde{\mathcal{D}}_s^2 f(x) = \begin{cases} \mathcal{L}_{\mathbb{Z}^4_+} f(x) \text{ if } x \in \mathbb{Z}^4_+, \\ \mathcal{L}_{\mathbb{Z}^4_-} f(x) \text{ if } x \in \mathbb{Z}^4_-. \end{cases}
$$

(Observe that these two Laplacians are *not* generalized Laplacians of $\mathbb{Z}^4$.)

2.1.9 Difference Operators on Line Graphs

In Sect. 2.2 we will introduce differential operators acting on spaces of functions defined on a metric graph's edges. As a warm up we consider some operators acting on functions defined on the edges of a (discrete) oriented graph G—that is, on edges seen as lumped, dimensionless objects. Equivalently, we can study corresponding operators defined on the nodes of the associated oriented line graph G_L, cf. Definition A.21.

Like in the case of any other graph, also the connectivity of the line graph G_L of G is of course determined by its degree matrix $\mathcal{D}_L$ and its adjacency matrix $\mathcal{A}_L$. The relevant function space is here $\mathbb{C}^{\mathsf{E}} = \mathbb{C}^{\mathsf{V}_L}$. We will see that $\mathcal{D}_L, \mathcal{A}_L$ can in turn be described in terms of matrices of the original graph G.

Matrix multiplication yields the following.

Proposition 2.33. *Let G_L be the weighted oriented line graph of G. Then the incidence matrices $\mathcal{I}^+, \mathcal{I}^-$ and the weight matrix $\mathcal{C}$ of G as well as the adjacency matrices $\mathcal{A}_L^{\mathrm{in}}, \mathcal{A}_L^{\mathrm{out}}$ of G_L satisfy*

$$\mathcal{I}^{+T}\mathcal{I}^-\mathcal{C} = \mathcal{A}_L^{\mathrm{out}} \quad \textit{and} \quad \mathcal{C}\mathcal{I}^{-T}\mathcal{I}^+ = \mathcal{A}_L^{\mathrm{in}}. \tag{2.30}$$

We will see in Sect. 2.2.2 that the following matrices, which have been introduced in [90], play an important role for the analysis of advection processes on metric graphs.

Definition 2.34. The *head* and *tail normalized adjacency matrices* of the line graph G_L of G are

$$\overrightarrow{\mathcal{B}} := \mathcal{C}\mathcal{I}^{-T}(\mathcal{D}^{\mathrm{out}})^{-1}\mathcal{I}^+ \quad \text{and} \quad \overleftarrow{\mathcal{B}} := \mathcal{I}^{-T}(\mathcal{D}^{\mathrm{in}})^{-1}\mathcal{I}^+\mathcal{C}, \tag{2.31}$$

respectively, provided G has no sinks or sources, respectively.

Remark 2.35. (1) The name we have given to $\overrightarrow{\mathcal{B}}, \overleftarrow{\mathcal{B}}$ can be explained observing that their $\mathsf{e}-\mathsf{f}$-entries are

$$\begin{cases} \frac{\gamma(\mathsf{e})}{\deg_\gamma^{\mathrm{out}}(\mathsf{e}_{\mathrm{init}})} & \text{if } \mathsf{f}_{\mathrm{term}} = \mathsf{e}_{\mathrm{init}}, \\ 0 & \text{otherwise,} \end{cases} \quad \text{and} \quad \begin{cases} \frac{\gamma(\mathsf{f})}{\deg_\gamma^{\mathrm{in}}(\mathsf{f}_{\mathrm{term}})} & \text{if } \mathsf{f}_{\mathrm{term}} = \mathsf{e}_{\mathrm{init}}, \\ 0 & \text{otherwise,} \end{cases} \tag{2.32}$$

respectively. That is, the $\mathsf{e}-\mathsf{f}$-entry of $\overrightarrow{\mathcal{B}}$ vanishes unless $(\mathsf{f},\mathsf{e}) \in \mathsf{E}_L$, in which case it agrees with the proportion of $\gamma(\mathsf{e})$ among the weights of all edges that *follow* f. Likewise, the $\mathsf{e}-\mathsf{f}$-entry of $\overleftarrow{\mathcal{B}}$ vanishes unless $(\mathsf{f},\mathsf{e}) \in \mathsf{E}_L$ (again!), in which case it agrees with the proportion of $\gamma(\mathsf{f})$ among the weights of all edges that *precede* e. This shows that $\overrightarrow{\mathcal{B}}$ is column stochastic, while $\overleftarrow{\mathcal{B}}$ is row stochastic.

(2) As suggested in Remark 2.31 for $\overrightarrow{\mathcal{T}}$, one can regard $\overrightarrow{\mathcal{B}}$ as a matrix that shifts some quantity from one edge to the following ones, splitting it in accordance with certain preference rules.
(3) Upon factorizing them as

$$\overrightarrow{\mathcal{B}} = \left(\mathcal{C}\mathcal{I}^{-T}(\mathcal{D}^{\text{out}})^{-1}\right)\mathcal{I}^{+} \qquad \text{and} \qquad \overleftarrow{\mathcal{B}} = \mathcal{I}^{-T}\left((\mathcal{D}^{\text{in}})^{-1}\mathcal{I}^{+}\mathcal{C}\right), \tag{2.33}$$

one is now motivated to consider the $\mathsf{V} \times \mathsf{V}$ matrices

$$\mathcal{I}^{+}\left(\mathcal{C}\mathcal{I}^{-T}(\mathcal{D}^{\text{out}})^{-1}\right) \qquad \text{and} \qquad \left((\mathcal{D}^{\text{in}})^{-1}\mathcal{I}^{+}\mathcal{C}\right)\mathcal{I}^{-T}, \tag{2.34}$$

which are in turn the matrices $\overrightarrow{\mathcal{T}}, \overleftarrow{\mathcal{T}}$ introduced in Sect. 2.1.7.3, respectively. By (2.33)–(2.34), a well-known result in linear algebra implies that the sets of nonzero eigenvalues of $\overrightarrow{\mathcal{T}}, \overrightarrow{\mathcal{B}}$ (resp., $\overleftarrow{\mathcal{T}}, \overleftarrow{\mathcal{B}}$) coincide.
(4) In view of Remark 2.4.(2), even if G is unweighted one should distinguish between the adjacency matrix $\mathcal{A}_L = \mathcal{A}_L^{\text{in}} + \mathcal{A}_L^{\text{out}}$ of the oriented line graph and the adjacency matrix $\tilde{\mathcal{A}}_L$ of its underlying simple graph. In the non-oriented case, the relation $\mathcal{J}^T\mathcal{J} = 2\,\text{Id} + \tilde{\mathcal{A}}_L$ between the signless incidence matrix introduced in (2.3) and the $\mathsf{E} \times \mathsf{E}$ adjacency matrix $\mathcal{A}_L$ of G_L is well-known, cf. [71, Lemma 8.2.2]. In the light of the above caveat, one should not be surprised by the differences between this formula and (2.30).
(5) By (2.14), $\lambda \neq 0$ is an eigenvalue of $\mathcal{Q}$ if and only if $\lambda \neq 0$ is an eigenvalue of $2\,\text{Id} + \tilde{\mathcal{A}}_L$.

The normalized adjacency matrices are at the basis of an object that play a fundamental role in scattering theory. To introduce it, we must for once abandon the setting of simple graphs.

If $\mathsf{e} \in \mathsf{E}$, then we denote as in Definition A.1 by $\bar{\mathsf{e}}$ its reversed edge. We consider the (non-simple) weighted digraph $\overline{\mathsf{G}} := (\mathsf{V}, \overline{\mathsf{E}}, \mu)$ introduced in Remark A.12. Recall that by Definition A.14 $\gamma(\mathsf{e}) = \mu(\bar{\mathsf{e}})$ for all $\mathsf{e} \in \overline{\mathsf{E}}$.

Definition 2.36. The *node scattering matrix* of G is the $\overline{\mathsf{E}} \times \overline{\mathsf{E}}$-matrix $\mathcal{S} := (\sigma_{\mathsf{ef}})$ defined by

$$\sigma_{\mathsf{ef}} := \begin{cases} 2\frac{\gamma(\mathsf{e})}{\deg_\gamma(\mathsf{e}_{\text{term}})} - 1, & \text{if } \mathsf{e} = \bar{\mathsf{f}}, \\ 2\frac{\gamma(\mathsf{e})}{\deg_\gamma(\mathsf{e}_{\text{term}})}, & \text{if } \mathsf{e}_{\text{term}} = \mathsf{f}_{\text{init}} \text{ but } \mathsf{e} \neq \bar{\mathsf{f}}, \\ 0, & \text{otherwise.} \end{cases}$$

Here we are denoting by $\deg_\gamma$ the degree function in G, which is clearly identical with both the indegree and the outdegree functions in $\overline{\mathsf{G}}$: this shows the relation to the above normalized adjacency matrices. Roughly speaking, σ_{ef} describes how likely it is that a particle hopping between edges of G, and currently on e, gets transmitted into a new edge ($\mathsf{f} \neq \mathsf{e}$) or reflected back into e ($\bar{\mathsf{f}} = \mathsf{e}$); these probability densities sum up to 0 in each node. Thus, the likelihood of a path $(\mathsf{v}_1, \ldots, \mathsf{v}_{n+1}), (\mathsf{e}_1, \ldots, \mathsf{e}_n)$ is encoded in the coefficient

$$\sigma(\mathsf{C}) := \sigma_{\mathsf{e}_1\mathsf{e}_2} \cdots \sigma_{\mathsf{e}_{n-1}\mathsf{e}_n}. \tag{2.35}$$

2.2 Differential Operators on Metric Graphs

We have already mentioned in Sect. 2.1.4.1 that in the classical matrix theory of electric networks γ, μ are interpreted as conductance as resistance, respectively. In particular, μ can be regarded as (proportional to) the length. Since we want to discuss differential properties, it seems therefore more natural to construct a

$$\boxed{\text{metric graph } \mathfrak{G} := (\mathsf{V}, \mathfrak{E}) \text{ over } \mathsf{G} = (\mathsf{V}, \mathsf{E}, \mu)}$$

(in the sense of Definition 3.12), rather than over $(\mathsf{V}, \mathsf{E}, \gamma)$. (It is clearly only a matter of notation, but in this way we can keep the role of the edge weight μ consistent with our electrostatic interpretation of circuits in Sect. 2.1.4.1, where γ was rather considered as *inversely* proportional to the length of an edge.)

In the following a function over a metric graph $\mathfrak{G}$ will be usually denoted by u. Its value in a point x of $\mathfrak{G}$ may be denoted by $u(x)$, but we will usually rather adopt the notation

$$u_{\mathsf{e}}(x) := u(\mathsf{e}, x) \qquad x \in (0, \mu(\mathsf{e})),$$

whenever x is a point of the metric edge associated with e, or

$$u(\mathsf{v}) := \begin{cases} u(\mathsf{e}, 0) & \text{if } \mathsf{v} = \mathsf{e}_{\text{init}}, \\ u(\mathsf{e}, 1) & \text{if } \mathsf{v} = \mathsf{e}_{\text{term}}. \end{cases}$$

Clearly, any operator that acts on functions on an interval determines an operator acting on functions over a metric graph $\mathfrak{G} = (\mathsf{V}, \mathfrak{E})$—this is done by extending the operator's definition edgewise.

Example 2.37. Given a function $p : \mathfrak{G} \to \mathbb{C}$, we can define a *multiplication operator M_p* by

$$(M_p u)_{\mathsf{e}}(x) := p_{\mathsf{e}}(x) u_{\mathsf{e}}(x) \qquad \text{for } u : \mathfrak{E} \to \mathbb{C},\ x \in (0, \mu(\mathsf{e})),\ \mathsf{e} \in \mathsf{E}. \tag{2.36}$$

(We can likewise extend M_p to an operator acting on functions defined also in the nodes, of course.)

Whenever we want to extend to the network setting a differential operator, however, boundary conditions have to be prescribed. The relevant issue in this section is to determine which are the most suitable boundary conditions—or rather, *node conditions*—for differential operators on metric graphs. It seems that there are essentially two possible approaches:

- The first one consists of choosing exactly those node conditions that enforce the behavior one expects on the basis of analogies with other known physical systems: e.g., finite speed of propagation in the case of an advective problem, parabolic maximum principle in the case of a diffusive one, etc.
- In the second one, to begin with we choose the usual discretization $\mathcal{A}$ of the relevant differential operator A (say, the discrete Laplacian or an advection matrix for the second or first derivative operator, respectively) and consider it on the node space of the given metric graph. Then, we start subdividing the edges, thus clearly obtaining more and more inessential nodes, i.e., of nodes with two neighbors that formerly belonged to the interior of the original metric edge. If the discretized version is well-behaved, in the limit the values $\mathcal{A}u(\mathsf{v})$ in the inessential nodes v should converge towards the values $Au(x)$ in the same points $x \equiv \mathsf{v}$. In the ramification nodes, however, we can usually recover a sequence of conditions whose limit as the subdivision becomes finer and finer delivers a natural node condition for the operator A.

We have begun Sect. 2.1 showing how to define a version of the Laplace operator for discrete graphs, and have subsequently discussed the advection matrix—the pendant of a first order differential operator. We have chosen to progress in this order since the latter operator is in a certain sense slightly less natural, as it relies upon the non-isotropic geometry of an oriented graph. We are going to follow a similar path in the case of differential operators on metric graphs, too, beginning with the second derivative.

Remark 2.38. When reading the following sections, one should bear in mind that metric graphs (in the sense of structures whose dimension is one) do not exist in the physical world, strictly speaking. Though, they are useful idealizations of higher-dimensional structures, if one size is largely predominant (think of a river or a wire, or perhaps of the axon inside a neuron, cf. Chap. 5) or if all but one dimensions are irrelevant for describing a certain system (say, a set of trains moving along a railway network). While it is comparatively easy to study a one-dimensional operator's behavior, as we are going to see, there is a priori no reason why this should give precise information about the real, higher-dimensional structure one is approximating. This natural question was studied in [47] and later on it got to become very popular after the further advances in [64, 92, 123], cf. the monograph [118] and the long list of references therein. However, naive but ingenious forerunners of one-dimensional approximation methods had been proposed already back in the 1940s by Kron and others, see e.g. [91]. Convergence of non-self-adjoint operators is more delicate and has been treated in [107].

2.2.1 *The Second Derivative*

Let us begin studying the paradigmatic case of a Laplacian on $\mathfrak{G}$: We can define a second derivative operator edgewise by

$$u_{\mathsf{e}} \mapsto \frac{d^2 u_{\mathsf{e}}}{dx^2} \qquad \text{for } u_{\mathsf{e}} : [0, \mu(\mathsf{e})] \to \mathbb{C},\ \mathsf{e} \in \mathsf{E}.$$

In the case of the second derivative, the node conditions typically prescribe the behavior of the normal derivative, and therefore depend on the metric of the graph. We choose to rescale the graph's edges by considering the isomorphism defined by

$$(\Psi u)_{\mathsf{e}}(x) := u_{\mathsf{e}}\left(\frac{x}{\mu(\mathsf{e})}\right), \qquad \text{for } u_{\mathsf{e}} : [0, 1] \to \mathbb{C},\ x \in [0, \mu(\mathsf{e})],\ \mathsf{e} \in \mathsf{E}. \tag{2.37}$$

In this way we have effectively stretched or shortened the network's edges in such a way that all of them have unit length. The price we have to pay is that we have to replace the second derivative by a more general elliptic operator. More explicitly, we obtain

$$\Psi \frac{d^2}{dx^2} \Psi^{-1} = \mathcal{C}^2 \frac{d^2}{dx^2}. \tag{2.38}$$

This operator $\mathcal{C}^2 \frac{d^2}{dx^2}$ acts by

$$u_{\mathsf{e}} \mapsto \frac{1}{\mu(\mathsf{e})^2} \frac{d^2 u_{\mathsf{e}}}{dx^2} \equiv \gamma(\mathsf{e})^2 \frac{d^2 u_{\mathsf{e}}}{dx^2} \qquad \text{for } u_{\mathsf{e}} : [0, 1] \to \mathbb{C},\ \mathsf{e} \in \mathsf{E}. \tag{2.39}$$

We can now follow again the *Ansatz* proposed in the introduction of Sect. 2.2 in order to determine correct boundary conditions for this operator. To do so, we make use of the discretization underlying the equations that describe the distribution of potential in an electric network sketched in Sect. 2.1.4.1. (Even if one is not specifically interested in diffusion of electric potential, one may regard that setting as prototypical for physical phenomena of diffusive type.)

In view of (2.39) the correct discrete version of the diffusion operator on $\mathfrak{G}$ is the Laplace–Beltrami matrix with coefficients γ^2, i.e.,

$$\mathcal{I}\mathcal{C}^2\mathcal{I}^T.$$

Given a node v and an incident edge e we will interpret the value $u_{\mathsf{e}}(x)$ of a function u as the electric potential at a point x of the edge e. We will need two linearly independent conditions for each edge, and by the Handshaking Lemma applied to the underlying unweighted graph this amounts to imposing in each node v a number of linearly independent conditions equal to the number of incident edges.

Ohm's law imposes a proportionality between current and voltage (i.e., difference of potential). Upon taking an infinitely fine subdivision of G, this can be interpreted as a requirement of proportionality

$$\frac{\partial_{\gamma^2} u_{\mathsf{e}}}{\partial \nu}(\mathsf{v}) = \gamma(\mathsf{e})^2 \iota_{\mathsf{ve}} u'_{\mathsf{e}}(\mathsf{v}). \tag{2.40}$$

between the current $\frac{\partial_{\gamma^2} u_{\mathsf{e}}}{\partial n}(\mathsf{v})$ flowing through e into v and the voltage $\iota_{\mathsf{ve}} u'_{\mathsf{e}}(\mathsf{v})$ along e evaluated at v by a factor given by the conductance $\gamma(\mathsf{e})^2$: (2.40) is nothing but the definition of *conormal* derivative $\frac{\partial_{\gamma^2} u_{\mathsf{e}}}{\partial n}$, cf. Definition B.2.

Kirchhoff's current law is just as easy to interpret: It states that in an electric network the total incoming current flowing into each node has to be equal to the total outgoing current flowing out of the same node, i.e., the *net current* has to vanish. In our context and in view of the previous rule, this means that

$$\partial_{\gamma^2} u(\mathsf{v}) = 0, \qquad \text{for all } \mathsf{v} \in \mathsf{V}, \tag{Kc}$$

where

$$\partial_{\gamma^2} u(\mathsf{v}) := \sum_{\mathsf{e} \in \mathsf{E}_{\mathsf{v}}} \frac{\partial_{\gamma^2} u_{\mathsf{e}}}{\partial \nu}(\mathsf{v}) \equiv \sum_{\mathsf{e} \in \mathsf{E}} \iota_{\mathsf{ve}} \gamma(\mathsf{e})^2 u'_{\mathsf{e}}(\mathsf{v}), \qquad \mathsf{v} \in \mathsf{V}. \tag{2.41}$$

For obvious reasons, this is usually referred to as *Kirchhoff condition* (on the normal derivative). Observe that the terms $\gamma(\mathsf{e})^2 u'(\mathsf{v})$ are added or subtracted depending on whether the node v is terminal or initial endpoint of e, hence this node condition seems to depend on the orientation of the graph—but, just like in the discrete case, this is only apparent, because re-orienting an edge forces the voltage to change sign.

In each v we have so far imposed only one boundary condition—using the notation in (2.4), we still need $\deg(\mathsf{v}) - 1$. In order to complete our task, we need a condition on the evaluations of u_{e} at their endpoints corresponding to v. In Kirchhoff's voltage law the potential function f is defined in the nodes: This suggests that we may look for a condition that reproduces the univocal node definition of discrete potential functions. Due to physical intuition (in the motivating example, potential does not make jumps in a node) we decide to impose a continuity condition in the nodes: More precisely, we assume equality of the values $u_{\mathsf{e}}(\mathsf{v})$ for all $|\mathsf{E}_{\mathsf{v}}|$ edges e that are incident at v,

$$u_{\mathsf{e}}(\mathsf{v}) = u_{\mathsf{f}}(\mathsf{v}) =: u(\mathsf{v}), \qquad \text{for all } \mathsf{e}, \mathsf{f} \in \mathsf{E}_{\mathsf{v}},\ \mathsf{v} \in \mathsf{V}, \tag{Cc}$$

with the notational convention

$$u_{\mathsf{e}}(\mathsf{v}) := \begin{cases} u_{\mathsf{e}}(0) & \text{if } \mathsf{v} = \mathsf{e}_{\mathrm{init}}, \\ u_{\mathsf{e}}(\mu(\mathsf{e})) & \text{if } \mathsf{v} = \mathsf{e}_{\mathrm{term}}. \end{cases} \tag{2.42}$$

(Denoting by $u(\mathsf{v})$ the joint value of all u_e at their respective endpoint corresponding to v seems appropriate and natural, but we should bear in mind that this is solely justified by condition (Cc).) Observe that (Cc) yields at each v the $|\mathsf{E}_\mathsf{v}|-1$ conditions we were looking for.

Remark 2.39. The Kirchhoff condition becomes a Neumann boundary condition in each leaf—recall Definition A.3. It amounts to prescribing continuity of the first derivative on each inessential node. Hence, we can add a new node in each interior point of a metric edge and find that each smooth function in the original graph defines a function in the new graph that automatically satisfies (Cc) and (Kc). This shows that the assumption of simplicity on the weighted graph underlying a metric graph is not restrictive at all, since we can always artificially subdivide an edge and thus produce a new simple graph that induces equivalent standard node conditions. Conversely, we can always remove an inessential node, thus obtaining a new function that is smooth on the new, longer metric edge.

Using the incidence matrices $\mathcal{I}^+, \mathcal{I}^-$ we can re-write these conditions in a more compact way. Condition (Cc) is equivalent to the existence of

$$u_{|\mathsf{V}} \in \mathbb{C}^\mathsf{V}$$

(which is then necessarily unique, by linearity) such that

$$\mathcal{I}^{-\top} u_{|\mathsf{V}} = u(0) \qquad \text{and} \qquad \mathcal{I}^{+\top} u_{|\mathsf{V}} = u(1)\,: \tag{2.43}$$

We denote in this case

$$u_{|\mathsf{V}} \equiv (u(\mathsf{v}))_{\mathsf{v}\in\mathsf{V}}.$$

Similarly, condition (Kc) is satisfied if and only if the weight and the incidence matrices satisfy

$$\mathcal{I}^+ C^2 u'(1) - \mathcal{I}^- C^2 u'(0) = 0. \tag{2.44}$$

This suggests the *generalized Kirchhoff conditions*

$$\partial_{\gamma^2} u(\mathsf{v}) + \mathcal{W} u_{|\mathsf{V}} = 0, \tag{KRc}$$

where $\mathcal{W}$ is some linear operator on $\mathbb{C}^\mathsf{V}$ and $\partial_{\gamma^2} u(\mathsf{v})$ has been introduced in (2.41).

Thus, we can finally introduce the following.

Definition 2.40. Let $\mathcal{W}$ be a linear operator on $\mathbb{C}^\mathsf{V}$. The *second derivative* (or *Laplacian) on $\mathfrak{G}$ with standard node conditions* is

$$\Delta u := \frac{d^2u}{dx^2} \quad \text{for } u : [0,1] \to \mathbb{C}^\mathsf{E} \text{ s.t. } \begin{array}{l} \exists u_{|\mathsf{V}} \in \mathbb{C}^\mathsf{V} \text{ s.t. } (\mathcal{I}^-)^\top u_{|\mathsf{V}} = u(0), \\ (\mathcal{I}^+)^\top u_{|\mathsf{V}} = u(1), \\ \text{and } \mathcal{I}^+ u'(1) - \mathcal{I}^- u'(0) + \mathcal{W} u_{|\mathsf{V}} = 0. \end{array}$$

A function u such that $\Delta u = 0$ is called *harmonic*.

More generally, for a given function $c : [0,1] \to \mathbb{C}^{\mathsf{E}}$ the *elliptic operator* (or sometimes *Laplace–Beltrami operator*) on $\mathfrak{G}$ with *elliptic coefficients* c^2 *and standard node conditions* is

$$\nabla(c^2\nabla u) := \frac{d}{dx}\left(c^2\frac{du}{dx}\right)$$

$$\text{for } u : [0,1] \to \mathbb{C}^{\mathsf{E}} \text{ s.t. } \begin{array}{l} \exists u_{|\mathsf{V}} \in \mathbb{C}^{\mathsf{V}} \text{ s.t. } (\mathcal{I}^-)^\top u_{|\mathsf{V}} = u(0),\ (\mathcal{I}^+)^\top u_{|\mathsf{V}} = u(1), \\ \text{and } \mathcal{I}^+c^2(1)u'(1) - \mathcal{I}^-c^2(0)u'(0) + \mathcal{W}u_{|\mathsf{V}} = 0. \end{array}$$

Example 2.41. Taking $\mathcal{W} = 0$ we of course recover the node conditions (Cc)–(Kc) (continuity and Kirchhoff).

If more generally $\mathcal{W}$ is diagonal, then (Cc)–(KRc) are often jointly referred to as *δ-coupling* in mathematical physics, where the expression "Kirchhoff conditions" is for some reasons deprecated. In analogy with the *Robin* boundary conditions, considered in the theory of partial differential equations, we will rather refer to them as *continuity and Kirchhoff–Robin node conditions* if $\mathcal{W} \neq 0$.

An especially interesting case is that of $\mathcal{W} = -\mathcal{L}$, where $\mathcal{L}$ is the Laplace–Beltrami matrix of the graph introduced in Definition 2.11. These conditions arise in the Friedrichs–Krein–von Neumann theory of extensions of self-adjoint operators. Also in view of Remark 2.48 below, Δ with conditions (Cc)–(KRc) for $\mathcal{W} = -\mathcal{L}$ turns out to be the so-called Krein–von Neumann extension (cf. [124, § 14.8], [11])—i.e., the largest self-adjoint, negative definite extension among those that respect the connectivity of the graph by including the continuity node conditions (Cc)—on $L^2\big((0,1);\ell^2_\mu(\mathsf{E})\big)$ of the second derivative defined on test functions over $\mathfrak{G}$. Some properties of this and related realizations have been studied in [104].

2.2.2 The First Derivative

Again, we follow the latter approach discussed at the beginning of Sect. 2.2 and use the ideas of Remark 2.35 in order to associate $\vec{\mathcal{B}}$ with a boundary condition for a differential operator that models advection, which can be defined edgewise in a natural way by

$$u_{\mathsf{e}} \mapsto k\frac{du_{\mathsf{e}}}{dx} \qquad \text{for } u_{\mathsf{e}} : [0,\mu(\mathsf{e})] \to \mathbb{C},\ \mathsf{e} \in \mathsf{E}, \tag{2.45}$$

for $k = +1$ or $k = -1$. First order differential operators are routinely used for describing physical phenomena of transport or advection type. One has to assign exactly one boundary condition to the first derivative on an interval, thus $|\mathsf{E}|$ boundary conditions in case of a metric graph with edge set E. It turns out that for the first derivative on an interval one cannot freely choose on which endpoint a condition must be imposed. Indeed, the correct endpoint is the right one if $k = +1$, and the left one if $k = -1$.

Let us for a moment restrict ourselves to the case of $k = +1$. On an interval $(0, 1)$ we may for instance impose the Dirichlet boundary condition $u(1) = 0$. But this choice is inappropriate on a network, as the operator would then effectively decay into a mere collection of decoupled first derivative operators on unrelated intervals.

Instead, thinking of $u(x)$ at each point x of the graph as a density, we would like to push $u(x)$ through a node into the outgoing edges. For each node v and each edge e whose terminal endpoint is v, the most natural choice is to split the value of $u_\mathsf{e}(\mathsf{v})$ into several parts, proportionally to the weights of the outgoing edges—that is, imposing the condition

$$u_\mathsf{e}(\mathsf{v}) = \sum_{\mathsf{f}\in\mathsf{E}_\mathsf{v}^+} \overleftarrow{\beta}_{\mathsf{ef}} u_\mathsf{f}(\mathsf{v}),$$

where $\overleftarrow{\mathcal{B}} = (\overleftarrow{\beta}_{\mathsf{ef}})$ is the $\mathsf{E} \times \mathsf{E}$ normalized adjacency matrix of the line graph of G defined in (2.31). Row stochasticity of $\overleftarrow{\mathcal{B}}$ ensures that there is no loss of mass.

As in the previous section we prefer to rescale the arguments of functions on a networks in order to obtain edges of unit length. This is done again by applying the unitary transformation Ψ defined in (2.37). In this way (plus or minus) the first derivative is transformed and we obtain

$$\Psi\left(\pm\frac{d}{dx}\right)\Psi^{-1} = \pm C\frac{d}{dx} :$$

This operator acts by

$$u_\mathsf{e} \mapsto \frac{1}{\mu(\mathsf{e})}\frac{du_\mathsf{e}}{dx} \equiv \gamma(\mathsf{e})\frac{du_\mathsf{e}}{dx} \qquad \text{for } u_\mathsf{e} : [0,1] \to \mathbb{C},\ \mathsf{e} \in \mathsf{E}. \tag{2.46}$$

Definition 2.42. The *first derivative on the metric graph $\mathfrak{G}$ with standard boundary conditions* is the operator $\overleftarrow{A}$ defined by

$$\overleftarrow{A}u := \frac{du}{dx} \qquad \text{for } u : [0,1] \to \mathbb{C}^\mathsf{E} \text{ s.t. } u(1) = \overleftarrow{\mathcal{B}}u(0).$$

Likewise, *(minus) the first derivative on $\mathfrak{G}$ with standard boundary conditions* is

$$\overrightarrow{A}u := -\frac{du}{dx} \qquad \text{for } u : [0,1] \to \mathbb{C}^\mathsf{E} \text{ s.t. } u(0) = \overrightarrow{\mathcal{B}}u(1).$$

We stress that in this definition the orientation of G does play a role. Observe also that the formal adjoint of $\overleftarrow{A}$ is instead given by

$$\overleftarrow{A}^* u = -\frac{du}{dx} \qquad \text{for } u : [0,1] \to \mathbb{C}^{\mathsf{E}} \text{ s.t. } u(0) = \overleftarrow{\mathcal{B}}^T u(1).$$

Remark 2.43. Different node conditions may also be justified if one drops the assumption of mass conservation—e.g., because one is modeling non-advective phenomena. For instance, it turns out that for the study of traveling waves in [139, § 16] the first derivative has to be equipped with continuity condition (Cc).

Besides the first derivative with continuity node conditions (Cc), which we denote by $\overleftarrow{A}_C$, one may also consider the first derivative $\overleftarrow{A}_K$ with a Kirchhoff-type condition analogous to (Kc)—this time not imposed on the normal derivatives, but rather on the values of the function, i.e.,

$$\sum_{\mathsf{e}\in\mathsf{E}} \iota_{\mathsf{ve}}\gamma(\mathsf{e})u_{\mathsf{e}}(\mathsf{v}) = 0, \qquad \text{for all } \mathsf{v} \in \mathsf{V}. \tag{Kc'}$$

A direct computation shows that the second derivative Δ with standard node conditions can be factorized as

$$-\Delta = \overleftarrow{A}_K \overleftarrow{A}_C = \overleftarrow{A}_C^* \overleftarrow{A}_C. \tag{2.47}$$

Indeed, one can check that A_C, A_K are mutually adjoint with respect to a natural inner product—more precisely, to the inner product of the Hilbert space $L^2(\mathfrak{G})$ defined in Sect. 3.2 below.

Also, one may consider a coefficient $c \in i\mathbb{R}$ in (2.45). In consideration of Example B.9 we introduce the following.

Definition 2.44. For any $\mathsf{E} \times \mathsf{E}$-matrix $\mathcal{U}$, a *momentum operator on* $\mathfrak{G}$ is defined by

$$\tilde{A}u := i\frac{du}{dx} \qquad \text{for } u : [0,1] \to \mathbb{C}^{\mathsf{E}} \text{ s.t. } u(0) = \mathcal{U}u(1).$$

It is convenient for later purposes to describe an alternative description of the same operators. Applying the isomorphism Ψ defined in (2.37) we namely obtain $\Psi \overleftarrow{A} \Psi^{-1} = C\frac{d}{dx}$, i.e., $\Psi \overleftarrow{A} \Psi^{-1}$ is the operator defined edgewise by

$$u_{\mathsf{e}} \mapsto \gamma(\mathsf{e})\frac{du_{\mathsf{e}}}{dx} \qquad \text{for } u_{\mathsf{e}} : [0,1] \to \mathbb{C},\ \mathsf{e} \in \mathsf{E},$$

with node conditions

$$u(1) = \overleftarrow{\mathcal{B}}u(0).$$

This corresponds simply to a change of measurement units that forces all edges to have unit length. Similar expressions can be obtained for $\overrightarrow{A}$ and $\tilde{A}$.

2.2.3 The Dirac Operator

The Dirac equation is the fundamental equation of relativistic quantum mechanics. Mathematically speaking, it is a hyperbolic system of first order partial differential equations whose unknown is a function with values in $\mathbb{C}^4$. Hence, the relevant Hamiltonian—the so-called *Dirac operator*—can be represented as an operator matrix that bears some formal similarity to the first derivative operators introduced in the previous section.

The Dirac operator can be also defined on metric graphs as follows. Up to some multiplicative constant (possibly different on each edge) we may as in the previous section assume without loss of generality that all edges have unit length.

Definition 2.45. For any $\mathsf{E} \times \mathsf{E}$-matrices $\mathcal{Z}_1, \mathcal{Z}_2$ a *Dirac operator on* $\mathfrak{G}$ *for two-component spinors* is defined by

$$Du := -i\hbar c \begin{pmatrix} 0 & -\frac{d}{dx} \\ \frac{d}{dx} & 0 \end{pmatrix} u + mc^2 \begin{pmatrix} \mathrm{Id} & 0 \\ 0 & -\mathrm{Id} \end{pmatrix} u \qquad \text{for } u : [0,1] \to \mathbb{C}^{\mathsf{E}} \times \mathbb{C}^{\mathsf{E}},$$

with node conditions

$$\mathcal{Z}_1 \underline{u} + \mathcal{Z}_2 \underline{\underline{u}} = 0$$

where

$$\underline{u} := \begin{pmatrix} u^1(0) \\ u^1(1) \end{pmatrix}, \quad \underline{\underline{u}} := \begin{pmatrix} -u^2(0) \\ u^2(1) \end{pmatrix}, \qquad \text{for } u \equiv \begin{pmatrix} u^1 \\ u^2 \end{pmatrix} : [0,1] \to \mathbb{C}^{\mathsf{E}} \times \mathbb{C}^{\mathsf{E}}.$$

We have already met a discretized version $\mathcal{D}_s$ of the Dirac operator in Sect. 2.1.8. Likewise, also D squares to an elliptic-type operator acting on vector-valued functions $u : [0,1] \to \mathbb{C}^{\mathsf{E}} \times \mathbb{C}^{\mathsf{E}}$ (we have neglected for simplicity the node condition). More precisely, one has

$$D^2 u = \hbar^2 c^2 \begin{pmatrix} \mathrm{Id} & 0 \\ 0 & \mathrm{Id} \end{pmatrix} u'' - 2i\hbar mc^3 \begin{pmatrix} 0 & \mathrm{Id} \\ \mathrm{Id} & 0 \end{pmatrix} u' + m^2 c^4 \begin{pmatrix} \mathrm{Id} & 0 \\ 0 & \mathrm{Id} \end{pmatrix} u, \tag{2.48}$$

and in particular D^2 is simply a Laplacian acting on $\mathbb{C}^2$-valued functions if $\hbar = c = 1$ and $m = 0$. Operator of this kind are sometimes called *Pauli operators*.

2.2.4 Higher Derivatives

Partial differential operators of fourth order have been crucial in the theory of elastic beams since the pioneering work of Euler. Of course, fourth order operators need four different boundary conditions that have to be carefully chosen to reflect model-oriented physical properties—in fact, most of them can be derived from a physically meaningful model. For the sake of simplicity we have chosen to present only the following simple one, see e.g. [55]. It is based on a common construction, that of *power* of an unbounded operator: If A is operator with domain $D(A)$ on a Banach space X, its square A^2 has domain

$$D(A^2) := \{u \in D(A) : Au \in D(A)\}.$$

Thus, each Au has to satisfy the same conditions imposed on u. This leads to the following.

Definition 2.46. For any $\mathsf{V} \times \mathsf{V}$-matrix $\mathcal{W}$ the *fourth derivative* (or *bi-Laplacian*) *on* $\mathfrak{G}$ *with standard node conditions* is the operator defined by

$$\Delta^2 u := \gamma(\mathsf{e})^4 \frac{d^4 u}{dx^4} \quad \text{for } u : [0,1] \to \mathbb{C}^{\mathsf{E}} \text{ s.t.}$$

$$\begin{aligned}
&\exists u_{|\mathsf{V}}, \tilde{u}_{|\mathsf{V}} \in \mathbb{C}^{\mathsf{V}} \quad \text{with} \\
&(\mathcal{I}^-)^\top u_{|\mathsf{V}} = u(0),\ (\mathcal{I}^+)^\top u_{|\mathsf{V}} = u(1), \\
&(\mathcal{I}^-)^\top \tilde{u}_{|\mathsf{V}} = u''(0),\ (\mathcal{I}^+)^\top \tilde{u}_{|\mathsf{V}} = u''(1), \\
&\mathcal{I}^+ \mathcal{C}^2 u'(1) - \mathcal{I}^- \mathcal{C}^2 u'(0) + \mathcal{W} u_{|\mathsf{V}} = 0, \\
&\text{and } \mathcal{I}^+ \mathcal{C}^4 u'''(1) - \mathcal{I}^- \mathcal{C}^4 u'''(0) + \mathcal{W} \tilde{u}_{|\mathsf{V}} = 0.
\end{aligned}$$

A function u such that $\Delta^2 u = 0$ is called *bi-harmonic*.

Elliptic operators of order 2^n, $n \geq 3$, may be of course defined likewise, but they do not appear often in models of applied mathematics.

2.3 Hybrid Operators on Metric Graphs

We have so far considered two classes of operators: On the one hand, those in Sect. 2.1 act on sequences, i.e., on functions defined on a discrete set; on the other hand, the operators treated in Sect. 2.2 deal with spatiality, as their arguments are functions defined on a continuum of points. In this section we present two operators that we regard as hybrid, as they combine both aspects. Neither of them is a differential operator, but both do build on the definition of the second derivative with standard node conditions.

2.3.1 *The Dirichlet-to-Neumann Operator*

Each time one considers a second order differential operator on a domain $U \subset \mathbb{R}^d$ with smooth boundary ∂U, one can also introduce a further operator that acts on functions defined on $\partial\Omega$. Indeed, choosing $f : \partial U \to \mathbb{C}$ in a suitable function space it is possible to solve uniquely (in a weak sense) the elliptic boundary value problem

$$\begin{cases} \nabla \cdot (c^2 \nabla u)(x) = 0, & x \in U, \\ u(z) = f(z), & z \in \partial U, \end{cases} \tag{2.49}$$

see e.g. [62, Chapter 6]. Then the *Dirichlet-to-Neumann* operator $\mathbb{DN}$ acts on f by taking the conormal derivative (with respect to c^2) of the solution u of (2.49), i.e.,

$$\mathbb{DN} f := \frac{\partial_{c^2} u}{\partial \nu}, \qquad \text{where } u \text{ is the weak solution of (2.49).}$$

By definition, $\mathbb{DN}$ is a linear operator whose arguments are functions defined on ∂U, rather than on U.

Defining a Dirichlet-to-Neumann operator on the metric graph $\mathfrak{G}$ does not require particular skills in partial differential equations. To begin with, we fix a node set $\mathsf{V}_0 \subset \mathsf{V}$ which we consider as *boundary* of the graph. We replace the space of functions over ∂U by $\mathbb{C}^{\mathsf{V}_0}$ and the second order differential operator $\nabla \cdot (c^2 \nabla)$ by the second derivative—with a suitable elliptic coefficient that allows us to renorm each interval $[0, \mu(\mathsf{e})]$ to $[0, 1]$.

Unlike in Sect. 2.2.1, we do not impose everywhere the same node conditions: (Cc) has to hold on each $\mathsf{v} \in \mathsf{V}$ (and hence the vector $u_{|\mathsf{V}}$ of node values is well-defined), but (Kc) is imposed only on

$$\mathsf{V}_0^C = \mathsf{V} \setminus \mathsf{V}_0.$$

Then, our aim is to impose a boundary value at each node in V_0, find a piecewise affine, twice weakly differentiable function u that additionally satisfies (Cc) in each node and (Kc) only in V_0^C, and finally read off its net current at each node that belongs to V_0.

More precisely, using the notation $\partial_{\gamma^2} u$ introduced in (2.41) we consider the following vector-valued system, a pendant of the boundary value problem (2.49):

$$\begin{cases} \gamma(\mathsf{e})^2 u_{\mathsf{e}}''(x) = 0, & x \in [0,1], \ \mathsf{e} \in \mathsf{E}, \\ \partial_{\gamma^2} u(\mathsf{v}) = 0, & \mathsf{v} \in \mathsf{V}_0^C, \\ \exists u_{|\mathsf{V}} \in \mathbb{C}^{\mathsf{V}} \text{ s.t. } (\mathcal{I}^-)^\top u_{|\mathsf{V}} = u(0), \ (\mathcal{I}^+)^\top u_{|\mathsf{V}} = u(1), & \\ u(\mathsf{v}) = f(\mathsf{v}), & \mathsf{v} \in \mathsf{V}_0, \end{cases} \tag{2.50}$$

for a given $f \in \mathbb{C}_0^{\mathsf{V}}$. This problem admits always a solution.

Definition 2.47. For some $\mathsf{V}_0 \subset \mathsf{V}$ the *Dirichlet-to-Neumann operator on* G *with respect to the boundary set* V_0 is the operator defined by

$$\mathbb{D}\mathbb{N} f := \partial_{\gamma} 2 u, \qquad \text{where } u \text{ is the unique weak solution of (2.50).}$$

The solution to (2.50) is necessarily a vector-valued affine function $u(s) := Ps + Q$, $s \in [0, 1]$, for some coefficients $P, Q \in \mathbb{C}^{\mathsf{E}}$ that have to satisfy

$$(\mathcal{I}^-)^T u_{|\mathsf{V}} = Q, \qquad (\mathcal{I}^+)^T u_{|\mathsf{V}} = P + Q \qquad \text{on } \mathsf{V},$$

as well as

$$\mathcal{I}^+ \mathcal{C}^2 P - \mathcal{I}^- \mathcal{C}^2 P = \mathbb{D}\mathbb{N} f \qquad \text{on } \mathsf{V}_0, \tag{2.51}$$

by (Cc) on V and (Kc) on V_0, respectively.

This amounts to a number $\deg(\mathsf{v})$ of conditions—recall the notational convention in (2.4)!—deriving from (Cc) in each $\mathsf{v} \in \mathsf{V}_0$, a number $\deg(\mathsf{v}) - 1$ of conditions deriving from (Cc) in each $\mathsf{v} \in \mathsf{V}_0^C$, and one condition deriving from (Kc) in each $\mathsf{v} \in \mathsf{V}_0^C$. All these conditions are linearly independent. We thus arrive at a total of

$$\begin{aligned}\sum_{\mathsf{v}\in\mathsf{V}_0} \deg(\mathsf{v}) - \sum_{\mathsf{v}\in\mathsf{V}_0^C} (\deg(\mathsf{v}) - 1) - |\mathsf{V}_0^C| &= \sum_{\mathsf{v}\in\mathsf{V}_0} \deg(\mathsf{v}) - \sum_{\mathsf{v}\in\mathsf{V}_0^C} \deg(\mathsf{v}) \\ &= \sum_{\mathsf{v}\in\mathsf{V}} \deg(\mathsf{v}) = 2|\mathsf{E}|\end{aligned}$$

linearly independent conditions, where the last identity follows by the Handshaking Lemma. It is therefore always possible to recover P, Q and hence to find a solution u of (2.50). In particular, an extension operator

$$E : \mathbb{C}^{\mathsf{V}_0} \ni f \mapsto u_{|\mathsf{V}} \in \mathbb{C}^{\mathsf{V}}$$

is well-defined and satisfies

$$\mathcal{I}^T E f = P.$$

Recalling that $\mathcal{I} = \mathcal{I}^+ - \mathcal{I}^-$ we conclude from (2.51) that

$$\mathbb{D}\mathbb{N} f(\mathsf{v}) := \partial_{\gamma^2} u(\mathsf{v}) = \mathcal{I}\mathcal{C}^2 P(\mathsf{v}) \qquad \text{for all } \mathsf{v} \in \mathsf{V}_0,$$

hence

$$\mathbb{D}\mathbb{N} = \mathcal{I}\mathcal{C}^2 \mathcal{I}^T E =: \mathcal{L} E,$$

where $\mathcal{L}$ is the Laplace–Beltrami matrix.

Remark 2.48. In particular, the Dirichlet-to-Neumann operator with respect to V_0 agrees with the Laplace–Beltrami matrix $\mathcal{L}$ if $\mathsf{V}_0 = \mathsf{V}$.

2.3.2 The Laplacian with Standard and Dynamic Node Conditions

Let $\mathsf{V}_0 \subset \mathsf{V}$. We consider the associated metric graph $\mathfrak{G}$ in order to work with functions spatially defined on the intervals associated with the edges, but we also want to take into particular account their values in the nodes belonging to V_0, which are regarded as massive enough to influence the behavior of the whole system. We define the operator matrix

$$\mathbb{A} := \begin{pmatrix} \Delta & 0 \\ -\partial_{\gamma^2} & -\mathcal{W}_1 \end{pmatrix},$$

which acts on vectors $\mathbb{u} := (u, f)$, for $u : [0,1] \to \mathbb{C}^{\mathsf{E}}$ and $f := u_{|\mathsf{V}_0} \in \mathbb{C}^{\mathsf{V}_0}$ (in the sense of (2.43)), just like a naive multiplication of a 2×2 matrix and a 2×1 matrix would—apart from the fact that some node conditions have to be imposed. Here $\mathcal{W}_1$ is some linear operator on $\mathbb{C}^{\mathsf{V}_0}$ that may or may not respect the structure of the graph: one may, e.g., think of a diagonal matrix, or rather of a Laplace–Beltrami matrix on the subgraph of G induced by V_0.

The operator Δ from Sect. 2.2.1 has too many node conditions for our current purposes. In particular we want ∂_{γ^2} to be a non-vanishing entry of $\mathbb{A}$. Therefore, we define $\mathbb{A}$ on pairs (u, f), where u is a functions that is merely continuous in the nodes (no Kirchhoff-type condition!), and f is exactly the value attained by u in the nodes. More precisely, we introduce the following.

Definition 2.49. Let $\mathsf{V}_0 \subset \mathsf{V}$ and let $\mathcal{W}_1, \mathcal{W}_2$ be a $\mathsf{V}_0 \times \mathsf{V}_0$-matrix and a $\mathsf{V}_0^C \times \mathsf{V}_0^C$-matrix, respectively. The *hybrid Laplacian on* $\mathfrak{G}$ *with standard node conditions on* V_0^C *and dynamic ones on* V_0 is

$$\mathbb{A}\mathbb{u} := \begin{pmatrix} \gamma^2 \frac{d^2}{dx^2} u \\ -\partial_{\gamma^2} u - \mathcal{W}_1 f \end{pmatrix}$$

$$\text{for } \mathbb{u} = \begin{pmatrix} u \\ f \end{pmatrix} \text{ s.t. } \begin{array}{l} u : [0,1] \to \mathbb{C}^{\mathsf{E}},\ f \in \mathbb{C}^{\mathsf{V}_0}, \\ \exists u_{|\mathsf{V}} \in \mathbb{C}^{\mathsf{V}} \text{ with} \\ (\mathcal{I}^-)^\top u_{|\mathsf{V}} = u(0),\ (\mathcal{I}^+)^\top u_{|\mathsf{V}} = u(1), \\ \mathcal{I}^+_{\mathsf{V}_0^C} C^2 u'(1) - \mathcal{I}^-_{\mathsf{V}_0^C} C^2 u'(0) + \mathcal{W}_2 u_{|\mathsf{V}_0^C} = 0, \\ \text{and } u_{|\mathsf{V}_0} = f. \end{array}$$

(Here $\mathcal{I}^+_{\mathsf{V}_0^C}, \mathcal{I}^-_{\mathsf{V}_0^C}$ are the $\mathsf{V}_0^C \times \mathsf{E}$ matrices consisting of those rows of $\mathcal{I}^+, \mathcal{I}^-$, respectively, that correspond to the nodes in V_0^C.)

The name in Definition 2.49 is justified by the evolution equation associated with $\mathbb{A}$, which we will meet again in Sect. 6.6.

The null space of the Laplacian with standard node conditions and the null space of $\mathbb{A}$ are isomorphic: Indeed, each harmonic function introduced in Definition 2.40 is the first component of some $\mathfrak{u} \in D(\mathbb{A})$ (and hence determines $\mathfrak{u}$ univocally) for which $\Delta u = 0$.

More generally, if one takes the eigenvalue equation associated with $\mathbb{A}$ one finds

$$\gamma^2 \frac{d^2}{dx^2} u = \lambda u, \qquad -\partial_{\gamma^2} u_{|\mathsf{V}_0} - \mathcal{W}_1 u_{|\mathsf{V}_0} = \lambda u_{|\mathsf{V}_0}$$

If we assume u to be smooth enough, evaluating the first expression at the nodes in V_0 and plugging into the second equation yields

$$\gamma^2 \frac{d^2}{dx^2} u_{|\mathsf{V}_0} + \partial_{\gamma^2} u_{|\mathsf{V}_0} + \mathcal{W}_1 u_{|\mathsf{V}_0} = 0. \tag{2.52}$$

This is a special instance of so-called Feller node conditions

$$\alpha \gamma^2 \frac{d^2}{dx^2} u_{|\mathsf{V}_0} + \beta \partial_{\gamma^2} u_{|\mathsf{V}_0} + \delta \mathcal{W}_1 u_{|\mathsf{V}_0} = 0, \tag{2.53}$$

where $\alpha, \beta, \delta : \mathsf{V} \to \mathbb{C}$ are not necessarily positive functions that satisfy certain compatibility conditions.

2.4 Nonlinear Operators

We have considered so far only linear operators, which in later chapters will be associated with linear Cauchy problems. However, in view of applications in several different fields, and most notably in theoretical neuroscience, nonlinear phenomena are often an important source of mathematical challenges. While discussing nonlinear evolution equations goes beyond the scope of this book, it seems appropriate to at least mention a few natural nonlinear extensions of our model.

The analysis of nonlinear operators on discrete graphs originates in the theory of nonlinear electric networks, and in particular in Minty's work [102]. Broadly speaking, a network is called *nonlinear* if Ohm's law is replaced by a more general nonlinear relation, given by function γ, between current flowing through an edge and voltage between the edge's endpoints, say

$$u(\mathsf{e}) = \gamma(\mathsf{e}) \left(\mathcal{I}^T f(e)\right) = \gamma(\mathsf{e})\big(f(\mathsf{e}_{\mathrm{term}}) - f(\mathsf{e}_{\mathrm{init}})\big).$$

The rougher γ, the harder the analysis of the network. Among benign nonlinearities we mention the two classes considered in [128, Appendix], where γ is either assumed to be continuous, odd, strictly monotone increasing, and satisfying $\lim_{x\to\infty}\gamma(x)=\infty$; or else to be a p-th power for some $p\in(1,\infty)$. The former class reflects Minty's approach but the latter—first introduced in [110]—has a more convenient variational structure that leads to introducing the *discrete p-Laplacian*

$$\mathcal{L}_p f(\mathsf{v}) := \sum_{\substack{\mathsf{w}\in\mathsf{V}\\ \mathsf{w}\sim\mathsf{v}}} |f(\mathsf{v})-f(\mathsf{w})|^{p-2}(f(\mathsf{v})-f(\mathsf{w})), \qquad \mathsf{v}\in\mathsf{V}. \tag{2.54}$$

Analogous classes of nonlinear operators play a role in the context of consensus problems, cf. Remark 4.60 below, if one allows for nonlinear protocols, cf. [15,113]. Also the model proposed by Kuramoto in [93] in his pioneering investigations on nonlinear oscillators has the same structure, with $\gamma=\sin$.

Also nonlinear Ohm-type laws of the form

$$u(e)=\mathcal{I}^T(\Phi f)(e)=\Phi(f(\mathsf{e}_{\mathrm{term}}))-\Phi(f(\mathsf{e}_{\mathrm{init}})).$$

are frequent in the literature. They lead to quasi-linear operators of the form $\mathcal{L}\circ\Phi$, where $\Phi:\mathbb{R}\to\mathbb{R}$ is suitably smooth and satisfies some appropriate additional assumption. The analysis of their properties goes back at least to [84]. Operators defined in a formally similar fashion arise in the theory of the *filtration equation* on domains and can be studied introducing suitable Sobolev spaces of negative order. In the case of graphs we mention the synchronization analysis performed in [82, 105] in the case of discrete and continuous dynamical systems, respectively. Yet other different but comparable nonlinear operators arise in some classes of neural models, like that in (5.6) below.

Semilinear operators on metric graphs are particularly interesting: On one hand they arise naturally in applications, in particular in neuronal models as well as in quantum mechanics; on the other hand they may have properties that are typically different from those of the corresponding operators on intervals. For instance, it is proved in [141] that the stationary solutions of a certain class of equations (related in particular of the Hodgkin–Huxley model presented in Chap. 5) display an *exceptional* behavior on five classes of graphs, one of which is the plain interval. This is in sharp contrast to the properties of *linear* differential equations, which typically extend with only minor changes from individual intervals to more general metric graphs—at least under standard node conditions. Semilinear problems on networks have also been studied, among many others, in [6, 34, 138] (parabolic equations), [4, 5] (hyperbolic equations), and [1, 7, 12, 25, 106, 122] (dispersive equations), cf. also [117] for an overview of the Soviet and Russian literature (often in Russian language) and several sections in the monograph [139].

Unlike the above ones, many common nonlinear network models do not have such an elementary variational structure: Let us mention e.g. the traffic flows studied in [45, 69], in which the relevant spatial operators are of type $\frac{d}{dx}\circ\Phi$—a nonlinear versions of the first derivative on an oriented graph with standard node conditions. Here Φ is some appropriate flux function, e.g. $\Phi(x) := x(1-x)$. The associated evolution equations are hyperbolic conservation laws and have to be studied by completely different methods from those of this book, e.g. along the lines of [62, Chapt. 11]. It is quite interesting that this, which is tightly related to the inviscid Burgers' equation, can be seen as the scaling limit of the *Asymmetric Simple Exclusion Process* (ASEP), see [119]. The ASEP is a favorite percolation-type model for transport on $\mathbb{Z}$ and some further more general graphs, cf. [109]. It displays most of the main features of the above differential model, in spite of being discrete in both time and space, and is therefore often regarded as a discrete traffic model. Indeed, a few rather connections between percolation and diffusion processes are known, especially on trees and lattices, cf. [88, 108, 116].

2.5 Notes and References

Section 2.1. A complete treatment of general graph theory can be found in several books, including [22, 26, 57]. We are particularly interested in algebraic graph theory: We thus especially refer to the classical monograph by Biggs and by Cvetković, Doob, and Sachs [18, 53], and to the more recent ones by Godsil and Royle, and by Brouwer and Haemers, cf. [31, 71], for the proofs of all results we have mentioned.

Adjacency is a very natural relation. The adjacency matrix frequently appears, possibly under different names, in applied sciences: it is e.g. dubbed "sociomatrix" in sociology and "connectivity matrix" in neuroscience.

The study of the Laplacian matrix was initiated, already in the weighted setting, by Kirchhoff (at least implicitly: matrices were to be formally defined in the same years by Sylvester). In [86] he introduced $\mathcal{L}$ and proved the following result.

Theorem 2.50. *Let* G *be a finite, connected, weighted graph. Consider an arbitrary orientation of* G*. Then all minors of* $\mathcal{L}$ *agree and their common value and their common value is*

$$\sum \mathsf{cap}(\mathsf{T})$$

where the sum is taken over the set of spanning trees T *of* G*.*

In particular, in the unweighted case of $\mu \equiv 1$ *each tree has capacity 1 and therefore then the common value of all minors of* $\mathcal{L}$ *is the number of spanning trees of* G*.*

This is known in graph theory as *Matrix-Tree Theorem* and was formally (if old-fashionedly) stated in [30]. Several related results can be found in [71, §§ 13.1–2].

It seems that the matrix $\mathcal{L}$ was referred to for the first time as *Laplacian* in [9]. In fact, besides the analogies with the Laplace equation of electrostatics sketched in Sect. 2.1.4.1, there are further several further reasons to think of $\mathcal{L}$ as a Laplacian-type operator.

- Harmonic functions for $\mathcal{L}$ satisfy appropriate versions of several equations that are known to hold for usual harmonic functions in the continuous case, e.g. the following mean value property: If $f \in \mathbb{C}^{\mathsf{V}}$ is harmonic, then by construction its value at a certain point v agrees with the average value of f on the unit sphere centered in v, i.e.,

$$f(\mathsf{v}) = \frac{1}{\deg(\mathsf{v})} \sum_{\mathsf{w} \sim \mathsf{v}} f(\mathsf{w}).$$

- Besides the above classical introduction of the discrete Laplacian on the basis of the theory of electric networks, one may also think of $\mathcal{L}$ as the discrete operator associated with a system of *second order* ordinary differential equations that govern an elastic system describing a network of springs. This observation goes back at least to [33] and is usually referred to as "electric circuit analogy" The study of spring networks has a biological motivation, cf. [21, Chapt. 5].
- A popular problem in machine learning theory consists in recovering the Riemannian metric of a manifold M, or rather the action of the Laplace–Beltrami operator Δ_M on M. Generally, the rule one is interested in consists in drawing a sample of points $(x_i)_{1 \le i \le n} \subset M$ and then suitably defining a weight for each pair of points x_i, x_j, thus turning the sample into the node set of a weighted graph G_x. The ultimate goal is to show that the Laplace–Beltrami matrix $\mathcal{L}$ of G_x, or perhaps $\mathcal{L}_{\text{norm}}$ or some other related difference operator on G_x, converge to Δ_M as $n \to \infty$ in a suitable, usually rather technical sense. Several results of this kind are known, cf. e.g. [16,46,77]. Conversely, it is known that in some cases spectral properties of Δ_M can be studied by spectral properties of the Laplace–Beltrami matrix $\mathcal{L}$ on some suitable associated graph, cf. [47].
- This is also related to the discretization scheme in (2.11). That heuristics is one of the main motivations for introducing $\mathcal{L}$: more generally, in the methods of finite differences one tries to approximate an (unknown) Green function for a partial differential equation by a sequence of "discrete" Green functions associated with certain difference problems. This is justified by the well-known (see e.g. [10, § 9.1]) fact that

$$|\Delta u(x) - \mathcal{L}u(x)| = O(a^2) \qquad \text{for all } u \in C^4(\overline{U}) \text{ and all } x \in (0,1),$$

where $\mathcal{L}$ is the discrete Laplacian of the lattice graph $\mathbb{Z}$, with lattice constant a, one of whose nodes is x. (We have seen a different but similar application of the same principle in Sect. 2.1.8.) If $U \subset \mathbb{R}$, then the condition $u \in C^4(\overline{U})$ is not very restrictive, but in higher dimensions non-convex, non-smooth domains U can easily lead to the case of $u \notin C^4(\overline{U})$, no matter how smooth Δu is. This poses severe restrictions to the application of the finite difference method in order to discretize domains of $\mathbb{R}^d$ for $d > 1$. Even if U is $\mathbb{R}^d$ or a smooth manifold (or perhaps a metric graph), the normalized Laplacian $\mathcal{L}_{\text{norm}}$ seems to yield often a more convenient and flexible discretization of the usual Laplace (or Laplace–Beltrami) operator than $\mathcal{L}$.

(On the other hand, Remark 2.48 seems to contradict this point of view. Indeed, whenever U is a smooth bounded domain of $\mathbb{R}^d$, $d \geq 2$, the Dirichlet-to-Neumann operator does not compare with the Laplace–Beltrami operator on ∂U, but rather with its square root—cf. [61] for details.)

Further cases of emergence of $\mathcal{L}$ and its generalizations are collected in [129], the speech delivered by Spielman on the occasion of the award of the Nevanlinna prize in 2010. Laplace–Beltrami matrices have proved a particularly efficient and handy replacement for Laplace–Beltrami (differential) operators on manifolds when it comes to produce examples of unusual spectral behaviors, see e.g. [127].

The signless Laplacian seems to have been introduced in [56] but its study has gained much momentum only in the last decade, cf. the survey [54]. An estimate on the largest eigenvalue of the normalized Laplace based on the smallest eigenvalue of the signless Laplacian has been discovered in [87]. The Laplacian and signless Laplacian are special cases of a class of operators defined taking $\sigma \in \mathbb{C}^{\mathsf{V}}$, considering $\mathcal{I}_\sigma = (\iota^\sigma_{\mathsf{ve}})$ where

$$\iota^\sigma_{\mathsf{ve}} := \iota^+_{\mathsf{ve}} + \sigma(\mathsf{v})\iota^-_{\mathsf{ve}}, \qquad \mathsf{v} \in \mathsf{V},\ \mathsf{e} \in \mathsf{E},$$

and introducing

$$\mathcal{L}_\sigma := \mathcal{I}_\sigma \mathcal{I}^*_\sigma. \tag{2.55}$$

These operators can be interpreted as discrete versions of the Schrödinger operators with a magnetic potential and as such have been thoroughly investigated in the series of papers [49, 50, 132]. Observe that $\mathcal{L}_\sigma$ depends on the orientation of G unless $\sigma(\mathsf{v}) \equiv e^{i\theta}$. The matrix $\mathcal{L}_{\text{norm}}$ was largely popularized by Chung, cf. [43].

The non-Hermitian matrices $\mathcal{K}^{\text{in}}, \mathcal{K}^{\text{out}}, \overrightarrow{\mathcal{N}}, \overleftarrow{\mathcal{N}}$ are less common than the different versions of the discrete Laplacians. While one of the fascinating aspects of the Matrix-Tree-Theorem is precisely the interplay of oriented and non-oriented graphs—that is, the possibility of finding an invariant of a graph by studying an arbitrary orientation of it—an oriented version of the Matrix-Tree Theorem exists, cf. [134, Thm. VI.27]. Indeed, all minors of $\mathcal{K}^{\text{in}}$ agree—and so do those of $\mathcal{K}^{\text{out}}$, by symmetry.

Advection matrices on (oriented versions of) $\mathbb{Z}$ act of course as left and right shifts, respectively, but on general graphs they have been seemingly first studied in [73, § 2.5.6]. Their name is due to their emergence in a certain discretized version of partial differential equations of advective time. Heuristically, the relation between $\overrightarrow{\mathcal{N}}$ and advection processes can be explained observing that if one applies $\overrightarrow{\mathcal{N}}$ to the constant function **1**, which can be thought of as a mass distribution, then the v-th entry of $\vec{\mathcal{N}}\mathbf{1}$ yields the quantity of mass gained or lost in transversing v.

Generalized Laplacians have been introduced by Colin de Verdière in [48] in the course of defining a graph invariant that is now known as the *Colin de Verdière number*, cf. [71, § 13.9]. Actually, due to his interest in their spectrum he assumed additionally the matrices to be symmetric. The connection between Google PageRank and advection matrices has been stressed already in [73, § 7.2.1.]. Random walks on suitable configurations, like that associated with the rate matrix, are ubiquitous in applied sciences: Many systems where the actual strength of interaction between configurations is unknown are for simplicity modeled by complete graphs, see e.g. [19, 28] for comparable (but much more complicated) settings in statistical physics.

Following a pattern that is similar to the one that leads to the introduction of the general Schrödinger matrix $\mathcal{L}_\sigma$, one can also consider general Dirac matrices by adding general coefficients in (2.29). The non-trivial spectral properties of such matrices have been studied in [66], cf. also [40].

A very complete overview of matrices related to random walks and further stochastic processes on graphs can be found in [20]. Also a huge amount of interesting applications of algebraic graph-theoretic methods for the analysis of examples from rather different fields of applied sciences is presented therein.

Section 2.2. Several quantum chemists began in the early 1950s to investigate our Δ, the second derivative operator on metric graphs, and were in particular able to conceive the correct node conditions—the same we have referred to as "standard". Among them we mention Ruedenberg and Scherr in [121], Griffith in [75], and Coulson in [52, § 4]. Especially Coulson suggested to use the Schrödinger equation associated with Δ to study the σ orbitals of a graphite monolayer—i.e., of what is nowadays called *graphene*—to complement the description of the π orbitals based on Hückel's theory. Montroll further developed these early investigations and studied in [103] the Hamiltonian $\Delta + M_V$ on a metric graph, where the potential term M_V is the multiplication by $V(x) := -\text{sech}^2\gamma x$. He went on to show that, depending on the values of γ, this Hamiltonian interpolates between Δ and suitable Hückel matrices $\mathcal{H}_{\alpha,\beta}$.

Lumer has been probably the first who performed a pure mathematical analysis of second order differential operators on metric graphs. In [98] he introduced the necessary functional setting, proposed a more general class of node conditions, and characterized reality of the associated spectra. In the 1980s several researchers extended Lumer's results considering more and more general node conditions, providing interesting descriptions of the spectrum, discussing nonlinear and/or higher dimensional operators, and elaborating on the connections with quantum mechanics, among others in [3, 65, 111, 120, 136] and subsequent papers. The latest wave of interest is connected to the theory of so-called *quantum graphs*, which we will briefly sketch in Sect. 7.3.

Lumer's general node conditions have been thoroughly studied in [89]. They can be compactly formulated in several equivalent ways: We mention the alternative descriptions given in [24,68,115,137]. The factorization (2.47) motivates in particular the study of $\overleftarrow{A}_C \overleftarrow{A}_C^*$, which is sometimes called the *Laplacian with δ'-coupling*, or *δ'_s-coupling* if lower order node terms are also allowed: See e.g. the historical discussion in [2, § 4]. Because of this factorization the spectra of $-\Delta$ and $\overleftarrow{A}_C \overleftarrow{A}_C^* = \overleftarrow{A}_C \overleftarrow{A}_K$ have to agree, possibly up to $\{0\}$.

Higher dimensional generalizations of Lumer's setting have been considered, among others, by Ali Mehmeti, von Below, Freidlin, Nicaise, and Wentzell, cf. [4,17,35,67,70,112,140].

To the best of our knowledge, the first derivative operator on metric graphs has been first studied in [36]. As this was interpreted as a *momentum operator*, hence as an observable of quantum mechanics, the attention was devoted to describing those node conditions that yield self-adjoint realizations. The peculiar nature of what we have called *the standard node conditions*, among all possible ones, was observed in [90]. The factorization in Remark 2.35 has been observed and exploited in [58, Chapter 3] in order to describe the spectrum of the operator A that has been introduced in Sect. 2.2.2, much simplifying the approach in [90]. More generally, over the last 10 years Nagel's school in Tübingen has been very active in the analysis of the first derivative with standard node conditions and various generalizations of it, in different directions and contexts. This group's results are conveniently surveyed in [59].

The attention for the self-adjoint case has been so far more limited. Momentum operators have been studied in [36] and in a few subsequent papers, including [63, 81, 125], where attention has been mostly devoted to self-adjoint realizations: However, it is not clear which of these infinitely many realizations has a "natural" physical interpretation—a difficulty of interpretation that is also found in connection with the Dirac operator. First derivatives on metric graphs have also been studied as part of mixed advection/diffusion phenomena in [79].

A standard introduction to the classical theory of Dirac equations in $\mathbb{R}^{3+1}$ is the monograph [131]. Dirac-type operators on metric graphs have been introduced in [32]. Apart from certain more or less heuristic discretization arguments, (2.48) is the defining property of all various Dirac-type operators and justifies the name of the matrix $\mathcal{D}_s$ in Sect. 2.1.8. Another possible way of introducing the Dirac operator is closer to the original three-dimensional definition and requires to work in spaces of $\mathbb{C}^4$-valued functions. However, the similarities between these two operators, at least under certain boundary conditions, suggest to opt for the lower dimensional approach, cf. [23, §§ 5–6].

Further operators—often with a natural interpretation in terms of stochastic processes on graphs—can be defined, but they are usually of non-differential nature, see e.g. the "averaging operator" in [38], which in [96] is shown to be related to the sine operator generated by the second derivative with standard node conditions, cf. Sect. 4.3.

Section 2.3. In higher dimensions the Dirichlet-to-Neumann operator is not a differential operator, but rather a *pseudo-differential* operator (of order 1). Luckily, in the case of metric graphs $\mathbb{DN}$ is of course simply a matrix. This significantly simplifies its study.

If one thinks of G as an electric network whose conductivities are given by γ, applying $\mathbb{DN}$ to some $f \in \mathbb{C}^{\mathsf{V}_0}$ amounts to applying electric potential $f(\mathsf{v})$ at each $\mathsf{v} \in \mathsf{V}_0$, letting the system reach equilibrium and finally reading off the current that flows through the same nodes. For this reason, the Dirichlet-to-Neumann operator is sometimes known as *voltage-to-current operator*. In different models, the trace and the conormal derivative of a function have different meanings, and so does the Dirichlet-to-Neumann operator: In elasticity, for example, the normal derivative of the displacement is interpreted as the strain at a certain point.

Yet another derivation of the Dirichlet-to-Neumann operator can be performed via Poiseuille's law (a fluid-dynamic analog of Ohm's law) for pressure in networks of pipes—say, in the lungs, cf. [74] or [99, Chapter 3]. In this case the potential at a node v is replaced by the pressure at that point and the current is replaced by the fluid flux (which, by Poiseuille's law, is proportional to the difference of pressure between adjacent nodes). The Dirichlet-to-Neumann operator can then be interpreted as an operator whose input is the vector of flux directed towards the nodes that belong to a certain node subset (in [74], the leaves of a tree). Just like the Laplace–Beltrami matrix, also the Dirichlet-to-Neumann is in general not invertible if G is finite, cf. Lemma 2.13; but upon factoring out its null space it can be inverted to yield the fluid flux at the endpoints of a tree on whose leaves a certain pressure was applied.

Remarkably, infinite graphs may develop topological features that strongly recall those of boundaries of manifolds in classical Riemannian geometry—the key notion in this context is that of *ends*, cf. [57, Chapter 8]. It turns out that difference (rather than differential) elliptic *end* value problems can also be studied, and therefore Dirichlet-to-Neumann-type operators can be defined in this setting, too, cf. [37, 51, 100].

The eigenvalue problem for the operator matrix $\mathbb{A}$ has a long history in mathematical physics: In this context one usually speaks of *energy* (or *eigenvalue*) *dependent boundary conditions*. While it is easy to derive evolution equations where the solution has to satisfy certain dynamic boundary conditions, it is less clear how to interpret the action of the operator, or perhaps of the associated Lagrangian. A physical derivation of the counterpart of this operator defined on domains (instead of metric graphs) has been attempted in [72]. An interesting interplay with the properties of a class of scattering problems has been pointed out in [94].

This kind of boundary conditions also appears in a different context—viz, approximation theory, as discussed in Remark 2.38. It has been observed both in two and three dimensional approximation schemes that the limiting system varies in dependence on the ratio between the area/volume of the neighborhoods (i.e., strips/cylinders) of *edges* and that of the neighborhoods (i.e., disks/balls) of *nodes*. In a certain critical case the sesquilinear form $\mathfrak{a}_{\mathcal{W}}$ considered in Lemma 6.92 turns out to be the natural limiting energy functional, cf. [92, § 3.3] and [64, § 7]. In the case of a metric graph, the same arguments as in [64, 92] also suggest that the operator $\mathbb{A}$ may be realized as a degenerate limit of second derivatives A_ϵ with standard node conditions on an auxiliary family of metric graphs $\mathfrak{G}_\epsilon$ defined replacing each node v of G with m incident edges by m new nodes $\mathsf{v}^{(1)}, \ldots, \mathsf{v}^{(n)}$, each adjacent to exactly one of the former neighbors of v as well as to all nodes $\mathsf{v}^{(1)}, \ldots, \mathsf{v}^{(n)}$; assigning to each of these new $m(m-1)/2$ edges the same weight $\mu_\mathsf{v} = \mu_\mathsf{v}(\epsilon^{-1})$; and letting ϵ tend to 0. This construction has been introduced in [42] and thoroughly deployed later on in [41] also in order to approximate a certain class of differential operators through Laplacians with standard node conditions.

Conversely, it has been shown in [8] that a special instance of $\mathfrak{a}_\mathcal{W}$ can be used to approximate the sesquilinear form associated with a certain Laplacian on the Sierpiński gasket, cf. also [13, 14]. In the construction in [8] $\mathcal{W}$ is taken to be a suitable discrete Laplacian defined on a certain network associated with the so-called Hanoi graph. This corresponds to allowing for a jump process on the boundary of the considered domain.

While the case of an extension of the setting in Sect. 2.2.1 appears the most interesting in view of applications, one can possibly consider dynamic node conditions for every differential operator of Sect. 2.2. Indeed, this has been done e.g. in [126] as well as in several papers including [39, 101] in the case of the first and the fourth derivative, respectively.

References

1. R. Adami, C. Cacciapuoti, D. Finco, D. Noja, Fast solitons on star graphs. Rev. Math. Phys. **23**, 409–451 (2011)
2. S. Albeverio, L. Dabrowski, P. Kurasov, Symmetries of schrödinger operator with point interactions. Lett. Math. Phys. **45**, 33–47 (1998)
3. F. Ali Mehmeti, A characterization of a generalized c^∞-notion on nets. Int. Equations Oper. Theory **9**, 753–766 (1986)
4. F. Ali Mehmeti, Regular solutions of transmission and interaction problems for wave equations. Math. Meth. Appl. Sci. **11**, 665–685 (1989)
5. F. Ali Mehmeti, *Nonlinear Waves in Networks*. Mathematical Research, vol. 80 (Akademie, Berlin, 1994)
6. F. Ali Mehmeti, S. Nicaise, Nonlinear interaction problems. Nonlinear Anal. Theory Methods Appl. **20**, 27–61 (1993)
7. F. Ali Mehmeti, V. Régnier, Splitting of energy of dispersive waves in a star-shaped network. ZAMM Z. Angew. Math. Mech. **83**, 105–118 (2003)
8. P. Alonso-Ruiz, U. Freiberg, Weyl asymptotics for Hanoi attractors. arXiv:1307.6719 (2013)
9. W.N. Anderson Jr., T.D. Morley, Eigenvalues of the Laplacian of a graph. Technical report TR-71-45, University of Maryland, 1971
10. W. Arendt, K. Urban, *Partielle Differenzialgleichungen* (Spektrum, Heidelberg, 2010)
11. M.S. Ashbaugh, F. Gesztesy, M. Mitrea, R. Shterenberg, G. Teschl, A survey on the Krein–von Neumann extension, the corresponding abstract buckling problem, and Weyl-type spectral asymptotics for perturbed Krein Laplacians in nonsmooth domains, in *Mathematical Physics, Spectral Theory and Stochastic Analysis*, ed. by M. Demuth, W. Kirsch (Springer, Berlin, 2013), pp. 1–106
12. V. Banica, L.I. Ignat, Dispersion for the Schrödinger equation on networks. J. Math. Phys. **52**, 083703 (2011)
13. M.T. Barlow, R.F. Bass, On the resistance of the Sierpinski carpet. Proc. Royal Soc. London A **431**, 345–360 (1990)

14. M.T. Barlow, E.A. Perkins, Brownian motion on the Sierpinski gasket. Prob. Theory Related Fields **79**, 543–623 (1988)
15. D. Bauso, L. Giarré, R. Pesenti, Non-linear protocols for optimal distributed consensus in networks of dynamic agents. Syst. Contr. Lett. **55**, 918–928 (2006)
16. M. Belkin, P. Niyogi, Laplacian eigenmaps for dimensionality reduction and data representation. Neural Comput. **15**, 1373–1396 (2003)
17. A. Bendikov, L. Saloff-Coste, M. Salvatori, W. Woess, The heat semigroup and brownian motion on strip complexes. Adv. Math. **226**, 992–1055 (2011)
18. N.L. Biggs, *Algebraic Graph Theory*. Cambridge Tracts in Mathematics, vol. 67 (Cambridge University Press, Cambridge, 1974)
19. N.L. Biggs, *Interaction Models*. London Mathematical Society Lecture Note Series, vol. 30 (Cambridge University Press, Cambridge, 1977)
20. P. Blanchard, D. Volchenkov, *Random Walks and Diffusions on Graphs and Databases: An Introduction*. Springer Series in Synergetics, vol. 10 (Springer, Berlin, 2011)
21. D. Boal, *Mechanics of the Cell* (Cambridge University Press, Cambridge, 2012)
22. B. Bollobás, *Modern Graph Theory* (Springer-Verlag, Berlin, 1998)
23. J. Bolte, J. Harrison, Spectral statistics for the Dirac operator on graphs. J. Phys. A **36**, 2747–2769 (2003)
24. J. Boman, P. Kurasov, Symmetries of quantum graphs and the inverse scattering problems. Adv. Appl. Maths. **35**, 58–70 (2005)
25. J.L. Bona, R.C. Cascaval, Nonlinear dispersive waves on trees. Canad. Appl. Math. Quart. **16**, 1–18 (2008)
26. J.A. Bondy, U.S.R. Murty, *Graph Theory*. Graduate Texts in Mathematics, vol. 244 (Springer, Berlin, 2008)
27. G. Boole, *A Treatise on the Calculus of Finite Differences* (Macmillan, Cambridge, 1860)
28. J.-P. Bouchaud, Weak ergodicity breaking and aging in disordered systems. J de Phys. I **2**, 1705–1713 (1992)
29. S. Brin, L. Page, The anatomy of a large-scale hypertextual web search engine. Comp Netw ISDN Syst. **30**, 107–117 (1998)
30. R.L. Brooks, C.A.B. Smith, A.H. Stone, W.T. Tutte, The dissection of rectangles into squares. Duke Math. J. **7**, 312–340 (1940)
31. A.E. Brouwer, W.H. Haemers, *Spectra of Graphs. Universitext* (Springer, Berlin, 2012)
32. W. Bulla, T. Trenkle, The free Dirac operator on compact and noncompact graphs. J. Math. Phys. **31**, 1157–1163 (1990)
33. V. Bush, Structural analysis by electric circuit analogies. J. Franklin Inst. **217**, 289–329 (1934)
34. S. Cardanobile, D. Mugnolo, Analysis of a FitzHugh–Nagumo–Rall model of a neuronal network. Math. Meth. Appl. Sci. **30** 2281–2308 (2007)
35. S. Cardanobile, D. Mugnolo, Parabolic systems with coupled boundary conditions. J. Differ. Equ. **247**, 1229–1248 (2009)
36. R. Carlson, Inverse eigenvalue problems on directed graphs. Trans. Amer. Math. Soc. **351**, 4069–4088 (1999)
37. R. Carlson, Dirichlet to Neumann maps for infinite quantum graphs. Netw. Het. Media **7**, 483–501 (2012)
38. D. Cartwright, W. Woess, The spectrum of the averaging operator on a network (metric graph). Illinois J. Math. **51**, 805–830 (2007)
39. C. Castro, E. Zuazua, Boundary controllability of a hybrid system consisting in two flexible beams connected by a point mass. SIAM J. Control Opt. **36**, 1576–1595 (1998)
40. S.N. Chandler-Wilde, E.B. Davies, Spectrum of a Feinberg–Zee random hopping matrix. arXiv:1110.0792 (2011)
41. T. Cheon, P. Exner, O. Turek, Approximation of a general singular vertex coupling in quantum graphs. Ann. Phys. **325**, 548–578 (2010)
42. T. Cheon, T. Shigehara, Realizing discontinuous wave functions with renormalized short-range potentials. Phys. Lett. A **243**, 111–116 (1998)

43. F.R.K. Chung, *Spectral Graph Theory*. Regional Conference Series in Mathematics, vol. 92 (American Mathematical Society, Providence, 1997)
44. F. Chung, The heat kernel as the pagerank of a graph. Proc. Nat. Acad. Sci. **104**, 19735–19740 (2007)
45. G.M. Coclite, M. Garavello, B. Piccoli, Traffic flow on a road network. SIAM J. Math. Anal. **36**, 1862–1886 (2005)
46. R.R. Coifman, S. Lafon, Diffusion maps. Appl. Comput. Harmon. Anal. **21**, 5–30 (2006)
47. Y. Colin de Verdière, Sur la mulitplicité de la première valeur propre non nulle du laplacien. Comment. Math. Helv. **61**, 254–270 (1986)
48. Y. Colin de Verdière, Sur un nouvel invariant des graphes et un critère de planarité. J. Comb. Theory. Ser. B **50**, 11–21 (1990)
49. Y. Colin de Verdière, N. Torki-Hamza, F. Truc, Essential self-adjointness for combinatorial Schrödinger operators II-Metrically non complete graphs. Math. Phys. Anal. Geom. **14**, 21–38 (2011)
50. Y. Colin de Verdière, N. Torki-Hamza, F. Truc, Essential self-adjointness for combinatorial Schrödinger operators. III: magnetic fields. Ann. Fac. Sci. Toulouse **20**, 599–611 (2011)
51. Y. Colin de Verdière, F. Truc, Scattering theory for graphs isomorphic to a homogeneous tree at infinity. J. Math. Phys. **54**, 063502 (2013)
52. C.A. Coulson, Note on the applicability of the free-electron network model to metals. Proc. Phys. Soc. Sect. A **67**, 608 (1954)
53. D.M. Cvetković, M. Doob, H. Sachs, *Spectra of Graphs: Theory and Applications*. Pure and Applied Mathematics (Academic, New York, 1979)
54. D. Cvetković, P. Rowlinson, S.K. Simić, Signless Laplacians of finite graphs. Lin. Algebra Appl. **423**, 155–171 (2007)
55. B. Dekoninck, S. Nicaise, The eigenvalue problem for networks of beams. Lin. Alg. Appl. **314**, 165–189 (2000)
56. M. Desai, V. Rao, A characterization of the smallest eigenvalue of a graph. J. Graph Theory **18**, 181–194 (1994)
57. R. Diestel, *Graph Theory*. Graduate Texts in Mathematics, vol. 173 (Springer, Berlin, 2005)
58. B. Dorn, Semigroups for flows on infinite networks, Master's thesis, Eberhard-Karls-Universität, Tübingen, 2005
59. B. Dorn, M. Kramar Fijavž, R. Nagel, A. Radl, The semigroup approach to transport processes in networks. Physica D **239**, 1416–1421 (2010)
60. P.G. Doyle, J.L. Snell, *Random Walks and Electric Networks*. Carus Mathematical Monographs, vol. 22 (Mathematical Association of America, Washington, DC, 1984)
61. J. Escher, The Dirichlet-Neumann operator on continuous functions. Ann. Sc. Norm. Super. Pisa Cl. Sci. **21**, 235–266 (1994)
62. L.C. Evans, *Partial Differential Equations*, 2nd edn. Graduate Studies in Mathematics, vol. 19 (American Mathematical Society, Providence, 2010)
63. P. Exner, Momentum operators on graphs, in *Spectral Analysis, Differential Equations and Mathematical Physics: A Festschrift in Honor of Fritz Gesztesy's 60th Birthday*, ed. by H. Holden, B. Simon, G. Teschl. Proceedings of Symposia in Pure Mathematics, vol. 87 (American Mathematical Society, Providence, 2013), pp. 105–118
64. P. Exner, O. Post, Convergence of spectra of graph-like thin manifolds. J. Geom. Phys. **54**, 77–115 (2005)
65. P. Exner, P. Šeba. Free quantum motion on a branching graph. Rep. Math. Phys. **28**, 7–26 (1989)
66. J. Feinberg, A. Zee, Non-Hermitian localization and delocalization. Phys. Rev. E **59**, 6433 (1999)
67. M.I. Freidlin, A.D. Wentzell, Diffusion processes on an open book and the averaging principle. Stochastic Processes Appl. **113**, 101–126 (2004)
68. S.A. Fulling, P. Kuchment, J.H. Wilson, Index theorems for quantum graphs. J. Phys. A **40**, 14165–14180 (2007)

69. M. Garavello, B. Piccoli, *Traffic Flow on Networks*. Applied Mathematics, vol. 1 (American Institute of Mathematical Sciences, Providence, 2006)
70. F. Gesztesy, M. Mitrea, R. Nichols, Heat kernel bounds for elliptic partial differential operators in divergence form with Robin-type boundary conditions. arXiv:1210.0667 (2012)
71. C. Godsil, G. Royle, *Algebraic Graph Theory*. Graduate Texts in Mathematics, vol. 207 (Springer, Berlin, 2001)
72. G.R. Goldstein, Derivation and physical interpretation of general boundary conditions. Adv. Diff. Equ. **11**, 457–480 (2006)
73. L.J. Grady, J.R. Polimeni, *Discrete Calculus: Applied Analysis on Graphs for Computational Science* (Springer, New York, 2010)
74. C. Grandmont, B. Maury, N. Meunier, A viscoelastic model with non-local damping application to the human lungs. ESAIM: Math. Modelling Numerical Anal. **40**, 201–224 (2006)
75. J.S. Griffith, A free-electron theory of conjugated molecules. part 1: polycyclic hydrocarbons. Trans. Faraday Soc. **49**, 345–351 (1953)
76. J.S. Ham, Viscoelastic theory of branched and cross-linked polymers. J. Chem. Phys. **26**, 625–633 (1957)
77. M. Hein, J.Y. Audibert, U. von Luxburg, Graph Laplacians and their convergence on random neighborhood graphs. J. Mach. Learning Res. **8**, 1325–1368 (2007)
78. E. Hückel, Quantentheoretische Beiträge zum Benzolproblem. Z. Physik A **70**, 204–286 (1931)
79. A. Hussein, D. Mugnolo, Quantum graphs with mixed dynamics: the transport/diffusion case. J. Phys. A **46**, 235202 (2013)
80. A. Isaev, *Introduction to Mathematical Methods in Bioinformatics. Universitext* (Springer, Berlin, 2006)
81. P. Jorgensen, S. Pedersen, F. Tiang, Momentum operators in two intervals: Spectra and phase transition. Compl. Anal. Oper. Theory **7**, 1735–1773 (2013)
82. J. Jost, M.P. Joy, Spectral properties and synchronization in coupled map lattices. Phys. Rev. E **65**, 016201 (2001)
83. T.H. Jukes, C.R. Cantor, Evolution of protein molecules, in *Mammalian Protein Metabolism*, vol. III, ed. by M.N. Munro (Academic, New York, 1969), pp. 21–132
84. K. Kaneko, Period-doubling of kink-antikink patterns, quasiperiodicity in antiferro-like structures and spatial intermittency in coupled logistic lattice. Progr. Theor. Phys. **72**, 480–486 (1984)
85. M. Kimura, Evolutionary rate at the molecular level. Nature **217**, 624–626 (1968)
86. G. Kirchhoff, Ueber die Auflösung der Gleichungen, auf welche man bei der Untersuchung der linearen Vertheilung galvanischer Ströme geführt wird. Ann. Physik **148**, 497–508 (1847)
87. S. Kirkland, D. Paul, Bipartite subgraphs and the signless Laplacian matrix. Applicable Anal Discrete Math **5**, 1–13 (2011)
88. W. Kirsch, P. Müller, Spectral properties of the Laplacian on bond-percolation graphs. Math. Z. **252**, 899–916 (2006)
89. V. Kostrykin, R. Schrader, Kirchhoff's rule for quantum wires. J. Phys. A **32**, 595–630 (1999)
90. M. Kramar, E. Sikolya, Spectral properties and asymptotic periodicity of flows in networks. Math. Z. **249**, 139–162 (2005)
91. G. Kron, Tensorial analysis and equivalent circuits of elastic structures. J. Franklin Inst. **238**, 399–442 (1944)
92. P. Kuchment, H. Zeng, Asymptotics of spectra of Neumann Laplacians in thin domains, in *Advances in Differential Equations and Mathematical Physics*, ed. by Y. Karpeshina, G. Stoltz, R. Weikard, Y. Zeng. Proceedings of the 9th UAB International Conference, University of Birmingham, 2002. Contemporary Mathematics, vol. 327 (American Mathematical Society, Providence, 2003), pp. 199–213
93. Y. Kuramoto, Self-entrainment of a population of coupled non-linear oscillators, in *International Symposium on Mathematical Problems in Theoretical Physics* (Springer, New York, 1975), pp. 420–422

94. P. Kurasov, Energy dependent boundary conditions and the few-body scattering problem. Rev. Math. Phys. **9**, 853–906 (1997)
95. A.N. Langville, C.D. Meyer, *Google's PageRank and Beyond: The Science of Search Engine Rankings* (Princeton University Press, Princeton, 2006)
96. D. Lenz, K. Pankrashkin, New relations between discrete and continuous transition operators on (metric) graphs. arXiv:1305.7491 (2013)
97. C.-K. Li, Lecture notes on numerical range (2005), people.wm.edu/~cklixx/nrnote.pdf
98. G. Lumer, Connecting of local operators and evolution equations on networks, in *Potential Theory (Proc. Copenhagen 1979)*, ed. by F. Hirsch (Springer, Berlin, 1980), pp. 230–243
99. B. Maury, *The Respiratory System in Equations*. Modeling, Simulations and Applications, vol. 7 (Springer, Milan, 2013)
100. B. Maury, D. Salort, C. Vannier, Trace theorems for trees and application to the human lungs. Netw. Het. Media **4**, 469–500 (2009)
101. D. Mercier, V. Régnier, Spectrum of a network of Euler–Bernoulli beams. J. Math. Anal. Appl. **337**, 174–196 (2008)
102. G.J. Minty, Monotone networks. Proc. R. Soc. Lond. Ser. A Math. Phys. Eng. Sci. **257**, 194–212 (1960)
103. E. Montroll, Quantum theory on a network I. J. Math. Phys. **11**, 635–648 (1970)
104. D. Mugnolo, Gaussian estimates for a heat equation on a network. Netw. Het. Media **2**, 55–79 (2007)
105. D. Mugnolo, Parabolic theory of the discrete p-Laplace operator. Nonlinear Anal. Theory Methods Appl. **87**, 33–60 (2013)
106. D. Mugnolo, J.-F. Rault, Construction of exact travelling waves for the Benjamin–Bona–Mahony equation on networks. Bull. Belg. Math. Soc. - Simon Stevin **21** (2014)
107. D. Mugnolo, R. Nittka, O. Post, Convergence of sectorial operators on varying Hilbert spaces. Oper. Matrices **7**, 955–995 (2013)
108. P. Müller, P. Stollmann, Spectral asymptotics of the Laplacian on supercritical bond-percolation graphs. J. Funct. Anal. **252**, 233–246 (2007)
109. K. Nagel, Particle hopping models and traffic flow theory. Phys. Rev. E, **53**, 4655–4672 (1996)
110. T. Nakamura, M. Yamasaki, Generalized extremal length of an infinite network. Hiroshima Math. J. **6**, 95–111 (1976)
111. S. Nicaise, Some results on spectral theory over networks, applied to nerve impulse transmission, in *Polynômes Orthogonaux et Applications*, ed. by C. Brezinski, A. Draux, A. P. Magnus, P. Maroni, A. Ronveaux. Proceedings, Bar-le-Duc 1984. Lecture Notes in Mathematics, vol. 1171 (Springer, Berlin, 1985), pp. 532–541
112. S. Nicaise, *Polygonal Interface Problems*. Methoden und Verfahren der Mathematischen Physik, vol. 39 (Peter Lang, Frankfurt/Main, 1993)
113. Olfati-Saber, R.M. Murray, Consensus protocols for networks of dynamic agents, in *Proceedings of the American Control Conference*, vol. 2, Denver, 4–6 June 2003, pp. 951–956
114. L. Page, S. Brin, R. Motwani, and T. Winograd. The PageRank citation ranking: bringing order to the web. Technical report 1999–66, Stanford InfoLab, November 1999
115. K. Pankrashkin, Reducible boundary conditions in coupled channels. J. Phys. A **38**, 8979–8992 (2005)
116. Y. Peres, Probability on trees: an introductory climb, in *Lectures on Probability Theory and Statistics*, Lecture Notes in Mathematics, vol. 1717 (Springer, Berlin, 1999), pp. 193–280
117. Y.V. Pokornyi, A.V. Borovskikh, Differential equations on networks (geometric graphs). J. Math. Sci. **119**, 691–718 (2004)
118. O. Post, *Spectral Analysis on Graph-Like Spaces*. Lecture Notes in Mathematics, vol. 2039 (Springer, Berlin, 2012)
119. F. Rezakhanlou, Hydrodynamic limit for attractive particle systems on $\mathbb{Z}^d$. Commun. Math. Phys. **140**, 417–448 (1991)
120. J.-P. Roth, Spectre du laplacien sur un graphe. C. R. Acad. Sci. Paris Sér. I Math. **296**, 793–795 (1983)

121. K. Ruedenberg, C.W. Scherr, Free-electron network model for conjugated systems. I. Theory. J. Chem. Phys. **21**, 1565–1581 (1953)
122. K.K. Sabirov, Z.A. Sobirov, D. Babajanov, D.U. Matrasulov, Stationary nonlinear Schrödinger equation on simplest graphs. Phys. Lett. A **377**, 860–865 (2013)
123. Y. Saito, The limiting equation for Neumann Laplacians on shrinking domains. Electronic J. Differ. Equ. **31**, 1–25 (2000)
124. K. Schmüdgen, *Unbounded Self-adjoint Operators on Hilbert Space*. Graduate Texts in Mathematics, vol. 265 (Springer, Berlin, 2012)
125. C. Schubert, C. Seifert, J. Voigt, M. Waurick, Boundary systems and (skew-) self-adjoint operators on infinite metric graphs. arXiv:1308.2635 (2013)
126. E. Sikolya, Flows in networks with dynamic ramification nodes. J. Evol. Equ. **5**, 441–463 (2005)
127. B. Simon, Operators with singular continuous spectrum, VI. Graph Laplacians and Laplace–Beltrami operators. Proc. Am. Math. Soc. **124**, 1177–1182 (1996)
128. P. Soardi, *Potential Theory on Infinite Networks*. Lecture Notes in Mathematics, vol. 1590 (Springer, Berlin, 1994)
129. D.A. Spielman, Algorithms, graph theory, and linear equations in Laplacian matrices, in *Proceedings of the International Congress of Mathematicians*, vol. 4, Hyderabad, 2010, pp. 2698–2722
130. J.J. Sylvester, Chemistry and algebra. Nature, **17**, 284 (1878)
131. B. Thaller, *The Dirac Equation* (Springer, New York, 1992)
132. N. Torki-Hamza, Essential self-adjointness for combinatorial Schrödinger operators I: metrically complete graphs. Confluentes Math. **2**, 333–350 (2010)
133. W.T. Tutte, The dissection of equilateral triangles into equilateral triangles. Math. Proc. Cambridge Phil. Soc. **44**, 463–482 (1948)
134. W.T. Tutte, *Graph Theory*. Encyclopedia of Mathematics and Its Applications, vol. 21 (Addison-Wesley, Reading, 1984)
135. W.T. Tutte, *Graph Theory As I Have Known It* (Clarendon Press, Oxford, 1998)
136. J. von Below, A characteristic equation associated with an eigenvalue problem on c^2-networks. Lin. Algebra Appl. **71**, 309–325 (1985)
137. J. von Below, Sturm–Liouville eigenvalue problems on networks. Math. Methods Appl. Sci. **10**, 383–395 (1988)
138. J. von Below, A maximum principle for semilinear parabolic network equations, in *Differential Equations with Applications in Biology, Physics, and Engineering (Proc. Leibnitz)*, ed. by F. Kappel J.A. Goldstein, W. Schappacher. Lecture Notes in Pure and Applied Mathematics, vol. 133 (Marcel Dékker, New York, 1991), pp. 37–45
139. J. von Below, *Parabolic Network Equations* (Tübinger Universitätsverlag, Tübingen, 1994)
140. J. von Below, S. Nicaise, Dynamical interface transition in ramified media with diffusion. Comm. Partial Differ. Equations **21**, 255–279 (1996)
141. E. Yanagida, Stability of nonconstant steady states in reaction-diffusion systems on graphs. Japan J. Indust. Appl. Math. **18**, 25–42 (2001)
142. Z. Yang, Estimating the pattern of nucleotide substitution. J. Mol. Evol. **39**, 105–111 (1994)

Chapter 3
Function Spaces on Networks

The first step in the study of evolution equations is the choice of suitable functions spaces that capture the structure of the problem. In this chapter we introduce some relevant spaces of functions on networks. We assume the theory of Lebesgue spaces to be known to the reader, whilst the fundamental aspects of the theory of Sobolev spaces are summarized in the Appendix B. Throughout this chapter

$$\boxed{\mathsf{G} = (\mathsf{V}, \mathsf{E}, \rho) \quad \text{is a weighted oriented graph.}}$$

In particular, all results of this section will be applicable to both the resistance network $(\mathsf{V}, \mathsf{E}, \mu)$ and the conductance network $(\mathsf{V}, \mathsf{E}, \gamma)$, cf. Remark 2.1.4.1.

3.1 The Discrete Setting

We begin by discussing some analytical properties of functions defined over a discrete graph. The basic framework for our investigations comprehends two classes of summable functions. The first arise in connection with the weights ρ associated to all edges.

Definition 3.1. For $p \in [1, \infty]$ we denote by $\ell^p_\rho(\mathsf{E})$ the space of all functions $u : \mathsf{E} \to \mathbb{C}$ such that

$$\|u\|_{\ell^p_\rho(\mathsf{E})} := \left(\sum_{\mathsf{e} \in \mathsf{E}} |u(\mathsf{e})|^p \rho(\mathsf{e}) \right)^{\frac{1}{p}} < \infty \quad \text{for } p \in [1, \infty),$$

or else

$$\|u\|_{\ell^\infty_\rho(\mathsf{E})} := \sup_{\mathsf{e} \in \mathsf{E}} |u(\mathsf{e})| \rho(\mathsf{e}) < \infty.$$

D. Mugnolo, *Semigroup Methods for Evolution Equations on Networks*,
Understanding Complex Systems, DOI 10.1007/978-3-319-04621-1_3,

We will occasionally need to assign weights not only to edges, but also to nodes. This leads to the following.

Definition 3.2. Let $\nu : \mathsf{V} \to (0,\infty)$. For $p \in [1,\infty]$ we denote by $\ell^p_\nu(\mathsf{V})$ the space of all functions $f : \mathsf{V} \to \mathbb{C}$ such that

$$\|f\|_{\ell^p_\nu(\mathsf{V})} := \left(\sum_{\mathsf{v}\in\mathsf{V}} |f(\mathsf{v})|^p \nu(\mathsf{v})\right)^{\frac{1}{p}} < \infty \quad \text{for } p \in [1,\infty),$$

or else

$$\|f\|_{\ell^\infty_\nu(\mathsf{V})} := \sup_{\mathsf{v}\in\mathsf{V}} |f(\mathsf{v})|\nu(\mathsf{v}) < \infty.$$

We simply write $\ell^p(\mathsf{V})$ if $\nu \equiv 1$.

Remark 3.3. If G is unweighted, or equivalently if ρ is the function of constant value 1, we write $\ell^p(\mathsf{E})$ for $\ell^p_\rho(\mathsf{E})$.

Furthermore, $\ell^p(\mathsf{E}) = \ell^p_\rho(\mathsf{E})$ with equivalence of norms if the counting measure weighted by ρ is equivalent to the usual counting measure, i.e.,

$$m_0 \le \rho(\mathsf{e}) \le m_1 \qquad \text{for some } m_0, m_1 > 0 \text{ and all } \mathsf{e} \in \mathsf{E}. \tag{3.1}$$

Likewise, $\ell^p(\mathsf{V}) \equiv \ell^p_1(\mathsf{V}) \simeq \ell^p_{\deg_\rho}(\mathsf{V})$ with equivalence of norms if

$$n_0 \le \deg_\rho(\mathsf{v}) \le n_1 \qquad \text{for some } n_0, n_1 > 0 \text{ and all } \mathsf{v} \in \mathsf{V}. \tag{3.2}$$

Likewise, $\ell^p(\mathsf{V}) = \ell^p_{\deg^{\text{in}}_\rho}(\mathsf{V})$ (resp., $\ell^p(\mathsf{V}) = \ell^p_{\deg^{\text{out}}_\rho}(\mathsf{V})$) with equivalence of norms if

$$n_0 \le \deg^{\text{in}}_\rho(\mathsf{v}) \le n_1 \qquad \text{for some } n_0, n_1 > 0 \text{ and all } \mathsf{v} \in \mathsf{V}, \tag{3.3}$$

$$(\text{resp., } n_0 \le \deg^{\text{out}}_\rho(\mathsf{v}) \le n_1 \qquad \text{for some } n_0, n_1 > 0 \text{ and all } \mathsf{v} \in \mathsf{V}). \tag{3.4}$$

Especially, in the unweighted case (3.2) (resp., (3.3), (3.4)) is satisfied if and only if $\deg_\rho \in \ell^\infty(\mathsf{V})$ (resp., $\deg^{\text{in}}_\rho \in \ell^\infty(\mathsf{V})$, $\deg^{\text{out}}_\rho \in \ell^\infty(\mathsf{V})$), and in particular a necessary condition is that $\rho \in \ell^\infty(\mathsf{E})$.

As the incidence matrix $\mathcal{I}$ is the discrete counterpart of the first derivative, it is tempting to introduce discrete versions of the Sobolev spaces discussed in the Appendix B.

Definition 3.4. Let $\nu \in \mathbb{C}^{\mathsf{V}}$. For $p \in [1,\infty]$ we define the *discrete Sobolev space of order 1* by

$$w^{1,p}_{\rho,\nu}(\mathsf{V}) := \left\{ f \in \ell^p_\nu(\mathsf{V}) : \mathcal{I}^T f \in \ell^p_\rho(\mathsf{E}) \right\},$$

or simply $w^{1,p}_\rho(\mathsf{V})$ if $\nu \equiv 1$.

Remark 3.5. Let G be connected and let $\nu \in \mathbb{C}^{\mathsf{V}}$. If $f \in \mathbb{C}^{\mathsf{V}} \setminus \{0\}$ satisfies $\mathcal{I}^T f = 0$, then f must be constant—thus, $f \notin \ell^p_\nu(\mathsf{V})$ unless $\|\mathbf{1}\|_{\ell^p_\nu} < \infty$, i.e., unless G has finite surface with respect to ν in the sense of Definition A.17 (for $p < \infty$) or unless ν is bounded (for $p = \infty$).

Discrete Sobolev spaces keep most of the nice features of usual ones. Functions with finite support are a discrete replacement of usual test functions on domains.

Lemma 3.6. *For all $p \in [1, \infty]$ $w^{1,p}_{\rho,\nu}(\mathsf{V})$ is a Banach space with respect to the norm defined by*

$$\|f\|^p_{w^{1,p}_{\rho,\nu}} := \|f\|^p_{\ell^p_\nu} + \|\mathcal{I}^T f\|^p_{\ell^p_\rho},$$

and a Hilbert space for $p = 2$. If $p \in [1, \infty)$, then $w^{1,p}_{\rho,\nu}(\mathsf{V})$ is separable in $\ell^p_\nu(\mathsf{V})$. If $p \in (1, \infty)$, then $w^{1,p}_{\rho,\nu}(\mathsf{V})$ is uniformly convex and hence reflexive.

Proof. Completeness of $w^{1,p}_{\rho,\nu}(\mathsf{V})$ holds because for all $p \in [1, \infty)$, $\ell^p_\nu(\mathsf{V})$ and $\ell^p_\rho(\mathsf{E})$ are Banach spaces. Also continuity of the embedding is clear, by definition of the norm of $w^{1,p}_{\rho,\nu}(\mathsf{V})$. Considering

$$T : w^{1,p}_{\rho,\nu}(\mathsf{V}) \ni f \mapsto (f, \mathcal{I}^T f) \in \ell^p_\nu(\mathsf{V}) \times \ell^p_\rho(\mathsf{E}),$$

which is an isometry if the Cartesian product on the right is endowed with the ℓ^1-norm, shows that $w^{1,p}_{\rho,\nu}$ is uniformly convex for all $p \in (1, \infty)$, since so is $\ell^p_\nu(\mathsf{V})$. Finally, since the space $c_{00}(\mathsf{V})$ of functions with finite support is dense in $\ell^p_\nu(\mathsf{V})$, so is $w^{1,p}_{\rho,\nu}(\mathsf{E})$. □

Definition 3.7. The *distance* $\mathrm{dist}_\rho(\mathsf{v}, \mathsf{w})$ of two nodes v, w is defined as the infimum of the lengths of all paths from v to w.

In this way G or, to be more precise, V becomes a metric space—n general not a complete one, unless ρ is uniformly bounded away from 0, i.e., unless $\rho^{-1} \in \ell^\infty(\mathsf{E})$. The ball of radius $r > 0$ and center v_0 with respect to dist_ρ will be denoted by

$$B_\rho(\mathsf{v}_0, r) := \left\{ \mathsf{w} \in \mathsf{V} : \mathrm{dist}_\rho(\mathsf{v}_0, \mathsf{w}) < r \right\}.$$

Proposition 3.8. *If G is connected, then the following assertions hold.*

(1) The space $w^{1,p}_{\rho,\nu}(\mathsf{V})$ is densely and continuously embedded in $\ell^p_\nu(\mathsf{V})$ for all $1 \le p \le \infty$.
(2) Let additionally $p < \infty$. Then this embedding is compact if for every $\epsilon > 0$ there are $\mathsf{v} \in \mathsf{V}$ and $r > 0$ such that

(i) $B_\rho(\mathsf{v}, r)$ *is a finite set and additionally*
(ii) there holds

$$\sum_{\mathsf{w} \notin B_\rho(\mathsf{v},r)} |f(\mathsf{w})|^p \nu(\mathsf{w}) < \epsilon^p$$

for all f *in the unit ball of* $w^{1,p}_{\rho,\nu}(\mathsf{V})$.

(3) Condition (i) on local finiteness of the neighborhoods is satisfied if ρ *is uniformly bounded from below away from 0, and in particular if* (3.1) *holds. For all* $\epsilon > 0$*, condition (ii) is satisfied for* $r > \mathrm{vol}_\rho(\mathsf{G})$ *if* $\mathrm{vol}_\rho(\mathsf{G})$ *is finite.*

Observe that both conditions in (3) can only be simultaneously satisfied if E and hence V are finite.

Proof. (1) For all $p \in [1, \infty]$, $w^{1,p}_{\rho,\nu}(\mathsf{V})$ contains the space $c_{00}(\mathsf{V})$ of functions with finite support. Since $c_{00}(\mathsf{V})$ is dense in $\ell^p_\nu(\mathsf{V})$, so is $w^{1,p}_{\rho,\nu}(\mathsf{V})$.

(2) If for a given $\varepsilon > 0$ we pick $\mathsf{v} \in \mathsf{V}$ and $r > 0$ in such a way that (ii) is satisfied and that $B_\rho(\mathsf{v}, r)$ only contains finitely many nodes, we can consider the finite-rank operator

$$\Phi : f \mapsto f_{|B_\rho(\mathsf{v},r)}.$$

Then, $\Phi(w^{1,p}_{\rho,\nu}(\mathsf{V}))$ is totally bounded in the metric space $\mathbb{C}^{B_\rho(\mathsf{v},r)}$ with respect to the metric induced by the ℓ^p_ν-norm. Moreover for all $f, g \in w^{1,p}_{\rho,\nu}(\mathsf{V})$ such that $\|f - g\|_{\mathbb{C}^{B_\rho(\mathsf{v},r)}} < \epsilon$ we have

$$\|f-g\|^p_{\ell^p_\nu} = \sum_{\mathsf{w} \in B_\rho(\mathsf{v},r)} |f(\mathsf{w})-g(\mathsf{w})|^p \nu(\mathsf{w}) + \sum_{\mathsf{w} \notin B_\rho(\mathsf{v},r)} |f(\mathsf{w})-g(\mathsf{w})|^p \nu(\mathsf{w}) \leq 3\epsilon.$$

The conditions of the Hanche-Olsen–Holden Lemma B.7 are thus satisfied.

□

Remark 3.9. It is known that in general $c_{00}(\mathsf{V})$ is not dense in the Banach space

$$D^p_\rho(\mathsf{V}) := \left\{ f : \mathsf{V} \to \mathbb{C} : \|\mathcal{I}^T f\|^p_{\ell^p_\rho} < \infty \right\} / \mathbb{C},$$

and the canonical basis $(\delta_\mathsf{v})_{\mathsf{v} \in \mathsf{V}}$ may not yield a total sequence of $w^{1,p}_{\rho,\nu}(\mathsf{V})$, cf. [25, Theorem 2.12].

In the classical theory of Sobolev spaces, for an open domain $U \subset \mathbb{R}^d$ the space $\mathring{W}^{1,p}(U)$ is defined as the closure of $C^\infty_c(U)$ in the norm of $W^{1,p}(U)$. We likewise consider the closure of the space $c_{00}(\mathsf{V})$ of finitely supported functions on V in the norm of $w_{\rho,\nu}^{1,p}(\mathsf{V})$: Let us denote it by $\overset{\circ}{w}_{\rho,\nu}^{1,p}(\mathsf{V})$, or $\overset{\circ}{w}_{\rho}^{1,p}(\mathsf{V})$ if $\nu \equiv 1$. Since it is a closed subspace of $w_{\rho,\nu}^{1,p}(\mathsf{V})$, the following is immediate in view of Lemma 3.6.

Corollary 3.10. *For all $p \in [1, \infty]$, $\overset{\circ}{w}_{\rho,\nu}^{1,p}(\mathsf{V})$ is a Banach space with respect to the norm of $w_{\rho,\nu}^{1,p}(\mathsf{V})$, and a Hilbert space for $p = 2$. For all $p \in [1, \infty]$, $\overset{\circ}{w}_{\rho,\nu}^{1,p}(\mathsf{V})$ is continuously and densely embedded into $\ell_\nu^p(\mathsf{V})$. If $p \in [1, \infty)$, then $\overset{\circ}{w}_{\rho,\nu}^{1,p}(\mathsf{V})$ is separable in $\ell_\nu^p(\mathsf{V})$. If $p \in (1, \infty)$, then $\overset{\circ}{w}_{\rho,\nu}^{1,p}(\mathsf{V})$ is uniformly convex, hence reflexive.*

By definition, $\overset{\circ}{w}_{\rho,\nu}^{1,p}(\mathsf{V})$ is a closed subspace of $w_{\rho,\nu}^{1,p}(\mathsf{V})$. In fact, the following counterpart of Lemma B.6 holds, cf. Definition B.5 for the notion of lattice ideal.

Lemma 3.11. *For all $p \in [1, \infty)$ $\overset{\circ}{w}_{\rho,\nu}^{1,p}(\mathsf{V})$ is a lattice ideal of $w_{\rho,\nu}^{1,p}(\mathsf{V})$.*

Proof. The first condition in Definition B.5 is clearly satisfied. In order to check the second condition, take $u \in \overset{\circ}{w}_{\rho,\nu}^{1,p}(\mathsf{V})$ and $v \in w_{\rho,\nu}^{1,p}(\mathsf{V})$ such that $|v| \leq |u|$. By definition of $\overset{\circ}{w}_{\rho,\nu}^{1,p}(\mathsf{V})$, $u = \lim_{n\to\infty} u_n$ in $w_{\rho,\nu}^{1,p}(\mathsf{V})$ for a sequence $(u_n)_{n\in\mathbb{N}}$ such that u_n has finite support for all n. Then, it suffices to consider the sequence $(v_n)_{n\in\mathbb{N}}$ with

$$v_n := v \cdot \mathbf{1}_{\mathrm{supp}\, u_n}, \qquad n \in \mathbb{N},$$

to deduce the claim. $\square$

3.2   The Continuous Setting

We know from Lemma A.19 that any weighted oriented graph can be embedded in $\mathbb{R}^3$ in a manner that respect some natural geometric rules—in a certain sense, the resulting embedded object should be a manifold that is singular, but not *too* singular. This is done by associating each edge with a simple arc in $\mathbb{R}^3$ and one sometimes speaks of a *geometric graph*. If the graph is weighted by a function $\rho : \mathsf{E} \to (0, \infty)$, then it seems natural to endow the arcs with a metric chosen in such a way that the arc corresponding to e has a length $\rho(\mathsf{e})$. This is the root of the notion of metric graph.

Definition 3.12. Let $(\mathsf{V}, \mathsf{E}, \rho)$ a weighted oriented graph. Denote

$$\mathfrak{E} := \prod_{\mathsf{e}\in\mathsf{E}} \{\mathsf{e}\} \times (0, \rho(\mathsf{e})).$$

Then the *metric graph over* G is the pair $\mathfrak{G} := (\mathsf{V}, \mathfrak{E})$. A *point* of $\mathfrak{G}$ is an element of either V or $\mathfrak{E}$. We refer to elements of $\mathfrak{E}$ as *metric edges*.

Observe that a weighted oriented graph uniquely determines a metric graph over it and vice versa, once we adopt for all edges e the convention of identifying $\mathsf{e}_{\mathrm{init}}$ with 0 and $\mathsf{e}_{\mathrm{term}}$ with $\rho(\mathsf{e})$.

Fig. 3.1 Subdivision of an edge

Throughout this section we are going to adopt the notational convention that

$$\boxed{\mathfrak{G} = (\mathsf{V}, \mathfrak{E}) \quad \text{is the metric graph over } \mathsf{G} = (\mathsf{V}, \mathsf{E}, \rho)}$$

as well as the shorthand in (2.42).

Definition 3.13. Let $x \in \mathfrak{E}$, i.e., $x \in (0, \rho(\mathsf{e}))$ for some $\mathsf{e} \in \mathsf{E}$. The *subdivision of* $\mathfrak{G}$ *at* x is the metric graph $\tilde{\mathfrak{G}} := (\tilde{V}, \tilde{\mathfrak{E}})$ over a new weighted oriented graph $\tilde{\mathsf{G}} := (\tilde{\mathsf{V}}, \tilde{\mathsf{E}}, \tilde{\rho})$ with node set $\tilde{\mathsf{V}} := \mathsf{V} \cup \{x\}$, edge set $\tilde{\mathsf{E}} := (\mathsf{E} \setminus \{\mathsf{e}\}) \cup \{(\mathsf{e}_{\text{init}}, x), (x, \mathsf{e}_{\text{term}})\}$, and weight $\tilde{\rho}$ defined by

$$\tilde{\rho}(\mathsf{f}) := \begin{cases} \rho(\mathsf{f}) & \text{if } \mathsf{f} \neq \mathsf{e}, \\ x\rho(\mathsf{e}) & \text{if } \mathsf{f} = (\mathsf{e}_{\text{init}}, x), \\ (1-x)\rho(\mathsf{e}) & \text{if } \mathsf{f} = (x, \mathsf{e}_{\text{term}}). \end{cases}$$

If instead $x \in \mathsf{V}$, then we stipulate that the subdivision of $\mathfrak{G}$ at x is again $\mathfrak{G}$.

Observe that subdividing $\mathfrak{G}$ first at x and then at y we find the same metric graph obtained subdividing $\mathfrak{G}$ first at y and then at x (Fig. 3.1).

The volume of each subdivision of G agrees with the volume of G. Because subdivisions do not affect measure properties, without loss of generality we will always assume all edges of a metric graph to have finite length—possibly upon subdividing it.

We can now generalize the notion of distance in Definition 3.7.

Definition 3.14. Let x, y be two points of $\mathfrak{G}$. Then the *distance* $\operatorname{dist}_\rho(x, y)$ of x, y is defined as their distance in the oriented weighted graph underlying the subdivision of $\mathfrak{G}$ at x and y.

Definition 3.15. We denote by $L^p(\mathfrak{G})$ the space of measurable functions $u : \mathfrak{E} \to \mathbb{C}$ such that

$$\|u\|_{L^p(\mathfrak{G})} := \left(\sum_{\mathsf{e} \in \mathsf{E}} \int_0^{\rho(\mathsf{e})} |u(\mathsf{e}, x)|^p dx \right)^{\frac{1}{p}} < \infty \quad \text{for } p \in [1, \infty),$$

or else

$$\|u\|_{L^\infty(\mathfrak{E})} := \inf \{c \in \mathbb{R} : |u(\mathsf{e}, x)| \leq c \ \text{ for a.e. } x \in (0, \rho(\mathsf{e})) \text{ and all } \mathsf{e} \in \mathsf{E}\}.$$

If $\rho \equiv 1$, then the space of functions from $\mathfrak{E}$ to $\mathbb{C}$ is isomorphic to the space of functions from $(0, 1)$ to $\mathbb{C}^{\mathsf{E}}$. Likewise, endowing $(0, 1)$ with the Lebesgue vector measure ρdx we see that

$$L^p(\mathfrak{G}) \simeq L^p\big((0,1);\ell^p_\rho(\mathsf{E})\big), \tag{3.5}$$

by means of the isomorphism introduced in (2.37), which is isometric for all $p \in [1,\infty)$ and even unitary for $p = 2$. Indeed, by Fubini's Theorem

$$\begin{aligned}
\|u\|^p_{L^p(\mathfrak{G})} &:= \sum_{\mathsf{e}\in\mathsf{E}} \int_0^{\rho(\mathsf{e})} |u(\mathsf{e},x)|^p\,dx \\
&= \sum_{\mathsf{e}\in\mathsf{E}} \int_0^1 |u(\mathsf{e},\tilde{x})|^p \rho(\mathsf{e}) d\tilde{x} \\
&= \int_0^1 \Big(\sum_{\mathsf{e}\in\mathsf{E}} |u(\mathsf{e},\tilde{x})|^p \rho(\mathsf{e})\Big) d\tilde{x} =: \|u\|^p_{L^p\left((0,1);\ell^p_\rho(\mathsf{E})\right)}.
\end{aligned}$$

That is, given a metric graph all of whose edges have unit length, endowing the edges with weights corresponds to considering a different unit of measurement on each metric—more explicitly,

$$x \to \tilde{x} := \frac{x}{\rho(\mathsf{e})}, \qquad x \in (0,\rho(\mathsf{e})),\ \mathsf{e} \in \mathsf{E}.$$

With an abuse of notation, we will occasionally tacitly identify the spaces in (3.5).

Following [23] we adopt the following definition.

Definition 3.16. A triple (M,d,m) is called a *metric measure space* whenever (M,d) is a complete separable metric space and m is a Borel measure such that for all $x \in M$ there is $r > 0$ small enough to have $m(B_r(x)) < \infty$.

We hence regard $\mathfrak{G}$ as the product measure space of $(0,1)$ and E, endowed with the Lebesgue measure and with the counting measure weighted by ρ, respectively: In this way we have built a new metric and measure structure upon G. If the measure ρ is equivalent to the usual counting measure, then the local boundedness condition in Definition 3.16 is also clearly satisfied and we obtain the following.

Lemma 3.17. *If* (3.1) *holds, then* $\mathfrak{G}$ *is a* metric measure space *in the sense of Definition 3.16.*

Remarks 3.18. (1) Observe that $\mathfrak{G}$ is a finite measure space if and only if G has finite volume.

(2) If G is finite, then the vector-valued Lebesgue space $L^p\big((0,1);\mathbb{C}^{\mathsf{E}}\big)$ is defined in a natural way and its dual is $L^{p'}\big((0,1);\mathbb{C}^{\mathsf{E}}\big)$ for all $p \in [1,\infty)$ where $\frac{1}{p} + \frac{1}{p'} = 1$. More generally, for all $p,q \in [1,\infty]$ $L^p\big((0,1);\ell^q_\rho(\mathsf{E})\big)$ is a *Bochner space*—a particular space of vector-valued Lebesgue measurable functions. We thus identify Lebesgue integrable functions on a metric graph and

(vector-valued) Bochner integrable functions on a single interval throughout. Observe that $\|\cdot\|_{L^p(\mathfrak{G})}$ is independent of the values of functions in the nodes.

(3) Working with Bochner spaces is delicate, but the theory becomes less tricky if the target Banach space—some $\ell^q_\rho(\mathsf{E})$-space in our case—is separable, cf. [2, Sect. 1.1]. For our purposes it suffices to recall that by [11, Theorem 4.1] if $p, q \in (1, \infty)$, then $L^p\big((0,1); \ell^q_\rho(\mathsf{E})\big)$ is reflexive and its dual is $L^{p'}\big((0,1); \ell^{q'}_\rho(\mathsf{E})\big)$, where $\frac{1}{p} + \frac{1}{p'} = 1$ and $\frac{1}{q} + \frac{1}{q'} = 1$. Furthermore, the dual of $L^1\big((0,1); \ell^q_\rho(\mathsf{E})\big)$ is known to agree with $L^\infty((0,1); \ell^{q'}_\rho(\mathsf{E})$ if and only if $\ell^{q'}_\rho(\mathsf{E}))$ has the Radon–NikodÃ¡m property—and unfortunately $\ell^{1'}_\rho(\mathsf{E}) = \ell^\infty_\rho(\mathsf{E})$ does not, unless E is finite. In particular, $(L^p(\mathfrak{G}))' = L^{p'}(\mathfrak{G})$ if and only if $p \in (1, \infty)$.

(4) For $p = 2$ the space $L^2(\mathfrak{G})$ is isomorphic to the tensor product space $L^2(0,1) \otimes \ell^2_\rho(\mathsf{E})$—this is not true for general p.

One important property of one-dimensional Bochner spaces is the following so-called *Aubin–Lions Lemma*.

Proposition 3.19. *Let E, F be Banach spaces. Let $I \subset \mathbb{R}$ be a bounded open interval. If E is compactly embedded in F, then $\{u \in L^p(I; E) : u' \in L^q(I; F)\}$ is compactly embedded in $L^p(I; F)$ for all $1 < p, q < \infty$.*

In particular, $W^{1,p}(I; E)$ is compactly embedded in $L^p(I; E)$ for all $p < \infty$ if E is finite dimensional. □

The following integration by parts formula is proved applying componentwise the relations that are known to hold for scalar-valued functions.

Lemma 3.20. *For all $f, h \in W^{1,2}\big((0,1); \ell^2_\rho(\mathsf{E})\big)$ there holds*

$$\int_0^1 \big(f'(x)|h(x)\big)_{\ell^2_\rho}\, dx = (f(1)|h(1))_{\ell^2_\rho} - (f(0)|h(0))_{\ell^2_\rho} - \int_0^1 \big(f(x)|h'(x)\big)_{\ell^2_\rho}\, dx.$$

In order to define differential operators acting on functions defined on $\mathfrak{G}$, one introduces the vector-valued Sobolev spaces $W^{1,p}\big((0,1); \ell^2_\rho(\mathsf{E})\big)$. The choice of $\ell^2_\rho(\mathsf{E})$ instead of $\ell^p_\rho(\mathsf{E})$ as target space may appear bizarre, but it often proves convenient in order to apply Hilbert space methods.

Lemma 3.21. *The following assertions hold for all $p \in [1, \infty]$.*

(1) $\mathring{W}^{1,p}\big((0,1); \ell^2_\rho(\mathsf{E})\big)$ *and hence* $W^{1,p}\big((0,1); \ell^2_\rho(\mathsf{E})\big)$ *are densely and continuously embedded in* $C\big([0,1]; \ell^2_\rho(\mathsf{E})\big)$ *and in* $L^q\big((0,1); \ell^2_\rho(\mathsf{E})\big)$ *for all* $q \in [1, \infty]$.

(2) $W^{1,p}\big((0,1); \ell^2_\rho(\mathsf{E})\big)$ *is a Banach space with respect to the norm defined by*

$$\|u\|^p_{W^{1,p}\left((0,1);\ell^2_\rho(\mathsf{E})\right)} := \|u\|^p_{L^p\left((0,1);\ell^2_\rho(\mathsf{E})\right)} + \|u'\|^p_{L^p\left((0,1);\ell^2_\rho(\mathsf{E})\right)}.$$

$W^{1,p}\big((0,1);\ell^2_\rho(\mathsf{E})\big)$ is separable in $L^p\big((0,1);\ell^2_\rho(\mathsf{E})\big)$ for $p \in [1,\infty)$. $W^{1,p}\big((0,1);\ell^2_\rho(\mathsf{E})\big)$ is uniformly convex and hence reflexive for $p \in (1,\infty)$. Finally, $W^{1,2}\big((0,1);\ell^2_\rho(\mathsf{E})\big)$ is a Hilbert space with respect to the inner products

$$(u|v)_{W^{1,2}\left((0,1);\ell^2_\rho(\mathsf{E})\right)} := (u|v)_{L^2(\mathfrak{G})} + (u'|v')_{L^2(\mathfrak{G})}. \tag{3.6}$$

(3) If E is finite, then for $p \in [1,\infty)$ both the embeddings of $W^{1,1}\big((0,1);\ell^2_\rho(\mathsf{E})\big)$ in $L^p\big((0,1);\ell^2_\rho(\mathsf{E})\big)$ and of $W^{1,p}\big((0,1);\ell^2_\rho(\mathsf{E})\big)$ in $C_b\big([0,1];\ell^2_\rho(\mathsf{E})\big)$ are compact, while the embedding of $W^{1,2}\big((0,1);\ell^2_\rho(\mathsf{E})\big)$ in $L^2(\mathfrak{G})$ is even a Hilbert–Schmidt operator.

(4) If $\rho \in \ell^\infty(\mathsf{E})$, then there exists $C > 0$ such that

$$\|u\|^3_{L^2(\mathfrak{G})} \le C\|u'\|_{L^2(\mathfrak{G})}\|u\|^2_{L^1\left((0,1);\ell^2_\rho(\mathsf{E})\right)}, \quad u \in W^{1,2}\big((0,1);\ell^2_\rho(\mathsf{E})\big) \cap L^1\big((0,1);\ell^2_\rho(\mathsf{E})\big). \tag{3.7}$$

(5) For all $p,q,r \in [1,\infty]$ such that $q \le p$ one has

$$\|u\|_{L^p\left((0,1);\ell^2_\rho(\mathsf{E})\right)} \le C\|u\|^a_{W^{1,r}\left((0,1);\ell^2_\rho(\mathsf{E})\right)}\|u\|^{1-a}_{L^q\left((0,1);\ell^2_\rho(\mathsf{E})\right)}, \quad u \in W^{1,r}\big((0,1);\ell^2_\rho(\mathsf{E})\big), \tag{3.8}$$

where

$$a := \frac{q^{-1} - p^{-1}}{(1 + q^{-1} - r^{-1})}. \tag{3.9}$$

Yet another possible identification is given by the following (Fig. 3.2).

Lemma 3.22. *For a fixed counting $\{\mathsf{e}_n : n = 1,2,\ldots\}$ of E and some $u : \mathfrak{E} \to \mathbb{C}$ define*

$$\Phi u(x) := u_{\mathsf{e}_{n+1}}\left(x - \sum_{m=0}^{n}\rho(\mathsf{e}_m)\right) \qquad \text{if } x \in \left(\sum_{m=0}^{n}\rho(\mathsf{e}_m), \sum_{m=0}^{n+1}\rho(\mathsf{e}_m)\right).$$

Then Φ is an isometric isomorphism from $L^p(\mathfrak{G})$ to $L^p(0,|\mathsf{E}|_\rho)$ for all $p \in [1,\infty]$.

But whenever we are interested in metric properties this fashion of identifying vector-valued and scalar-valued functions does not perform well on graphs with infinite volume, because points that are close in $\mathfrak{G}$ might become very distant in $(0,|\mathsf{E}|_\rho) \equiv \mathbb{R}$. Furthermore, Φu will in general not be continuous any more, even if u is with respect to the natural metric on $\mathfrak{G}$.

Thus, in order to enjoy some of the most useful properties of classical Sobolev spaces a finer construction is needed.

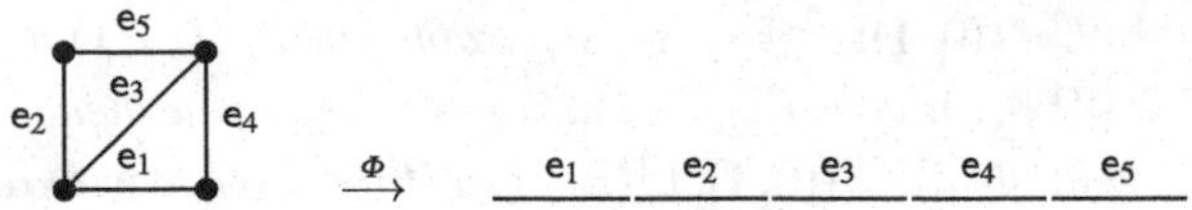

Fig. 3.2 Transformation of a graph under the isomorphism Φ: Observe that Φ ignores the nodes

Definition 3.23. For $p \in [1, \infty)$, we consider the closed subspace

$$\left\{u \in W^{1,p}\big((0,1); \ell^p_\rho(\mathsf{E})\big) : \exists u_{|\mathsf{V}} \in \mathbb{C}^{\mathsf{V}} \text{ s.t. } (\mathcal{I}^+)^\top u_{|\mathsf{V}} = u(0),\ (\mathcal{I}^-)^\top u_{|\mathsf{V}} = u(1)\right\}$$

of $W^{1,p}\big((0,1); \ell^p_\rho(\mathsf{E})\big)$ and define *Sobolev space of order* 1 *over* $\mathfrak{G}$ its isomorphic image under the operator Ψ in (2.37): We denote it by

$$W^{1,p}(\mathfrak{G}).$$

Similarly, we consider the closed subspace

$$\left\{u \in C^k([0,1]; \ell^\infty_\rho(\mathsf{E})) : \exists u_{|\mathsf{V}} \in \mathbb{C}^{\mathsf{V}} \text{ s.t. } (\mathcal{I}^+)^\top u_{|\mathsf{V}} = u(0),\ (\mathcal{I}^-)^\top u_{|\mathsf{V}} = u(1)\right\}$$

of $C^1([0,1]; \ell^\infty_\rho(\mathsf{E}))$ and define *space of bounded, k-times continuously differentiable functions over* $\mathfrak{G}$ its isomorphic image under the same Ψ: We denote it by

$$C^k_b(\overline{\mathfrak{G}}).$$

Finally, the *space of Radon measures over* $\mathfrak{G}$ is

$$M(\mathfrak{G}) := \Psi^{-1} M\big((0,1); \ell^\infty_\rho(\mathsf{E})\big)\Psi.$$

Roughly speaking, $W^{1,p}(\mathfrak{G})$ consists of those elements of $L^p(\mathfrak{G})$ that are weakly differentiable with weak derivative in $L^p(\mathfrak{G})$ and such that their boundary values are glued in accordance with the adjacency relations of G.

Remarks 3.24. (1) If G has finite volume, then $\mathfrak{G}$ is a compact Hausdorff space, each continuous function on $\mathfrak{G}$ is automatically bounded, and we simply write $C^0_b(\overline{\mathfrak{G}}) = C(\overline{\mathfrak{G}})$. Then, $M(\mathfrak{G})$ is the dual of $C(\overline{\mathfrak{G}})$.

(2) If G is a connected graph in the sense of Definition A.3, then the metric space $\mathfrak{G}$ is connected.

(3) By definition $L^2\big((0,1); \ell^2_\rho(\mathsf{E})\big) \simeq L^2(\mathfrak{G})$, while $W^{1,p}(\mathfrak{G})$ is in general only (isomorphic to) a closed subspace of $W^{1,2}\big((0,1); \ell^2_\rho(\mathsf{E})\big)$.

(4) If G is finite, then $L^p\big((0,1); \ell^2_\rho(\mathsf{E})\big) \simeq L^p(\mathfrak{G})$ and $W^{1,p}(\mathfrak{G})$ is (isomorphic to) a closed subspace of $W^{1,p}\big((0,1); \ell^2_\rho(\mathsf{E})\big)$.

Furthermore, we can mimic the construction of $\mathring{w}^{1,p}(\mathsf{V})$ and introduce the following.

Definition 3.25. For $p \in [1,\infty)$,

$$\mathring{W}^{1,p}(\mathfrak{G})$$

is defined as the closure in $W^{1,p}(\mathfrak{G})$ of the set

$$\left\{u : \mathfrak{G} \to \mathbb{C} : u \in C_b^k(\overline{\mathfrak{G}}) \text{ for all } k \in \mathbb{N} \text{ and } u \text{ has compact support}\right\}.$$

One can easily prove the following analog of Lemma 3.11.

Lemma 3.26. *For all $p \in [1,\infty)$ $\mathring{W}^{1,p}(\mathfrak{G})$ is a lattice ideal of $W^{1,p}(\mathfrak{G})$.*

In the following we denote by $C_0(\mathfrak{G})$ the vector space of all functions on the locally compact space $\mathfrak{G}$ such that for all $\epsilon > 0$ $\{x \in \mathfrak{G} : |f(x)| \geq \epsilon\}$ is compact.

Lemma 3.27. *The following assertions hold for all $p \in [1,\infty]$.*

(1) $W^{1,p}(\mathfrak{G})$ is densely and continuously embedded in $C_0(\mathfrak{G})$ and in $L^q(\mathfrak{G})$ for all $q \in [1,\infty]$.
(2) $W^{1,p}(\mathfrak{G})$ is a Banach space with respect to the norm defined by

$$\|u\|^p_{W^{1,p}(\mathfrak{G})} := \|u\|^p_{L^p(\mathfrak{G})} + \|u'\|^p_{L^p(\mathfrak{G})}.$$

$W^{1,p}(\mathfrak{G})$ is separable in $L^p(\mathfrak{G})$ for $p \in [1,\infty)$. $W^{1,p}(\mathfrak{G})$ is uniformly convex and hence reflexive for $p \in (1,\infty)$. Finally, $W^{1,2}(\mathfrak{G})$ is a Hilbert space.
(3) Let V be finite. Then both the embeddings of $W^{1,1}(\mathfrak{G})$ in $L^q(\mathsf{G})$ and of $W^{1,p}(\mathfrak{G})$ in $C_b(\overline{\mathfrak{G}})$ are compact for all $p,q < \infty$. The embedding of $W^{1,2}(\mathfrak{G})$ in $L^2(\mathfrak{G})$ is a Hilbert–Schmidt operator.
(4) If G only contains a finite number of connected components, then there exists $C > 0$ such that

$$\|u\|^3_{L^2(\mathfrak{G})} \leq C\|u'\|_{L^2(\mathfrak{G})}\|u\|^2_{L^1(\mathfrak{G})}, \qquad u \in W^{1,2}(\mathfrak{G}) \cap L^1(\mathfrak{G}). \tag{3.10}$$

(5) If G has finite volume and only contains a finite number of connected components, then there exists $C > 0$ such that

$$\|u\|_{L^p(\mathfrak{G})} \leq C\|u\|^a_{W^{1,r}(\mathfrak{G})}\|u\|^{1-a}_{L^q(\mathfrak{G})}, \qquad u \in W^{1,r}(\mathfrak{G}), \tag{3.11}$$

for all $p,q,r \in [1,\infty]$ such that $q \leq p$, where a is defined as in (3.9).
(6) If V is finite and G is connected, then for all $\mathsf{v} \in \mathsf{V}$ there exists $C > 0$ such that

$$\|u\|_{L^\infty} \leq C\left(\|u'\|_{L^1} + |u(\mathsf{v})|\right), \qquad u \in W^{1,1}(\mathfrak{G}). \tag{3.12}$$

In order to prove Lemma 3.27.(4) one would of course like to apply the one-dimensional Nash inequality (B.3), but it is a priori not clear why this holds for functions defined on a set that is not an interval. Naively mapping $\mathfrak{G}$ to an interval via the isomorphism Φ of Lemma 3.22 is not a solution, as Φu is in general not a $W^{1,2}$-function any more, so that the Nash inequality cannot be applied to it. The doubling trick presented in our proof below has been suggested in [20].

Proof. (1) and (2) are direct consequences of Lemma 3.21.(1) and (2).

(3) The assertion follows directly from Proposition 3.19.

(4) Let us consider the doubling $\mathsf{G}^{\|}$ of G, cf. Definition A.12, and define a metric graph over it. In particular, $\rho(\bar{\mathsf{e}}) = \rho(\mathsf{e})$ whenever e and hence $\bar{\mathsf{e}}$ are edges of G. Because each pair of twin edges has same length, one can define a new metric graph $\mathfrak{G}^{\|} = (\mathsf{V}, \mathfrak{E}^{\|})$ over $\mathsf{G}^{\|}$.

Now, take $u^{\|} \in W^{1,2}(\mathfrak{G})$ and a define a function $u^{\|} : \mathfrak{G}^{\|} \to \mathbb{C}$ by $u_{\bar{\mathsf{e}}}(x) = u_{\mathsf{e}}(1-x)$. Because e and its twin $\bar{\mathsf{e}}$ are directed in opposite directions, it follows from $u_{\bar{\mathsf{e}}}(0) = u_{\mathsf{e}}(1)$ that $u_{\bar{\mathsf{e}}}(\mathsf{v}) = u_{\mathsf{e}}(\mathsf{v})$ whenever v is either endpoint of e, and in particular $u \in W^{1,2}(\mathfrak{G}^{\|})$. Now, $\mathsf{G}^{\|}$ is Eulerian by Theorem A.10 and we can "unfold" $u^{\|}$ along the Eulerian tour, representing it as $u^{\|} \in W^{1,p}(\mathbb{R};\mathbb{C})$. One can thus apply (B.3) to this function obtaining

$$\|u^{\|}\|^3_{L^2(\mathfrak{G}^{\|})} \le C\,\|(u^{\|})'\|_{L^2(\mathfrak{G}^{\|})}\|u^{\|}\|^2_{L^1(\mathfrak{G}^{\|})}. \tag{3.13}$$

Accordingly,

$$\|u\|^3_{L^2(\mathfrak{G})} = 2^3\|u^{\|}\|^3_{L^2(\mathfrak{G}^{\|})} \le 2^3 C\,\|(u^{\|})'\|_{L^2(\mathfrak{G}^{\|})}\|u^{\|}\|^2_{L^1(\mathfrak{G}^{\|})} \le 2^8 C\,\|u'\|_{L^2(\mathfrak{G})}\|u\|^2_{L^1(\mathfrak{G})},$$

as $\mathsf{G}^{\|}$ has twice as many edges as G. This procedure yields the sought-after estimate on each connected component. If there are only finitely many connected components, then one can certainly find a global bound for the whole graph.

(5) The assertion can be proven likewise, applying again the doubling trick and the Gagliardo–Nirenberg inequality (B.6).

□

In the scalar case (i.e., if E is a singleton) it is known that lattice ideals of $L^2\big((0,1);\mathbb{C}\big)$ are exactly those spaces of the form $L^2(\omega;\mathbb{C})$ for a Borel set $\omega \subset (0,1)$. Invariance of such subspaces under the flow of a dynamical system is usually referred to as *reducibility* of the system and is known to have interesting consequences (among others, its failure implies a parabolic strong maximum principle). The following shows how the characterization of closed lattice ideals of scalar-valued function spaces can be generalized to the vector-valued case?

Theorem 3.28. *Every orthogonal projection $\mathcal{P}_L$ onto a closed lattice ideal L of $L^2\big((0,1);\ell^2_\rho(\mathsf{E})\big)$ is of the form*

$$(\mathcal{P}_L f)(x) = P_x(f(x)) \qquad \textit{for a.e. } x \in (0,1) \textit{ and all } f \in L^2(0,1;\ell^2_\rho(\mathsf{E})), \tag{3.14}$$

for some family $(P_x)_{x\in(0,1)}$ *of orthogonal projections onto closed lattice ideals of* $\ell^2_\rho(\mathsf{E})$ *such that* $x \mapsto P_x$ *is strongly measurable.*

Conversely, consider a family $(P_x)_{x\in(0,1)}$ *of orthogonal projections onto closed lattice ideals of* $\ell^2_\rho(\mathsf{E})$ *such that* $x \mapsto P_x$ *is strongly measurable. Then the bounded linear operator* $\mathcal{P}_L$ *defined via (3.14) is an orthogonal projection onto the closed lattice ideal*

$$L := \{f \in L^2(0,1;\ell^2_\rho(\mathsf{E})) : f(x) \in \operatorname{Rg} P_x \text{ for a.e. } x \in (0,1)\}.$$

3.3 Notes and References

Section 3.1. Although spaces of sequences are obviously a special class of Lebesgue spaces, Sobolev spaces on discrete graphs seem to have been considered in the literature only seldom—this has certainly to do with the fact that in fact, as we will see later, operator theory on infinite graphs first began to be systematically developed only at the end of the 1970s. Actually, apart from the introductory study by M. Yamasaki and his coauthors begun in [17, 26, 27] the only explicit study we are aware of appears in [19]. However, many inequalities typical of Sobolev spaces can be seen as properties of certain relevant functionals, or rather as geometric issues for singular manifolds. In these contexts, much attention has been devoted in particular to Hardy- and Poincaré-type inequalities that hold for certain Sobolev-type seminorms, cf. e.g. [4, 9, 10, 13, 25].

Among the most interesting uses of discrete Sobolev-type spaces there is surely the possibility of producing discrete versions of some of the most important results of vector analysis: We refer the reader to [21, Sect. 1] for a concise exposition of fundamental results, to [15] for a recent detailed collection of applications, and to [24, Chap. 5] for an overview of the more general interplays between graph theory and combinatorial topology.

Throughout [15], the basic rule of thumb is to replace scalar functions and vector fields on a domain U by functions on V and E, respectively. In particular, integral equations over U turn into summed equations over V. There is a rationale for this choice, but in our context it collides with the convention at the root of Definition A.17.

Also the metric properties of discrete graphs are very well-known. Certain classes of graphs, however, support distance functions with much finer properties: A typical example is that of trees, which yield a typical example of *ultrametric* spaces, cf. [6].

Section 3.2. Metric graphs are natural objects and they have been considered for many years in applied sciences, as we have seen in the historical notes on Sect. 2.2 at the end of Chap. 2. They have also occasionally appeared in pure mathematics: R. Diestel points out in [12, Sect. 8.5] that the construction of the *Freudenthal compactification* of an infinite graph G relies essentially upon the passage from G to the associated metric graph $\mathfrak{G}$, and in particular from dimensionless edges e to what he calls *topological edges* $(0,1)$—or rather $(0,\rho(\mathsf{e})$ in the weighted case.

(Freudenthal's procedure, presented in [14], has the goal of embedding discrete objects (like locally finite graphs) into compact Hausdorff spaces in a way that is overall more natural and descriptive than under either the Alexandroff or the Stone–Čech compactification.) Indeed Freudenthal's compactification of G, denoted $|\mathsf{G}|$, has an interesting structure: One has $\mathfrak{G} \subset |\mathsf{G}|$ and the "points at infinity" of G are the elements of $|\mathsf{G}| \setminus \mathfrak{G}$, called *ends*. In fact, one can regard the set of ends also as the boundary of G, and unsurprisingly this paves the road for the development of a rich potential theory on graphs, cf. [25, Chap. 4] and [22].

Probably because of the natural interpretation of a resistive electric network as a graph whose edges are assigned a length (recall that resistance of a wire is proportional to its length), metric graphs $\mathfrak{G}$ have been often dubbed *networks*, in particular during the 1980s. Indeed, a favorite notion in the early investigations of differential equations on graph-like ramified structures was that of c^k*-networks*, i.e., of geometric graphs whose edges are identified with simple arcs parametrized by k-times continuously differentiable functions, see e.g. [1,5,16,18].

A first version of the Aubin–Lions Lemma has been proved in [3]. Fully general versions are much more technical. It is known to be sharp in several senses, cf. [8]. Theorem 3.28 is taken from [7].

References

1. F. Ali Mehmeti, A characterization of a generalized c^∞-notion on nets. Int. Equ. Oper. Theory **9**, 753–766 (1986)
2. W. Arendt, C.J.K. Batty, M. Hieber, F. Neubrander, *Vector-Valued Laplace Transforms and Cauchy Problems*. Monographs in Mathematics, vol. 96 (Birkhäuser, Basel, 2001)
3. J.P. Aubin, Un théorème de compacité. C.R. Acad. Sci. Paris Sér. A **256**, 5042–5044 (1963)
4. P. Auscher, T. Coulhon, A. Grigoryan (eds.), *Heat Kernels and Analysis on Manifolds, Graphs, and Metric Spaces*. Contemporary Mathematics, vol. 338 (American Mathematical Society, Providence, 2003)
5. J. von Below, A characteristic equation associated with an eigenvalue problem on c^2-networks. Lin. Algebra Appl. **71**, 309–325 (1985)
6. A. Bendikov, A. Grigor'yan, C. Pittet, On a class of Markov semigroups on discrete ultra-metric spaces. Potential Anal. **37**, 125–169 (2012)
7. S. Cardanobile, D. Mugnolo, Towards a gauge theory for evolution equations on vector-valued spaces. J. Math. Phys. **50**, 103520 (2009)
8. X. Chen, A. Jüngel, J.-G. Liu, A note on Aubin-Lions-Dubinskii lemmas. Acta Appl. Math. (2013). doi:10.1007/s00233-013-9552-1
9. F.R.K. Chung, *Spectral Graph Theory*. Regional Conference Series in Mathematics, vol. 92 (American Mathematical Society, Providence, 1997)
10. P. Diaconis, L. Saloff-Coste, Comparison theorems for reversible Markov chains. Ann. Appl. Probab. **3**, 696–730 (1993)
11. J. Diestel, J.J. Uhl, *Vector Measures*. Mathematical Surveys, vol. 15 (American Mathematical Society, Providence, 1977)
12. R. Diestel, *Graph Theory*. Graduate Texts in Mathematics, vol. 173 (Springer, Berlin, 2005)
13. T. Ekholm, R.L. Frank, H. Kovarik, Remarks about Hardy inequalities on metric trees, in *Analysis on Graphs and its Applications*, ed. by P. Exner, J. Keating, P. Kuchment, T. Sunada, A. Teplyaev. Proceedings of the Symposium on Pure Mathematics, vol. 77 (American Mathematical Society, Providence, 2008), pp. 369–379

14. H. Freudenthal, Über die Enden diskreter Räume und Gruppen. Comment. Math. Helv. **17**, 1–38 (1944)
15. L.J. Grady, J.R. Polimeni, *Discrete Calculus: Applied Analysis on Graphs for Computational Science*. (Springer, New York, 2010)
16. G. Lumer, Connecting of local operators and evolution equations on networks, in *Potential Theory (Proc. Copenhagen 1979)*, ed. by F. Hirsch (Springer, Berlin, 1980), pp. 230–243
17. T. Nakamura, M. Yamasaki, Generalized extremal length of an infinite network. Hiroshima Math. J. **6**, 95–111 (1976)
18. S. Nicaise, Some results on spectral theory over networks, applied to nerve impulse transmission, in *Polynômes Orthogonaux et Applications (Proc. Bar-le-Duc 1984)*, ed. by C. Brezinski, A. Draux, A. P. Magnus, P. Maroni, A. Ronveaux. Lecture Notes on Mathematics, vol. 1171 (Springer, Berlin, 1985), pp. 532–541
19. M.I. Ostrovskii, Sobolev spaces on graphs. Quaest. Math. **28**, 501–523 (2005)
20. R. Pröpper, Heat kernel bounds for the Laplacian on metric graphs of polygonal tilings. Semigroup Forum **86**, 262–271 (2013)
21. M. Rigoli, M. Salvatori, M. Vignati, Subharmonic functions on graphs. Israel J. Math. **99**, 1–27 (1997)
22. P.M. Soardi, Rough isometries and Dirichlet finite harmonic functions on graphs. Proc. Am. Math. Soc. **119**, 1239–1248 (1993)
23. K.-T. Sturm, On the geometry of metric measure spaces. Acta Math. **196**, 65–131 (2006)
24. W.T. Tutte, *Graph Theory As I Have Known It* (Clarendon Press, Oxford, 1998)
25. W. Woess, *Random Walks on Infinite Graphs and Groups*. Cambridge Tracts in Mathematics, vol. 138 (Cambridge University Press, Cambridge, 2000)
26. M. Yamasaki, Extremum problems on an infinite network. Hiroshima Math. J. **5**, 223–250 (1975)
27. M. Yamasaki, Parabolic and hyperbolic infinite networks. Hiroshima Math. J. **7**, 135–146 (1977)

Chapter 4
Operator Semigroups

It is well-known from elementary linear algebra that the solution of the linear Cauchy problem

$$\begin{cases} \frac{dx}{dt}(t) = Ax(t), & t \in \mathbb{R}, \\ x(0) = x_0 \in \mathbb{C}^n, \end{cases}$$

associated with an $n \times n$ matrix A is given by $x(t) := e^{tA}x_0$, where the exponential matrix e^{tA} is

$$e^{zA} := \sum_{k=0}^{\infty} \frac{z^k}{k!} A^k, \qquad z \in \mathbb{C}. \tag{4.1}$$

This series converges in norm also if A is, more generally, a bounded linear operator on a Banach space. Furthermore, (4.1) also satisfies the group law $e^{tA}e^{sA} = e^{(t+s)A}$ for all $t, s \in \mathbb{C}$.

Example 4.1. (1) If A is a diagonal matrix on $\mathbb{C}^N$, say $A = \operatorname{diag}(a_n)_{1 \le n \le N}$, then

$$e^{zA} = \operatorname{diag}(e^{za_n})_{1 \le n \le N}, \qquad z \in \mathbb{C}. \tag{4.2}$$

(2) If P is a *projector* on a Banach space X, i.e., a bounded linear operator on X such that $P^2 = P$, then

$$e^{zP} = \sum_{k=0}^{\infty} \frac{z^k}{k!} P^k = P^0 + \sum_{k=1}^{\infty} \frac{z^k}{k!} P = \mathrm{Id} + (e^z - 1)P, \qquad z \in \mathbb{C}.$$

If a weighted oriented graph $\mathsf{G} = (\mathsf{V}, \mathsf{E}, \rho)$ is finite, then all matrices introduced in Sect. 2.1 represent of course bounded operators on $\ell^2_\rho(\mathsf{V})$, and we will see in Sect. 4.1 that the same holds under suitable connectivity conditions for infinite graphs, too. Hence, (4.1) can be applied to yield a solution of the associated linear

D. Mugnolo, *Semigroup Methods for Evolution Equations on Networks*, Understanding Complex Systems, DOI 10.1007/978-3-319-04621-1_4,

dynamical system. But the differential operators introduced in Sect. 2.2 are typically unbounded, though; and even matrices can give rise to unbounded operators—it has e.g. been proved by M. Fiedler in [14] in the unweighted case that $\mathcal{L}$ is unbounded on $\ell^2(\mathsf{V})$ if G is not uniformly locally finite. Besides, even for matrices of finite but large size the exponentiation can easily become a hard task, cf. [32].

How can existence and uniqueness of a solution to the above Cauchy problem be shown in the unbounded case? And can we deduce relevant properties of (ACP) if we cannot or do not want take the hassle of actually computing the exponential of a large matrix? These questions can be answered in the framework of the classical theory of operator semigroups, which we survey in the next section.

4.1 Matrix Semigroups

We assume throughout this section that

$\mathsf{G} = (\mathsf{V}, \mathsf{E}, \gamma)$ is a locally finite, weighted oriented graph without isolated nodes.

A direct computation yields the following and suggests a convenient weighting of the node space.

Lemma 4.2. *For all $p \in [1, \infty)$ the matrix $\mathcal{D}^{\frac{1}{p}} := \operatorname{diag}\left(\deg_\gamma(\mathsf{v})^{\frac{1}{p}}\right)$ is an isometric isomorphism from $\ell^p_{\deg_\gamma}(\mathsf{V})$ to $\ell^p(\mathsf{V})$, and hence on $\ell^p(\mathsf{V})$ if (3.2) is satisfied.*

Lemma 4.3. *The following assertions hold for $p \in [1, \infty]$.*

(1) $\mathcal{I}^+$ (resp., $\mathcal{I}^-$, $\mathcal{I}$) is an isometry from $\ell^p_\gamma(\mathsf{E})$ to $\ell^p_{\deg^{\text{in}}_\gamma}(\mathsf{V})$ (resp., $\ell^p_{\deg^{\text{out}}_\gamma}(\mathsf{V})$, $\ell^p_{\deg_\gamma}(\mathsf{V})$). In particular, all these operators have closed range.

(2) Thus, if G is outward uniformly (resp., inward uniformly, uniformly) locally finite, then $\mathcal{I}^+$ (resp., $\mathcal{I}^-, \mathcal{I}$) is bounded from $\ell^p_\gamma(\mathsf{E})$ to $\ell^p(\mathsf{V})$, and in particular $w^{1,p}_\gamma(\mathsf{V}) = \ell^p(\mathsf{V})$.

(3) The converse implication holds for $p \in [1, \infty)$. It also holds for $p = \infty$ if additionally $\gamma \in \ell^\infty(\mathsf{E})$.

Proof. Take $f : \mathsf{V} \to \mathbb{C}$ and observe that for all $\mathsf{e} \in \mathsf{E}$ there exists exactly one $\mathsf{v} \in \mathsf{V}$ s.t. $\iota^+_{\mathsf{ve}} \neq 0$, hence

$$\sum_{\mathsf{v} \in \mathsf{V}} \iota^+_{\mathsf{ve}} |f(\mathsf{v})|^p = |f(\mathsf{e}_{\text{term}})|^p = \sup_{\mathsf{v} \in \mathsf{V}} \iota^+_{\mathsf{ve}} |f(\mathsf{v})|^p \quad \text{for all } \mathsf{e} \in \mathsf{E} \text{ and all } p \in [1, \infty),$$

Take first $p \in [1, \infty)$. By Fubini's theorem

$$\|\mathcal{I}^{+T} f\|^p_{\ell^p_\gamma} = \sum_{\mathsf{e} \in \mathsf{E}} |f(\mathsf{e}_{\text{term}})|^p \gamma(\mathsf{e})$$

$$
\begin{aligned}
&= \sum_{\mathsf{e}\in\mathsf{E}} \left(\sum_{\mathsf{v}\in\mathsf{V}} |f(\mathsf{v})|^p \iota^+_{\mathsf{ve}} \right) \gamma(\mathsf{e}) \\
&= \sum_{\mathsf{v}\in\mathsf{V}} |f(\mathsf{v})|^p \sum_{\mathsf{e}\in\mathsf{E}} \iota^+_{\mathsf{ve}} \gamma(\mathsf{e}) \\
&= \sum_{\mathsf{v}\in\mathsf{V}} |f(\mathsf{v})|^p \deg^{\text{in}}_\gamma(\mathsf{v}) = \|f\|^p_{\ell^p_{\deg^{\text{in}}_\gamma}} .
\end{aligned}
$$

This shows that $\mathcal{I}^{+T}$ is an isometry from $\ell^p_{\deg^{\text{in}}}(\mathsf{V})$ to $\ell^p_\gamma(\mathsf{E})$, hence it is bounded from $\ell^p(\mathsf{V})$ to $\ell^p_\gamma(\mathsf{E})$ if G is inward uniformly locally finite.

For $p = \infty$ one has

$$
\begin{aligned}
\|\mathcal{I}^{+T} f\|_{\ell^\infty_\gamma} &= \sup_{\mathsf{e}\in\mathsf{E}} |f(\mathsf{e}_+)| \gamma(\mathsf{e}) \\
&= \sup_{\mathsf{e}\in\mathsf{E}} \sup_{\mathsf{v}\in\mathsf{V}} \iota^+_{\mathsf{ve}} |f(\mathsf{v})| \gamma(\mathsf{e}) \\
&\le \sum_{\mathsf{e}\in\mathsf{E}} \sup_{\mathsf{v}\in\mathsf{V}} \iota^+_{\mathsf{ve}} |f(\mathsf{v})| \gamma(\mathsf{e}) \\
&= \sup_{\mathsf{v}\in\mathsf{V}} |f(\mathsf{v})| \sum_{\mathsf{e}\in\mathsf{E}} \iota^+_{\mathsf{ve}} \gamma(\mathsf{e}) \\
&= \sup_{\mathsf{v}\in\mathsf{V}} |f(\mathsf{v})| \deg^{\text{in}}_\gamma(\mathsf{v}) = \|f\|_{\ell^\infty_{\deg^{\text{in}}_\gamma}} .
\end{aligned}
$$

Again, if G is inward uniformly locally finite this inequality suffices to say that $\mathcal{I}^{+T}$ is bounded from $\ell^\infty(\mathsf{V})$ to $\ell^\infty_\gamma(\mathsf{E})$.

(3) Take a sequence $(\mathsf{v}_n)_{n\in\mathbb{N}} \subset \mathsf{V}$ s.t.

$$
n \le \deg^{\text{in}}_\gamma(\mathsf{v}_n)
$$

and consider the functions $u_n : \mathsf{E} \to \mathbb{C}$ defined for all $n \in \mathbb{N}$ by

$$
u_n(\mathsf{e}) := \mathbf{1}_{\mathsf{E}^+_{\mathsf{v}_n}} ,
$$

where as usual $\mathsf{E}^+_{\mathsf{v}_n} = \{\mathsf{e} \in \mathsf{E} : \iota^+_{\mathsf{v}_n\mathsf{e}} \neq 0\}$, the set of edges outgoing from v_n. Then $\|u_n\|_{\ell^\infty_\gamma} \le \|\gamma\|_{\ell^\infty}$ for all $n \in \mathbb{N}$, but

$$
\|\mathcal{I}^+ u_n\|_{\ell^\infty} = \sup_{\mathsf{v}\in\mathsf{V}} \left| \sum_{\mathsf{e}\in\mathsf{E}} \iota^+_{\mathsf{ve}} u_n(\mathsf{e}) \right|
$$

$$\geq \sum_{\mathsf{e}\in \mathsf{E}^+_{\mathsf{v}_n}} (\iota^+_{\mathsf{v}_n\mathsf{e}}\gamma(\mathsf{e})) \frac{1}{\|\gamma\|_{\ell^\infty}}$$

$$= \deg(\mathsf{v}_n)\frac{1}{\|\gamma\|_{\ell^\infty}} \geq \frac{n}{\|\gamma\|_{\ell^\infty}} \stackrel{n\to\infty}{\to} +\infty.$$

The remaining assertions can be proved likewise.

□

Applying Lemmas 4.2–4.3 to (2.7) and the further "variational" definitions of the various operators introduced Sect. 2.1 we obtain the following.

Lemma 4.4. *Let $\gamma \in \ell^\infty(\mathsf{E})$ and $p \in [1,\infty]$. Then the following hold.*

(1) $\mathcal{L}, \mathcal{Q}, \mathcal{K}^{\text{in}}, \mathcal{K}^{\text{out}}, \overrightarrow{\mathcal{N}}, \overleftarrow{\mathcal{N}}$ are bounded on $\ell^p_{\deg_\gamma}(\mathsf{V})$.
(2) $\mathcal{L}_{\text{norm}}, \mathcal{Q}_{\text{norm}}$ are bounded on $\ell^2(\mathsf{V})$.
(3) $\mathcal{T}$ is bounded on $\ell^2_{\deg_\gamma}(\mathsf{V})$.
(4) If (3.3) holds, then both $\mathcal{K}^{\text{in}}, \overleftarrow{\mathcal{N}}$ are bounded from $w^{1,p}_\gamma(\mathsf{V})$ to $\ell^p(\mathsf{V})$.
(5) If (3.4) holds, then both $\mathcal{K}^{\text{out}}, \overrightarrow{\mathcal{N}}$ are bounded from $w^{1,p}_\gamma(\mathsf{V})$ to $\ell^p(\mathsf{V})$.
(6) If (3.2) holds, then $\mathcal{D}, \mathcal{A}, \mathcal{L}, \mathcal{Q}, \mathcal{K}^{\text{in}}, \mathcal{K}^{\text{out}}, \overrightarrow{\mathcal{N}}, \overleftarrow{\mathcal{N}}$ are bounded also on $\ell^p(\mathsf{V})$.
(7) If (3.2) holds, then $\overleftarrow{\mathcal{B}}, \overrightarrow{\mathcal{B}}$ are bounded on $\ell^1_\gamma(\mathsf{E})$ while $\mathcal{T}, \overleftarrow{\mathcal{T}}, \overrightarrow{\mathcal{T}}$ are bounded on $\ell^1(\mathsf{V})$.

Proof. (1) By its definition (cf. (2.9)) we can represent $\mathcal{L}$ as the composition

$$\ell^p_{\deg_\gamma}(\mathsf{V}) \stackrel{\mathcal{I}^T}{\longrightarrow} \ell^p_\gamma(\mathsf{E}) \stackrel{\mathcal{C}}{\longrightarrow} \ell^p_\gamma(\mathsf{E}) \stackrel{\mathcal{I}}{\longrightarrow} \ell^p_{\deg_\gamma}(\mathsf{V}) \quad \text{or rather}$$

$$\ell^p(\mathsf{V}) \stackrel{\mathcal{I}^T}{\longrightarrow} \ell^p_\gamma(\mathsf{E}) \stackrel{\mathcal{C}}{\longrightarrow} \ell^p_\gamma(\mathsf{E}) \stackrel{\mathcal{I}}{\longrightarrow} \ell^p(\mathsf{V}),$$

and analogous relations hold for the remaining matrices replacing $\mathcal{I}$ by $\mathcal{J}$ or $\mathcal{I}^\pm$.

(2) Boundedness of $\mathcal{L}_{\text{norm}}$ follows from (2.22), Lemma 4.2, and (1) representing $\mathcal{L}_{\text{norm}}$ as the composition

$$\ell^2(\mathsf{V}) \stackrel{\mathcal{D}^{-\frac{1}{2}}}{\longrightarrow} \ell^2_{\deg_\gamma}(\mathsf{V}) \stackrel{\mathcal{L}}{\longrightarrow} \ell^2_{\deg_\gamma}(\mathsf{V}) \stackrel{\mathcal{D}^{\frac{1}{2}}}{\longrightarrow} \ell^2(\mathsf{V}),$$

and likewise for $\mathcal{Q}_{\text{norm}}$.

The remaining assertions can be proved likewise.

□

Remarks 4.5. (1) A result similar to Lemma 4.4.(2)–(3) can be proved replacing $\ell^2_{\deg_\gamma}(\mathsf{V})$ by $\ell^p_{\deg_\gamma}(\mathsf{V})$, but this would require defining a new normalized Laplacian with $\mathcal{D}^{-\frac{1}{p}}$ in lieu of $\mathcal{D}^{-\frac{1}{2}}$.

(2) Lemma 4.4 shows that even in the case of an infinite Hückel graph, $\mathcal{H}_{\alpha,\beta}$ is bounded on $\ell^2(\mathsf{V})$ provided that (3.1) holds—e.g., if G is unweighted and regular, like in the case of graphene, and so is the Kogut–Susskind Dirac operator $\tilde{\mathcal{D}}_S$.
(3) The in/outdegree of $\mathsf{e} = (\mathsf{v}, \mathsf{w})$ in G_L is $\deg^{\text{in}}(\mathsf{v})$ and $\deg^{\text{out}}(\mathsf{w})$, respectively. Hence, G_L is uniformly locally finite if so is G. Combining this observation and (2) one deduces that the degree matrix of the line graph G_L is a bounded operator on $\ell^p(\mathsf{V}_L)$ for all $p \in [1, \infty]$ whenever G is uniformly locally finite. If (3.1) holds and $\gamma \in \ell^\infty(\mathsf{E})$, then by (2.30) $\mathcal{A}_L$ is a bounded linear operator on $\ell^p_\gamma(\mathsf{E})$ for all $p \in [1, \infty]$.

4.2 First Order Problems

Our approach to the study of difference and differential equations on networks is based on strongly continuous semigroup (shortly: *C_0-semigroup*) of operators. Whenever we want to discuss an evolution equation by semigroup methods, the first step is to hide the spacial dependence by transforming this partial differential equation into an *abstract Cauchy problem*, i.e., into a Cauchy problem

$$\begin{cases} \frac{d\xi}{dt}(t) = A\xi(t), & t \geq 0, \\ \xi(0) = x_0 \in X, \end{cases} \tag{ACP}$$

whose unknown ξ is a function taking values in an appropriate Banach space X. Here, A is typically an unbounded operator. If boundary conditions were imposed on the evolution equation, they have typically to be incorporated into the domain $D(A)$ of A.

Definition 4.6. A *classical solution* of (ACP) is a solution $x \in C^1(\mathbb{R}_+, X)$ such that $x(t) \in D(A)$ for all $t \geq 0$ and such that (ACP) is satisfied.

Definition 4.7. A *C_0-semigroup* is a family $(T(t))_{t\geq 0}$ of bounded linear operators on a Banach space X such that the *semigroup law*

$$T(t)T(s) = T(t+s), \qquad t, s \geq 0, \qquad \text{and} \qquad T(0) = \mathrm{Id},$$

is satisfied and moreover

$$\lim_{t \to 0+} T(t)x = x \qquad \text{for all } x \in X.$$

Its *generator* is the operator A with domain $D(A)$ defined by

$$D(A) := \left\{ x \in X : \lim_{t \to 0+} \frac{T(t)x - x}{t} \text{ exists} \right\},$$

$$Ax := \lim_{t \to 0+} \frac{T(t)x - x}{t}.$$

In the bounded case we already know that $(e^{tA})_{t \geq 0}$ is given by (4.1). *Analytic vectors*, i.e., those

$$x \in \bigcap_{k=0}^{\infty} D(A^k) \qquad \text{such that} \qquad \sum_{k=0}^{\infty} \frac{t^k}{k!} A^k x \text{ has positive radius of convergence}$$

form however only a proper subset of X in the case of general operators A.

Definition 4.8. Let X be a Banach space. A linear operator $T : D(T) \to X$ is called *closed* if for any $(f_n)_{n \in \mathbb{N}} \subset D(T)$ such that both $(f_n)_{n \in \mathbb{N}}$ and $(Tf_n)_{n \in \mathbb{N}}$ converge—say, towards some f and g, respectively—there holds $f \in D(T)$ and $Tf = g$.

Equivalently, a linear operator T on X is closed if

$$\text{Graph}(T) := \{(x, Tx) \in X \times X : x \in D(T)\}$$

is a closed subset of $X \times X$.

Remark 4.9. If A is a closed operator on X, then its domain $D(A)$ becomes a Banach space when endowed with the graph norm

$$\|x\|_A := \|x\|_X + \|Ax\|_X :$$

we denote it by $[D(A)]$.

Definition 4.10. Let A be a closed operator on a Banach space X, $\lambda \in \mathbb{C}$. If $\lambda \operatorname{Id} - A$ is an invertible operator, then one says that λ is in the *resolvent set* $\rho(A)$ of A. The inverse of $\lambda \operatorname{Id} - A$ is denoted by $R(\lambda, A)$, the *resolvent operator* of A at λ. The *spectrum* $\sigma(A)$ of A is $\mathbb{C} \setminus \rho(A)$. The *point spectrum* $\sigma_p(A)$ of A is the subset consisting of all $\lambda \in \sigma(A)$ such that $\lambda \operatorname{Id} - A$ is not injective. The *spectral bound* $s(A)$ of A is $\sup\{\operatorname{Re} \lambda : \lambda \in \sigma(A)\}$.

A direct computation shows that the resolvent operators of a closed operator satisfy

$$R(\lambda, A) = \big(\operatorname{Id} - (\lambda - \mu) R(\lambda, A)\big) R(\mu, A), \qquad \lambda, \mu \in \rho(A).$$

By the Closed Graph Theorem, resolvent operators of closed operators are always bounded. If $\rho(A) \neq \emptyset$, then in view of the ideal property of compact operators $R(\lambda_0, A)$ is compact for some $\lambda_0 \in \rho(A)$ if and only if $R(\lambda, A)$ is compact for any $\lambda \in \rho(A)$, and in this case A is said to have *compact resolvent*.

The relation between C_0-semigroups and abstract Cauchy problems is shown in the following.

Proposition 4.11. *Let A be a closed operator on a Banach space X. The following are equivalent.*

(a) A generates a C_0-semigroup $(T(t))_{t\geq 0}$.
(b) A has nonempty resolvent set and for all $x_0 \in D(A)$ the abstract Cauchy problem (ACP) *has a unique solution $\xi \in C^1(\mathbb{R}_+; X) \cap C(\mathbb{R}_+; [D(A)])$.*
(c) A is densely defined, for all $x_0 \in D(A)$ (ACP) *has a unique solution $\xi \in C^1(\mathbb{R}_+; X) \cap C(\mathbb{R}_+; [D(A)])$, and furthermore for each sequence $(x_{0n})_{n\in\mathbb{N}} \subset D(A)$ that tends to 0 with respect to $\|\cdot\|_X$ also the sequence $(\xi_n)_{n\in\mathbb{N}}$ of solutions to the corresponding* (ACP) *tends to 0 with respect to $\|\cdot\|_X$, uniformly in compact intervals of $\mathbb{R}_+$.*

If any of these conditions hold, then the solution to (ACP) *is given by $\xi(t) := T(t)x_0$, $t \geq 0$, and* (ACP) *is said to be* well-posed..

One can see that a linear operator cannot generate more than one C_0-semigroup. This suggests the notation $(e^{tA})_{t\geq 0}$ for the semigroup generated by A, which we adopt throughout.

One of the main motivations to study C_0-semigroups is given by the following.

Proposition 4.12. *Let a linear operator A generate a C_0-semigroup on a Banach space X. If $x_0 \in D(A)$, then*

$$\xi : \mathbb{R}_+ \ni t \mapsto e^{tA}x_0 \in X$$

defines a classical solution to (ACP), *i.e.,*

$$\frac{d}{dt}e^{tA}x_0 = Ae^{tA}x_0, \qquad t \geq 0,$$

and in fact $\xi(t) \in D(A)$ for all $t \geq 0$.

Remark 4.13. Among the important consequences of Proposition 4.12 we mention that

$$Ae^{tA}x_0 = e^{tA}Ax_0 \qquad \text{for all } t \geq 0,\ x_0 \in D(A),$$

that at most one C_0-semigroup can be generated by a given operator, and that if A, B are C_0-semigroup generators with $D(A) \subset D(B)$ and $Ax = Bx$ for all $x \in D(A)$—then necessarily $A = B$. In other words, a *strict* restriction of a C_0-semigroup generator cannot yield a further C_0-semigroup generator.

This explains the reason why adding further boundary conditions to a well-posed evolution equation can only result in an overdetermined problem.

Several interesting properties follow already from the semigroup law. For example, all C_0-semigroups $(e^{tA})_{t\geq 0}$ are *exponentially bounded*, i.e., there exist constants $M \geq 1$ and $\omega \in \mathbb{R}$ such that

$$\|e^{tA}\|_{\mathcal{L}(X)} \leq Me^{\omega t} \qquad \text{for all } t \geq 0. \tag{4.3}$$

This justifies the introduction of the *growth bound* $\omega_0(A)$ of $(e^{tA})_{t\geq 0}$

$$\omega_0(A) := \inf\{\omega \in \mathbb{R} : \exists M \geq 1 \text{ s.t. (4.3) } \textit{holds}\}.$$

Example 4.14. In order to study the one-dimensional heat equation

$$\frac{\partial u}{\partial t}(t,x) = \frac{\partial^2 u}{\partial x^2}(t,x), \qquad t \geq 0,\ x \in I, \tag{4.4}$$

for $I = (0,1)$ with Dirichlet boundary conditions

$$u(t,0) = u(t,1) = 0, \qquad t \geq 0,$$

we transform it into an abstract Cauchy problem

$$\dot{u}(t) = \Delta^D u(t), \qquad t \geq 0,$$

on the complex Hilbert space $L^p(0,1)$, where the operator Δ^D is defined by

$$D(\Delta^D) := W^{2,p}(0,1) \cap W_0^{1,p}(0,1),$$
$$\Delta^D u := u''.$$

We will show in Examples 4.24 and 6.34 that Δ^D generates a C_0-semigroup on $L^p(0,1)$.

Example 4.15. Let $d \in \mathbb{N}$ and $p \in [1,\infty)$. Consider the *d-dimensional Gaussian kernel* $G_d : (0,\infty) \times \mathbb{R}^d \to (0,\infty)$ defined by

$$G_d : (t,x) \mapsto \frac{1}{(4\pi t)^{\frac{d}{2}}} e^{-\frac{\|x\|^2}{4t}} \tag{4.5}$$

Then for $d = 1$

$$e^{tA}u(x) := (G_1(t,\cdot) * u)(x) = \frac{1}{(4\pi t)^{\frac{1}{2}}} \int_{\mathbb{R}} e^{-\frac{\|x-y\|^2}{4t}} u(y)dy, \qquad t > 0,\ x \in \mathbb{R}, \tag{4.6}$$

can be extended by the identity to a family indexed in $\mathbb{R}_+$ which defines a C_0-semigroup on $L^p(\mathbb{R})$—the so called *Gaussian semigroup*—whose generator is

$$D(A) := W^{2,p}(\mathbb{R}),$$
$$Au := u''.$$

This C_0-semigroup yields the solution of (4.4) for $I = \mathbb{R}$. Likewise, the convolution with G_d defines a C_0-semigroup on $L^p(\mathbb{R}^d)$ whose generator is the d-dimensional Laplacian.

Example 4.16. Let $c > 0$. The family $(T(t))_{t\geq 0}$ of *right shift operators* defined by

$$T(t)f(x) := f(x - c^{-1}t), \qquad t \geq 0,\ x \in \mathbb{R},$$

defines a C_0-semigroup on $L^p(\mathbb{R})$ for all $p \in [1, \infty)$—in fact, even a *C_0-group*: By this we mean that $(T(t))_{t\geq 0}$ embeds in a family of bounded linear operators $(T(t))_{t\in\mathbb{R}}$ that additionally satisfies

$$T(t)T(s) = T(t + s), \qquad t, s \in \mathbb{R}.$$

(It is easy to see that a linear operator A generates a C_0-group if and only if both A and $-A$ are C_0-semigroup generators.)

The generator is

$$\begin{aligned} D(A) &:= W^{1,p}(\mathbb{R}), \\ Au &:= -cu'. \end{aligned}$$

Example 4.17. Let $(U, \tilde{\mu})$ be a σ-finite measure space and let $q : U \to \mathbb{C}$ be a measurable function such that

$$|q(x)| \leq M \quad \text{for some } M \in \mathbb{R} \text{ and a.e. } x \in U.$$

Then the family $(e^{tM_q})_{t\geq 0}$ defined by

$$e^{tM_q} f(x) := e^{tq(x)} f(x), \qquad t \geq 0,\ x \in U,$$

defines a C_0-semigroup on $L^p(U, \tilde{\mu})$ for all $p \in [1, \infty)$. Its generator is the *multiplication operator*

$$\begin{aligned} D(M_q) &:= \{u \in L^p(U, \tilde{\mu}) : q \cdot u \in L^p(U, \tilde{\mu})\}, \\ M_q u &:= q \cdot u. \end{aligned}$$

By the Hölder inequality, this operator is bounded if and only if $q \in L^\infty(U, \tilde{\mu})$.

Corollary 4.18. *Let A generate a C_0-semigroup on a Banach space X. If $x_0 \in D(A)$ and $f \in W^{1,1}(\mathbb{R}_+; X)$ or $f \in C_0(\mathbb{R}_+; [D(A)])$, then*

$$\xi(t) := e^{tA}x_0 + \int_0^t e^{(t-s)A} f(s)ds, \qquad t \geq 0,$$

defines the unique classical solution *of the* inhomogeneous abstract Cauchy problem

$$\begin{cases} \frac{d\xi}{dt}(t) = A\xi(t) + f(t), & t \geq 0, \\ \xi(0) = x_0, \end{cases} \tag{iACP}$$

i.e., the unique $\xi \in C^1(\mathbb{R}_+; X)$ *with* $\xi(t) \in D(A)$ *for all* $t \geq 0$ *that satisfies* (iACP).

If in particular A is bounded, then $(e^{tA})_{t\geq 0}$ agrees with (or, more precisely, *embeds in*) the C_0-semigroup given by (4.1).

Definition 4.19. A C_0-semigroup $(T(t))_{t\geq 0}$ on a Banach space X is said to be

- *bounded* if for some $M > 0$

$$\|T(t)\|_{\mathcal{L}(X)} \leq M \qquad \text{for all } t \geq 0,$$

  and *contractive* if M can be taken to be 1.
- *ω-quasi-contractive* if there exists $\omega \in \mathbb{R}$ such that

$$\|T(t)\|_{\mathcal{L}(X)} \leq e^{\omega t} \qquad \text{for all } t \geq 0$$

  and *quasi-contractive* if $(T(t))_{t\geq 0}$ is ω-quasi-contractive for some $\omega \in \mathbb{R}$;
- *uniformly exponentially stable* if there exist $M, \epsilon > 0$ such that

$$\|T(t)\|_{\mathcal{L}(X)} \leq Me^{-\epsilon t} \qquad \text{for all } t \geq 0.$$

Since $\|e^{tA}x\|_X$ usually represents some relevant quantity of a physical system with initial data $x \in X$ at time t (e.g., total mass of a diffusive system if $\|\cdot\|_X$ is an L^1-norm, or total energy of an elastic system at time t if $\|\cdot\|_X$ is a $W^{1,2}$-norm), in many cases one actually expects that it is non-increasing in time. Generators of contractive C_0-semigroups are characterized by the following *Hille–Yosida Theorem*.

Theorem 4.20. *Let A be a linear operator on a Banach space X. Then the following are equivalent.*

(a) A generates a contractive C_0-semigroup $(e^{tA})_{t\geq 0}$.
(b) A is closed, densely defined, and furthermore $\lambda \in \rho(A)$ with

$$\lambda \|R(\lambda, A)\|_{\mathcal{L}(X)} \leq 1 \qquad \textit{for all real } \lambda > 0.$$

A few years after Theorem 4.20 was proved, it became clear that an ingenious rescaling argument allows for a generalization to the general, non-contractive case as follows.

Theorem 4.21. *Let $M, \omega \in \mathbb{R}$. Let A be a linear operator on a Banach space X. Then the following are equivalent.*

(a) A generates a C_0-semigroup such that $\|e^{tA}\|_{\mathcal{L}(X)} \le Me^{\omega t}$ for all $t \ge 0$.
(b) A is closed, densely defined, and furthermore $\lambda \in \rho(A)$ and

$$\left\| \Big((\lambda - \omega) R(\lambda, A) \Big)^n \right\|_{\mathcal{L}(X)} \le M \qquad \text{for all real } \lambda > \omega \text{ and all } n \in \mathbb{N}.$$

If X is a Banach space, then for each $x \in X$ there exists by the Hahn–Banach Theorem at least one element $J(x)$ of the dual space X' such that $\|x\|^2 = \langle x, J(x) \rangle = \|J(x)\|^2$, where $\langle \cdot, \cdot \rangle$ denotes the duality between X and X'. Then $\mathfrak{J} : X \ni x \mapsto \{J(x)\} \in 2^{X'}$ is called *duality mapping*: $\mathfrak{J}$ is in general nonlinear and indeed even multi-valued, but it is single-valued if X has strictly convex dual.

Definition 4.22. A linear operator A on a Banach space X is called *dissipative* if

$$\operatorname{Re}\langle Ax, J(x) \rangle \le 0 \qquad \text{for all } x \in D(A) \text{ and all } J(x) \in \mathfrak{J}(x). \tag{4.7}$$

It is called *ω-quasi-dissipative* if $A - \omega\,\mathrm{Id}$ is dissipative, and *ω-quasi-m-dissipative* if additionally $\operatorname{Rg}(\lambda\,\mathrm{Id} - A) = X$ for all $\lambda > \omega$. It is called *quasi-dissipative* (resp., quasi-m-dissipative) if A is ω-quasi-dissipative (resp., ω-quasi-m-dissipative) for some $\omega \in \mathbb{R}$.

On Hilbert spaces there holds even $\mathfrak{J}(x) = \{J(x)\} = \{x\}$, hence A is dissipative if and only if

$$\operatorname{Re}(Ax|x) \le 0 \qquad \text{for all } x \in D(A) :$$

The following *Lumer–Phillips Theorem* relates dissipativity and generation property.

Theorem 4.23. *Let A be a closed, densely defined operator on a Banach space X. Then for $\omega \in \mathbb{R}$ the following are equivalent.*

(a) A generates an ω-quasi-contractive C_0-semigroup.
(b) A is ω-quasi-m-dissipative.
(c) Both A and its adjoint A^* are ω-quasi-dissipative.

If in particular a C_0-semigroup is ω-quasi-contractive for some $\omega < 0$, then it is uniformly exponentially stable. Because of the similarities between (4.7) and the definition of numerical range, the Lumer–Phillips theorem can be regarded as a stability theorem of Lyapunov type.

Example 4.24. We review the setting of Example 4.14 for $p = 2$. For $u \in D(A)$

$$\operatorname{Re}(Au|u)_{L^2} = \operatorname{Re} \int_0^1 u''(x)\overline{u(x)}dx$$

$$= \mathrm{Re}\left(u'(1)\overline{u(1)} - u'(0)\overline{u(0)}\right) - \int_0^1 |u'(x)|^2 dx$$
$$= -\int_0^1 |u'(x)|^2 dx \le 0$$

using the boundary conditions. Hence, A is dissipative. One checks that $A = A^*$ and generation of a contractive C_0-semigroup thus follows from the Lumer–Phillips Theorem.

Example 4.25. Let us consider the first derivative on $L^2(0, 1)$, and more precisely

$$D(A) := \{u \in W^{1,2}(0,1) : u(1) = \beta u(0)\},$$
$$Au := u'$$

and its adjoint

$$D(A^*) := \{u \in W^{1,2}(0,1) : u(0) = \bar{\beta} u(1)\},$$
$$A^* u := -u'.$$

Then for all $u \in D(A)$

$$\mathrm{Re}(Au|u)_{L^2} = \mathrm{Re} \int_0^1 u'(x)\overline{u(x)} dx$$
$$= |u(1)|^2 - |u(0)|^2 - \mathrm{Re} \int_0^1 u'(x)\overline{u(x)} dx$$
$$= \left(|\beta|^2 - 1\right) |u(0)|^2 - \mathrm{Re} \int_0^1 u'(x)\overline{u(x)} dx.$$

A similar computation can be performed for A^* and one concludes that both A, A^* with boundary conditions $u(1) = \beta u(0)$ are dissipative provided that $|\beta| \le 1$, hence A generates a dissipative C_0-semigroup on $L^2(0, 1)$, and so does A^*. More generally, one sees that they generate a C_0-semigroup on $L^p(0, 1)$ which for $\beta = 0$ are the *left* and *right shift* semigroup

$$e^{tA} f(x) := \begin{cases} f(x+t) & \text{if } x + t \in [0,1], \\ 0 & \text{otherwise,} \end{cases} \qquad e^{tA^*} f(x) := \begin{cases} f(x-t) & \text{if } x - t \in [0,1], \\ 0 & \text{otherwise,} \end{cases}$$

respectively.

What about characterization of those operators that generate bounded, but not contractive, C_0-semigroups? A.M. Gomilko has provided the following sufficient condition.

Theorem 4.26. *Let A be a densely defined closed linear operator on a Banach space X. If $\sigma(A) \subset \{z \in \mathbb{C} : \mathrm{Re} z \leq 0\}$ and*

$$\sup_{\delta>0} \delta \int_{\delta-i\infty}^{\delta+i\infty} \left| \left\langle \frac{d}{d\lambda} R(\lambda, A)x, y \right\rangle \right| |d\lambda| < \infty \qquad \text{for all } x \in X,\ y \in X',$$

then A generates a bounded C_0-semigroup. If X is a Hilbert space, then also the converse is true.

Observe that boundedness of a C_0-semigroup is in principle already fully characterized by Theorem 4.21. However, that characterization involves infinitely many conditions on derivatives of the resolvent operators. Thus, before [17] virtually all concrete applications of the Hille–Yosida Theorem dealt with the quasi-contractive case, when the estimate in Theorem 4.21.(b) has to be checked for $n = 1$ only. In practice, even after [17], instead of checking Gomilko's condition it is often easier to introduce an equivalent norm with respect to which the Lumer–Phillips condition is satisfied: An example is given in Sect. 4.5.

It is often convenient to investigate whether a linear operator generates a C_0-semigroup by representing it as a sum of several well-behaved terms. The following is the most elementary result in the *perturbation theory* of generators of C_0-semigroups.

Lemma 4.27. *Let X be a Banach space and let B be a bounded linear operator on X. If A is the generator of a C_0-semigroup $(e^{tA})_{t\geq 0}$ on X, then so is $A + B$, with same domain as A. Define recursively a sequence $(S_n(t))_{n\in\mathbb{N}}$ of bounded linear operators on X by*

$$S_0(t) := e^{tA} \qquad \text{and} \qquad S_{n+1}(t) := \int_0^t e^{(t-s)A} B S_n(s) ds, \qquad t \geq 0,\ n \in \mathbb{N}.$$

Then the C_0-semigroup generated by $A + B$ is given by the Dyson–Phillips series

$$e^{t(A+B)} = \sum_{n=0}^{\infty} S_n(t), \qquad t \geq 0,$$

which converges in operator norm, uniformly in t on compact intervals of $\mathbb{R}_+$. Alternatively, the same semigroup is given by the Lie–Trotter product formula

$$e^{t(A+B)} = \lim_{n\to\infty} \left(e^{\frac{t}{n}A} e^{\frac{t}{n}B} \right)^n, \qquad t \geq 0,$$

with strong convergence.

In particular, A is a generator if and only if so is $A + \omega\,\mathrm{Id}$ for some/all $\omega \in \mathbb{C}$, and the semigroup generated by $A + \omega\,\mathrm{Id}$ satisfies

$$e^{t(A+\omega\,\mathrm{Id})} = e^{t\omega} e^{tA}, \qquad t \geq 0.$$

C_0-semigroups are tightly related to resolvent of their generators by means of the Laplace transform and of the backward Euler scheme.

Proposition 4.28. *Let A be the generator of a C_0-semigroup on a Banach space X. Then the following hold.*

(1) There is $\omega_0 > 0$ such that for $\lambda \in \mathbb{C}$ with $\operatorname{Re}\lambda > \omega_0$ one has $\lambda \in \rho(A)$ and

$$R(\lambda, A)x = \lim_{t\to\infty} \int_0^t e^{-\lambda s} e^{sA} x \, ds, \qquad x \in X$$

(2) For $t \geq 0$ one has

$$e^{tA}x = \lim_{n\to\infty} R\left(1, \frac{t}{n}A\right)^n x, \qquad x \in X.$$

(3) If additionally X is reflexive and $A - \omega_0 \operatorname{Id}$ is dissipative for some $\omega_0 \in \mathbb{R}$, then for $t \geq 0$ and $\omega > \omega_0$

$$e^{tA}x = \lim_{k\to\infty} \int_{-k}^{k} e^{(\omega+is)t} R(\omega + is, A)x \, ds, \qquad x \in X.$$

As a consequence of Proposition 4.28, the following holds.

Proposition 4.29. *Let $(e^{tA})_{t\geq 0}$ be a contractive C_0-semigroup on a reflexive Banach space X with generator A. Let C be a closed convex subset of X. Then the following are equivalent.*

(a) C is invariant under $(e^{tA})_{t\geq 0}$, i.e., $e^{tA}C \subset C$ for all $t > 0$;
(b) $\lambda R(\lambda, A)C \subset C$ for all $\lambda > 0$.

If C is a closed subspace, *then the above conditions are also equivalent to the following.*

(c) $R(\lambda, A)C \subset C$ for some $\lambda > 0$.

While a C_0-semigroup yields in principle a solution of an evolution equation, finding an explicit expression for its operators is usually hopeless. Though, their qualitative properties can be often discussed in a rather simple way by means of Proposition 4.29.

We assume in the remainder of this section that

$(U, \tilde{\mu})$ is a σ-finite measure space.

Definition 4.30. A bounded linear operator T on $H := L^2(U, \tilde{\mu})$ is called

- *real* if $Tf(x) \in \mathbb{R}$ for $\tilde{\mu}$-a.e. $x \in U$, whenever $f(x) \in \mathbb{R}$ for $\tilde{\mu}$-a.e. $x \in U$;
- *positive* (or sometimes *positivity preserving*) if T is real and $Tf(x) \geq 0$ for $\tilde{\mu}$-a.e. $x \in U$, whenever $f(x) \geq 0$ for $\tilde{\mu}$-a.e. $x \in U$;

- *stochastic* if T is positive and $\int_U Tf(x)d\tilde{\mu} = \int_U f(x)d\tilde{\mu}$ for all $f \in H$;
- *L^∞-contractive* if $\|Tf\|_{L^\infty} \leq \|f\|_{L^\infty}$ for all $f \in H \cap L^\infty(U, \tilde{\mu})$, or equivalently if $|Tf(x)| \leq 1$ for $\tilde{\mu}$-a.e. $x \in U$, whenever $|f(x)| \leq 1$ for $\tilde{\mu}$-a.e. $x \in U$;
- *sub-Markovian* if T is positive and L^∞-contractive; and
- *Markovian* if T is positive and an isometry for the L^∞-norm; and
- *irreducible* if it does not leave invariant any non-trivial closed ideal.

For a further bounded linear operator S on H one says that

- *S dominates T (in the sense of positive operators)* if $|Tf| \leq S|f|$ for all $f \in H$.

A C_0-semigroup $(T(t))_{t\geq 0}$ on $L^2(U, \tilde{\mu})$ is called *real* (resp., *positive, stochastic, L^∞-contractive, sub-Markovian, irreducible*) if so is each operator $T(t)$, $t \geq 0$.

All above properties can be described in terms of invariance (or non-invariance) of suitable closed convex subsets of $L^2(U, \tilde{\mu})$ under all operators of a C_0-semigroup.

Proposition 4.31. *Let $C \subset H$ be a closed convex subset of a Hilbert space H and P_C be the orthogonal projector onto C. Let A be an ω-quasi-m-dissipative operator on H. Then C is invariant under $(e^{tA})_{t\geq 0}$ if and only if*

$$\operatorname{Re}(Au|u - P_C u)_H \leq \omega \|u - P_C u\|_H^2 \qquad \textit{for all } u \in D(A). \tag{4.8}$$

Corollary 4.32. *Let $C \subset H$ be a closed convex subset of a Hilbert space H and P_C be the orthogonal projector onto C. Let A be an ω-quasi-m-dissipative operator on H. Then the following assertions hold.*

(1) Under the assumptions of Proposition 4.31, let $D(A)$ be invariant under P_C, i.e., $P_C u \in D(A)$ whenever $u \in D(A)$. If

$$\operatorname{Re}(AP_C u|u - P_C u)_H \leq 0 \qquad \textit{for all } u \in D(A),$$

then C is invariant under $(e^{tA})_{t\geq 0}$.

(2) If additionally C is a subspace of H and

$$\operatorname{Re}(AP_C u|u - P_C u)_H = 0 \qquad \textit{for all } u \in D(A), \tag{4.9}$$

then C is invariant under $(e^{tA})_{t\geq 0}$, and the converse is true if A is dissipative.

The observation in Corollary 4.32.(1) is valuable because it shows that (4.8) can be checked without having to determine ω.

Proof. (1) The assertion follows because

$$\operatorname{Re}(Au|u-P_C u)_H = \operatorname{Re}(AP_C u|u-P_C u)_H + \operatorname{Re}(A(u-P_C u)|u-P_C u)_H \leq \omega\|u-P_C u\|_H^2$$

for all $u \in D(A)$.

□

In order to appreciate the power of Proposition 4.31, one clearly needs both to express some relevant qualitative property in term of invariance of some subset C; and to show that the orthogonal projector P_C onto this subset, which is uniquely determined by the condition

$$\mathrm{Re}(x - P_C x | y - P_C x)_H \leq 0 \qquad \text{for all } y \in C,$$

can be explicitly expressed.

Lemma 4.33. *A bounded linear operator T on $H := L^2(U, \tilde{\mu})$ satisfies the following.*

(1) T is real if and only if $C := \{f \in H : f(x) \in \mathbb{R}$ for a.e. $x \in U\}$ is invariant under T; the orthogonal projector onto C is given by $P_C f :=$ Re f.
(2) Let T be real. Then T is positive if and only if $C := \{f \in H : f(x) \geq 0$ for a.e. $x \in U\}$ is invariant under T; the orthogonal projector onto C is given by $P_C f :=$ Re f^+.
(3) Let $\tilde{\mu}(U) < \infty$ and T be positive. Then T is stochastic if and only if $\int_U Tf(x)d\tilde{\mu} = 0$ whenever $\int_U f(x)d\tilde{\mu} = 0$, i.e., if and only if $C := \{f \in H : (f|\mathbf{1})_H = 0\}$ is invariant under T; the orthogonal projector onto C is given by $P_C f := f - (f|\mathbf{1})_H\mathbf{1}$.
(4) T is $L^\infty(U, \tilde{\mu})$-contractive if and only if $C := \{f \in H : |f(x)| \leq 1$ for $\tilde{\mu}$-a.e.$x \in U\}$ is invariant under T; the orthogonal projector onto C is given by $P_C f :=$ min$\{|f|, 1\}$ sgn f, where

$$\mathrm{sgn}\, y := \begin{cases} \frac{y}{|y|} & \text{if } y \neq 0, \\ 0 & \text{otherwise.} \end{cases}$$

(5) T is irreducible if and only if it does not *leave invariant $C_{U_0} := \{f \in H : f(x) = 0$ for $\tilde{\mu}$-a.e. $x \notin U_0\}$ for any measurable subset U_0 of U with $U_0 \neq \emptyset$ and $U_0 \neq U$; the orthogonal projector onto C_{U_0} is given by $P_C f := \mathbf{1}_{U_0} f$, where $\mathbf{1}_{U_0}$ denotes the characteristic function of U_0.*

Example 4.34. Consider the operator A on $L^2(0, 1)$ of Example 4.25 with $\beta = 0$, i.e., (minus) the first derivative with Dirichlet boundary condition in 1. By Lemma B.12.(2), $D(A)$ is invariant under P_C, where $C := \{w \in L^2(0, 1) : w(x) \geq 0$ for a.e. $x \in (0, 1)\}$, so that $P_C u = u^+$. Furthermore,

$$\mathrm{Re}(AP_C, u - P_C u) = -\mathrm{Re} \int_0^1 (u^+)'(x)\overline{u^-(x)}dx = 0,$$

as the intersection of the supports of u^+, u^- and hence of $(u^+)', u^-$ has zero Lebesgue measure. We conclude by Corollary 4.32 that the semigroup generated by A is positive.

Growth bound of a semigroup and spectral bound of its generator do in general differ, but are related by

$$-\infty \le s(A) \le \omega_0(A) < \infty. \tag{4.10}$$

This is unpleasant, because $\omega_0(A)$ is the quantity one is usually interested in but—unlike $s(A)$—it can be seldom determined. Indeed, $s(A)$ and $\omega_0(A)$ may sometimes agree—this is e.g. always the case if A is bounded—but it is known that the inequality may as well be strict, see e.g. [11, Sect. 4.2.7]. The following enhancement of (4.10), due to L. Weis, is therefore quite remarkable.

Theorem 4.35. *Let A generate a positive C_0-semigroup on $L^p(U, \tilde{\mu})$ for some $p \in [1, \infty)$. Then*

$$s(A) = \omega_0(A),$$

and in particular $(e^{tA})_{t \ge 0}$ is uniformly exponentially stable if and only if $s(A) < 0$.

Definition 4.36. For $p \in (1, \infty)$ we call a projector P on $L^p(U, \tilde{\mu})$ *rank-1* if it is of the form

$$Pu := \int_U \phi(y)u(y)\, d\tilde{\mu}(y) \cdot v$$

for some $v \in L^p(U, \tilde{\mu})$ and $\phi \in L^{\frac{p}{p-1}}(U, \tilde{\mu})$ such that $\int_U \phi(x)v(x)\, dx = 1$. Additionally, P is called *strictly positive* if $\phi > 0$ $\tilde{\mu}$-a.e.

Proposition 4.37. *Let A generate a positive, irreducible C_0-semigroup on $L^p(U, \tilde{\mu})$ for some $p \in [1, \infty)$. If $e^{t_0 A}$ is compact for some $t_0 > 0$, then the spectrum of A is nonempty, hence $s(A) > -\infty$, and there exists a strictly positive rank-1-projector P and constants $M \ge 0$ and $\epsilon > 0$ such that*

$$\|e^{-s(A)t} e^{tA} - P\|_{\mathcal{L}(L^p)} \le Me^{-\epsilon t} \qquad \textit{for all } t \ge 0.$$

The following is another typical example of the results yielded by the Perron–Frobenius theory, this time combined with the *Jacobs–Deleeuw–Glicksberg decomposition* from ergodic theory.

Proposition 4.38. *Let A generate a positive, irreducible C_0-semigroup on $L^p(U, \tilde{\mu})$ for some $p \in [1, \infty)$. If A has compact resolvent, then $(e^{tA})_{t \ge 0}$ converges to a periodic group $(U(t))_{t \ge 0}$; this means that there exist spaces L^p_0, L^p_{ap} such that*

- *$L^p(U, \tilde{\mu}) = L^p_0 \oplus L^p_{\text{ap}}$ and both L^p_0 and L^p_{ap} are invariant under $(e^{tA})_{t \ge 0}$;*
- *there exists a periodic group $(U(t))_{t \ge 0}$ on L^p_{ap} such that $e^{tA} \equiv U(t)$ on L^p_{ap}, $t \ge 0$;*
- *$\left(e^{tA}{}_{|L^p_0}\right)_{t \ge 0}$ is uniformly exponentially stable.*

When one thinks of diffusive processes, it is more natural to consider the L^1- and L^∞-norms (corresponding to the system's total mass and maximal pointwise density, respectively) are more natural to consider than the L^2-norm.

Definition 4.39. Let $p, q \in [1, \infty)$. A C_0-semigroup $(T_p(t))_{t\geq 0}$ on $L^p(U, \tilde{\mu})$ is said to *extrapolate* to $L^q(U, \tilde{\mu})$ if there is a (so-called *extrapolated*) C_0-semigroup $(T_q(t))_{t\geq 0}$ on $L^q(U, \tilde{\mu})$ that is *consistent*, i.e.,

$$T_p(t)f = T_q(t)f \quad \text{for all } f \in L^p(U, \tilde{\mu}) \cap L^q(U, \tilde{\mu}) \text{ and all } t > 0.$$

Then, L^∞-contractivity is particularly important because, jointly with L^2-(quasi)-contractivity, it yields by the interpolation theorem of Riesz–Thorin the following.

Proposition 4.40. *Let $(T(t))_{t\geq 0}$ be a C_0-semigroup that is quasi-contractive with respect to the norms of $L^2(U, \tilde{\mu})$ and $L^\infty(U, \tilde{\mu})$. Then $(T(t))_{t\geq 0}$ extrapolates to $L^p(U, \tilde{\mu})$ for all $p \in [2, \infty]$ and all $t \geq 0$. All such extrapolated C_0-semigroups are positive if so is $(T(t))_{t\geq 0}$. If moreover $(T(t))_{t\geq 0}$ is also quasi-contractive with respect to the norm of $L^1(U, \tilde{\mu})$ (i.e., if $(T(t)^*)_{t\geq 0}$ is L^∞-quasi-contractive), then it also extrapolates to $L^p(U, \tilde{\mu})$ for all $p \in [1, 2]$.*

All these semigroups are strongly continuous, with the only exceptions of $(T_\infty(t))_{t\geq 0}$ (never strongly continuous) and $(T_1(t))_{t\geq 0}$ (which by [49] is indeed strongly continuous if e.g. $(T(t))_{t\geq 0}$ is contractive with respect to all the L^p-norms, if it is positive, or if $\tilde{\mu}(U) < \infty$).

Especially, one is sometimes interested in invariance of some (either dense or closed) subspace $(e^{tA})_{t\geq 0}$. One may then restrict the semigroup to this space and study the Cauchy problem solved by the restricted semigroup: Does it have anything to do with the original Cauchy problem associated with A? The next result shows that this is often the case. Recall that given a linear operator A on a Banach space X, its *part* in a subspace Y of X is

$$D(A_|) := \{x \in D(A) : Ax \in Y\},$$
$$A_| x := Ax.$$

Lemma 4.41. *Let A generate a C_0-semigroup on a Banach space X. If Y is a subspace of X that is left invariant under $(e^{tA})_{t\geq 0}$ and such that the restriction of $(e^{tA})_{t\geq 0}$ to Y is strongly continuous, then its generator is the part of A in Y.*

This can be applied to identify the generator of an extrapolated semigroup.

Corollary 4.42. *Let $(T(t))_{t\geq 0}$ be a C_0-semigroup on $L^p(U)$ that extrapolates to $L^q(U)$, in the sense of Definition 4.39, for some $p < q < \infty$. Then the generator of the extrapolated semigroup is the part in $L^q(U)$ of the generator of $(T(t))_{t\geq 0}$ whenever it is known that there exists a Banach space V, with respect to whose norm the restriction of $(T(t))_{t\geq 0}$ is strongly continuous, and such that*

$$\|f\|_{L^q} \le M_\alpha \|f\|_V^\alpha \|f\|_{L^p}^{1-\alpha} \qquad f \in V,$$

for some $M > 0$ and some $\alpha \in (0,1)$.

Let us also observe that if A, B are generators, then the operator matrix

$$\operatorname{diag}(A,B) := \begin{pmatrix} A & 0 \\ 0 & B \end{pmatrix}$$

generates the C_0-semigroup defined by

$$\operatorname{diag}(e^{tA}, e^{tB}) := \begin{pmatrix} e^{tA} & 0 \\ 0 & e^{tB} \end{pmatrix}, \qquad t \ge 0.$$

Lemma 4.43. *Given two bounded linear operators S, T on a Hilbert space $H = L^2(U, \tilde{\mu})$, S dominates T if and only if $\mathbf{C} := \{(f,g) \in H \times H : |f(x)| \le g(x) \text{ for a.e. } x \in U\}$ is invariant under $\operatorname{diag}(T,S)$, if $H := L^2(U,\tilde{\mu})$; the orthogonal projector onto $\mathbf{C}$ is given by*

$$P_{\mathbf{C}}(f,g) := \begin{cases} (f,g) & \text{if } |f| \le g, \\ \frac{1}{2}\left((|f| + \min\{|f|, \operatorname{Re} g\})^+ \operatorname{sgn} f, (\max\{|f|, \operatorname{Re} g\} + \operatorname{Re} g)^+ \right) & \text{otherwise.} \end{cases}$$

Let us propose some classical applications of Proposition 4.31 to the case of a finite matrix.

Example 4.44. Let S be a finite set and $\mathcal{W} = (\omega_{ij})$ be an $S \times S$-matrix, which is of course always $\|S\|$-quasi-dissipative.

(1) According to Corollary 4.32, each $e^{-t\mathcal{W}}$ is a real and positive operator if and only if

$$0 \ge \operatorname{Re}(-\mathcal{W} \operatorname{Re} x | i \operatorname{Im} x)_{\mathbb{C}^S} = -\operatorname{Im} \sum_{i,j \in S} \omega_{ij} \operatorname{Re} x_j \operatorname{Im} x_i \qquad \text{for all } x \in \mathbb{C}^S$$

along with

$$0 \ge \operatorname{Re}(-\mathcal{W} x^+ | -x^-)_{\mathbb{C}^S} = \operatorname{Re} \sum_{i,j \in S} \omega_{ij} x_j^+ x_i^- \qquad \text{for all } x \in \mathbb{C}^S$$

or rather, considering the canonical basis vectors, if and only if all entries of $\mathcal{W}$ are real (for reality) and additionally the off-diagonal entries of $\mathcal{W}$ are negative (for positivity). (This criterion also holds for matrices $\mathcal{W}$ that are merely quasi-dissipative, because $e^{t\omega} e^{t\mathcal{W}}$ is positive if and only if $e^{t\mathcal{W}}$ is.) In particular, $\mathcal{W}$ and $-\mathcal{W}$ cannot both generate a positive semigroup unless $\mathcal{W}$ is diagonal.

(2) By Corollary 4.32.(2), the space C in Lemma 4.33.(3) is invariant under $(e^{-t\mathcal{W}})_{t\geq 0}$ if

$$0 = \mathrm{Re}(-\mathcal{W}x|\mathbf{1})_{\mathbb{C}^{\mathsf{S}}} = -\mathrm{Re}(x|\mathcal{W}^*\mathbf{1})_{\mathbb{C}^{\mathsf{S}}} \qquad \text{for all } x \in \mathbb{C}^{\mathsf{S}},$$

i.e., if $\mathcal{W}^*\mathbf{1} = 0$. Hence, if $(e^{-t\mathcal{W}})_{t\geq 0}$ is positive, then it is stochastic if $\mathrm{Id} - \mathcal{W}$ is column stochastic; and by Proposition 4.31, the converse implication provided $\mathcal{W}$ is dissipative.

(3) Likewise, $(e^{-t\mathcal{W}})_{t\geq 0}$ is sub-Markovian if and only if it is positive and $\mathcal{W}\mathbf{1} \geq 0$. In particular, $(e^{-t\mathcal{W}})_{t\geq 0}$ is Markovian *and* stochastic if and only if $\mathrm{Id} - \mathcal{W}$ is doubly stochastic.

(4) Define a new matrix $\mathcal{W}_\sharp = (\omega_{ij}^\sharp)$ by

$$\omega_{ij}^\sharp := \begin{cases} \mathrm{Re}\omega_{ii} & \text{if } i = j, \\ -|\omega_{ij}| & \text{if } i \neq j. \end{cases}$$

By (1), $(e^{-t\mathcal{W}_\sharp})_{t\geq 0}$ is positive: it is called the *modulus semigroup* of $(e^{-t\mathcal{W}})_{t\geq 0}$ and one can check that $(e^{-t\mathcal{W}_\sharp})_{t\geq 0}$ dominates $(e^{-t\mathcal{W}})_{t\geq 0}$ and is dominated by any other positive semigroup that dominates $(e^{-t\mathcal{W}})_{t\geq 0}$.

(5) Likewise, $(e^{-t\mathcal{W}})_{t\geq 0}$ is $\ell^\infty(S)$-contractive if and only if

$$\mathrm{Re}\omega_{ii} \geq \sum_{j\neq i} |\omega_{ij}| \qquad \text{for all } i \in S. \tag{4.11}$$

Indeed, this condition states that $\mathcal{W}_\sharp\mathbf{1} \geq 0$, and hence the modulus semigroup $(e^{-t\mathcal{W}_\sharp})_{t\geq 0}$ is sub-Markovian—and in particular ℓ^∞-contractive. But then also $(e^{-t\mathcal{W}})_{t\geq 0}$, which is dominated by $(e^{-t\mathcal{W}_\sharp})_{t\geq 0}$, has to be ℓ^∞-contractive. Conversely, if $(e^{-t\mathcal{W}})_{t\geq 0}$ is ℓ^∞-contractive, and hence its adjoint $(e^{-t\mathcal{W}^*})_{t\geq 0}$ is ℓ^1-contractive, then also the dominating semigroup $(e^{-t\mathcal{W}_\sharp^*})_{t\geq 0}$ has to be ℓ^1-contractive, hence its adjoint $(e^{-t\mathcal{W}_\sharp})_{t\geq 0}$ is ℓ^∞-contractive, i.e. $\mathcal{W}_\sharp\mathbf{1} \geq 0$.

In particular, $\mathcal{W}$ and $-\mathcal{W}$ cannot both generate an $\ell^\infty(S)$-contractive semigroup unless $\mathcal{W}$ is diagonal and its (diagonal) entries are purely imaginary.

(6) One shows analogously that irreducibility as in Definition 4.30 is equivalent to the condition that $\mathcal{W}$ is not similar via a permutation to a block upper triangular matrix—a formulation of the notion of irreducibility that is more usual for matrices.

It seems to be unknown whether Proposition 4.31 has a Banach space pendant. However, at least the following *Phillips' Theorem* holds, with the notation of the Lumer–Phillips Theorem.

Definition 4.45. A linear operator A on $L^p(U, \tilde{\mu})$ is called *dispersive* if it maps real-valued functions in real-valued functions and

$$\mathrm{Re}\langle Ax, y\rangle \leq 0 \qquad \text{for all } x \in D(A) \text{ and some } y \in \mathfrak{J}(x^+).$$

It is called *m-dispersive* if it is dispersive and the range $\mathrm{Rg}(\lambda\,\mathrm{Id} - A)$ of $\lambda\,\mathrm{Id} - A$ agrees with $L^p(U, \tilde{\mu})$ for all $\lambda > 0$.

Proposition 4.46. *Let A be a closed, densely defined operator on $L^p(U, \tilde{\mu})$. Then the following are equivalent.*

(a) A generates a contractive, real positive C_0-semigroup.
(b) A is m-dissipative and dispersive.
(c) A is m-dispersive.

4.3 Second Order Problems

We denote again by $[D(A)]$ the Banach space obtained by endowing the domain of closed, densely defined operator A on a Banach space X with the graph norm. In order to investigate well-posedness of the second-order abstract Cauchy problem

$$\begin{cases} \frac{d^2\xi}{dt^2} = A\xi(t), & t \geq 0, \\ \xi(0) = x_0, & \frac{d\xi}{dt}(0) = x_1, \end{cases} \tag{ACP2}$$

a classical approach is to introduce a *reduction operator matrix* along with an auxiliary unknown

$$\mathbf{A} := \begin{pmatrix} 0 & \mathrm{Id} \\ A & 0 \end{pmatrix} \quad \text{and} \quad \mathbf{u} := \begin{pmatrix} \xi \\ \frac{d\xi}{dt} \end{pmatrix}, \tag{4.12}$$

respectively, and then re-write the problem as

$$\begin{cases} \dot{\mathbf{u}}(t) = \mathbf{A}\mathbf{u}(t), & t \geq 0, \\ \mathbf{u}(0) = \mathbf{u}_0 := \begin{pmatrix} x_0 \\ x_1 \end{pmatrix}. \end{cases} \tag{4.13}$$

(The operator matrix $\mathbf{A}$ is thus used to *reduce* the second order equation to a system of first order equations.)

If $\mathbf{A}$ generates a C_0-semigroup, then its first coordinate yields the unique *classical solution* of (ACP2), i.e., a function $u \in C^2(\mathbb{R}_+, X) \cap C(\mathbb{R}_+, [D(A)])$ that satisfies (ACP2). If A is a matrix on a finite dimensional space, or more generally a bounded linear operator on X, then in analogy with the second order linear ordinary differential equations one considers the operator families

$$C(t, A) := \sum_{k=0}^{\infty} \frac{t^{2k}}{(2k)!} A^k, \qquad S(t, A) := \int_0^t C(s, A)ds, \qquad t \geq 0, \tag{4.14}$$

respectively, so that the solution of (ACP2) is given by

$$\xi(t) := C(t,A)x_0 + S(t,A)x_1, \qquad t \geq 0. \tag{4.15}$$

If especially A is a diagonal matrix all of whose eigenvalues are negative, then the power series $C(t,A)$ agrees with $\cos(t\sqrt{-A})$. This suggests to introduce the following.

Definition 4.47. A C_0*-cosine operator function* is a family $(C(t))_{t\geq 0}$ of bounded linear operators on a Banach space X such that

$$2C(t)C(s) = C(t+s) + C(t-s)), \qquad t,s \geq 0, \qquad \text{and} \qquad C(0) = \mathrm{Id},$$

and moreover

$$\lim_{t\to 0+} C(t)x = x \qquad \text{for all } x \in X.$$

Its *generator* is

$$D(A) := \left\{ x \in X : \lim_{t\to 0+} \frac{C(t)x - x}{t^2} \text{ exists} \right\},$$
$$Ax := 2 \lim_{t\to 0+} \frac{C(t)x - x}{t^2}.$$

Each C_0-cosine operator function has exactly one generator. We adopt the notation $(C(t,A))_{t\geq 0}$ throughout. The following relates all these objects.

Lemma 4.48. *Let A be a closed operator on a Banach space X. The following are equivalent.*

(a) A generates a C_0-cosine operator family on X.
(b) A is densely defined, for all $x_0, x_1 \in D(A)$ (ACP2) *has a unique solution $\xi \in C^2(\mathbb{R}_+, X) \cap C(\mathbb{R}_+, [D(A)])$, and furthermore for each pair of sequences $(x_{0n})_{n\in\mathbb{N}}, (x_{1n})_{n\in\mathbb{N}} \subset D(A)$ that tend to 0 with respect to $\|\cdot\|_X$ also the sequence $(u_n)_{n\in\mathbb{N}}$ of solutions to the corresponding* (ACP) *tends to 0 with respect to $\|\cdot\|_X$, uniformly in compact intervals of $\mathbb{R}_+$.*
(c) There exists a Banach space V, with $[D(A)] \hookrightarrow V \hookrightarrow X$, such that the operator matrix

$$\mathbf{A} := \begin{pmatrix} 0 & \mathrm{Id} \\ A & 0 \end{pmatrix}, \qquad D(\mathbf{A}) := D(A) \times V,$$

generates a C_0-semigroup $(e^{t\mathbf{A}})_{t\geq 0}$ on $V \times X$.

In this case there holds

$$e^{t\mathbf{A}} = \begin{pmatrix} C(t,A) & S(t,A) \\ AS(t,A) & C(t,A) \end{pmatrix}, \qquad t \geq 0. \tag{4.16}$$

If any of these conditions hold, then the solution to (ACP2) *is given by* (4.15) *and* (ACP2) *is said to be* well-posed..

If such a space V exists, then it is unique and is called *Kisyński space* associated with A, as its fundamental properties were first proved in [24]. The Kisyński space does not agree with X unless A is bounded.

Remarks 4.49. (1) A direct computation shows that $\lambda \in \rho(\pm\mathbf{A})$ and the resolvent operator of the reduction matrix $\mathbf{A}$ is given by

$$R(\lambda, \pm\mathbf{A}) = \begin{pmatrix} \lambda R(\lambda^2, A) & \pm R(\lambda^2, A) \\ \pm AR(\lambda^2, A) & \lambda R(\lambda^2, A) \end{pmatrix} \tag{4.17}$$

whenever λ is a complex number such that $\lambda^2 \in \rho(A)$.

(2) By an application of Lemma 4.27 one sees that the each of the conditions in Lemma 4.48 is also equivalent to the following one.

(c') There exists a Banach space V, with $[D(A)] \hookrightarrow V \hookrightarrow X$, such that the operator matrix

$$\tilde{\mathbf{A}} := \begin{pmatrix} 0 & I_V \\ A+B & C \end{pmatrix}, \qquad D(\tilde{\mathbf{A}}) := D(A) \times V, \tag{4.18}$$

generates a C_0-semigroup $(e^{t\tilde{\mathbf{A}}})_{t\geq 0}$ on $V \times X$ for all bounded linear operators B, C from V to X and on X, respectively.

The first order abstract Cauchy problem associated with $\tilde{\mathbf{A}}$ on $V \times X$ is equivalent to the *damped* second order abstract Cauchy problem

$$\begin{cases} \frac{d^2\xi}{dt^2} = (A+B)\xi(t) + C\frac{d\xi}{dt}(t), & t \geq 0, \\ \xi(0) = x_0, \quad \frac{d\xi}{dt}(0) = x_1. \end{cases} \tag{4.19}$$

Example 4.50. For all $c > 0$ the family $(C(t))_{t\geq 0}$ given by

$$C(t)f(x) := \frac{1}{2}\left(f(x+ct) + f(x-ct)\right), \qquad t \geq 0,\ x \in \mathbb{R}, \tag{4.20}$$

defines a C_0-cosine operator function on $L^p(\mathbb{R})$ for all $p \in [1, \infty)$. Its generator is

$$\begin{aligned} D(C) &:= W^{2,p}(\mathbb{R}), \\ Cu &:= c^2 u'', \end{aligned}$$

the square of A introduced in Example 4.16.

The following shows that the setting of Example 4.50 is just a special case of a typical behavior.

Proposition 4.51. *Let B generate a C_0-group on a Banach space X. Then $A := B^2$ generates a C_0-cosine operator function on X with KisyÅĎski space $[D(B)]$. This is given by*

$$C(t, A) = \frac{1}{2}\left(e^{itB} + e^{-itB}\right), \qquad t \geq 0.$$

Furthermore, if $\lambda^2 \in \rho(A)$, then $\lambda \in \rho(B) \cap \rho(-B)$ and the resolvent operators of $\pm B$ are given by

$$R(\lambda, B) = (\lambda \operatorname{Id} + B)R(\lambda^2, A), \qquad R(\lambda, -B) = (\lambda \operatorname{Id} - B)R(\lambda^2, A).$$

Clearly, one expects from a solution of a wave equation different properties than from solutions of a heat equation. We only mention two.

Definition 4.52. A C_0-cosine operator function $(C(t))_{t\geq 0}$ on a Banach space X is called *bounded* if there exists $M > 0$ such that $\|C(t)\|_{\mathcal{L}(X)} \leq M$ for all $t \geq 0$. It is called *periodic* if there is a *period* $\tau > 0$ such that $C(t + \tau) = C(t)$ for all $t \geq 0$.

Proposition 4.53. *Let A generate a bounded C_0-cosine operator function on a Banach space X. Then $C(t, A)_{t\geq 0}$ is periodic of period τ if and only if its spectrum $\sigma(A)$ consists of simple poles of the $R(\cdot, A)$, the set of eigenvectors of A is total in X, and finally*

$$\sigma(A) \subset \left\{-\frac{4k^2\pi^2}{\tau^2} : k \in \mathbb{Z}\right\}. \tag{4.21}$$

The *D'Alembert formula* (4.20) for the solution of the one dimensional wave equation is the prototypical cosine operator function and suggests the properties one expects from a well-behaved second order Cauchy problem.

Definition 4.54. Let A generate a C_0-cosine operator function on $X = L^p(U, \tilde{\mu})$, $p \in (1, \infty)$, and hence (ACP2) be well-posed. Then (ACP2) is said to enjoy *finite speed of propagation* if there exists $c_0 > 0$ such that

$$\langle C(t, A)u, v\rangle = 0$$

for all $\omega_1, \omega_2 \subset U$, all $u \in L^p(U, \tilde{\mu})$ and $v \in L^q(U, \tilde{\mu})$ $(p^{-1} + q^{-1} = 1)$ with $\operatorname{supp} u \subset \omega_1$, $\operatorname{supp} v \subset \omega_2$, and all $t \in (0, \frac{1}{c_0}\operatorname{dist}(\omega_1, \omega_2))$.

4.4 Semigroups on Discrete Graphs

Besides the analogy with electric networks there are further reasons for regarding $\mathcal{L}$ as a Laplacian. The following justification was suggested in [18, Sect. 2.5.5] and turns out to be particularly compelling in the viewpoint of this book.

When modeling diffusive phenomena for an isolated system in an open domain U, in the continuous case the conservation law

$$\frac{d}{dt}\int_U \varphi(t,x)dx = -\int_U \operatorname{div} j(t,x)dx$$

has to be enforced. Likewise, in the discrete case we have the conservation law

$$\frac{d}{dt}\sum_{\mathsf{v}\in\mathsf{V}} \varphi(t,\mathsf{v}) = -\sum_{\mathsf{v}\in\mathsf{V}} (\mathcal{I}j)(t,\mathsf{v}),$$

where φ represents e.g. a temperature or a density of a chemical substance and j is the flux function. Then, Fick's law of diffusion can be written in a discrete form as

$$j(t,\mathsf{e}) = c\left(\mathcal{I}^T\varphi(t,\mathsf{e})\right),$$

for a suitable function $c : \mathbb{R} \to \mathbb{R}$. It can in fact be used to derive a differential equation governing a flow on G: Choosing c to be the identity we finally obtain the *discrete diffusion equation*

$$\frac{d}{dt}\sum_{\mathsf{v}\in\mathsf{V}} \varphi(t,\mathsf{v}) = -\sum_{\mathsf{v}\in\mathsf{V}} \mathcal{I}\mathcal{I}^T\varphi(t,\mathsf{v}). \tag{4.22}$$

In view of the above derivation, we regard (4.22) as a linear, homogeneous heat equation—and hence think of $\mathcal{I}\mathcal{I}^T$ as a discrete version of the Laplacian. (Observe that, unlike its spatially continuous counterpart, (4.22) is a *backward* evolution equation.) If c is more generically a linear mapping we recover instead a general Laplace–Beltrami matrix.

Let us consider again the operators introduced in Sect. 2.1. The case of hermitian matrices is easy to treat.

Proposition 4.55. *Let* G *be a finite weighted graph. Then* $-\mathcal{L}, -\mathcal{Q}, -\mathcal{L}_{\mathrm{norm}}, -\mathcal{Q}_{\mathrm{norm}}$ *generate contractive* C_0*-semigroups.*

Proof. Because each bounded operator generates a group by (4.1), it suffices to check that these matrices are dissipative: Indeed, $-\mathcal{L}, -\mathcal{Q}, -\mathcal{L}_{\mathrm{norm}}, -\mathcal{Q}_{\mathrm{norm}}$ are dissipative by their definition. □

Also in the infinite case, in order to show that these matrices are C_0-semigroup generators one could apply the Lumer–Phillips theorem by showing that they are in fact m-dissipative, i.e., checking a condition on their range. This boils down to solve an infinite system of algebraic equations. However, in Sect. 6.4.1 we will prove a sharper result with less effort applying the theory of elliptic forms.

What about the non-Hermitian matrices in Sect. 2.1? We have seen in Example 2.24 that the advection matrices $-\overrightarrow{\mathcal{N}}, -\overleftarrow{\mathcal{N}}$ need not be dissipative, in sharp contrast to the first derivative on an interval, cf. Example 4.25. In which sense, then, the semigroups generated by $-\overrightarrow{\mathcal{N}}$ and $-\overleftarrow{\mathcal{N}}$ can be considered discrete pendants of the shift semigroups? We can make sense of this analogy at least in a very special case.

Lemma 4.56. *Let* G *a be a finite unweighted graph. If* G *is an oriented cycle, then* $(e^{-t\overrightarrow{\mathcal{N}}})_{t\geq 0}$ *is contractive. If* G *is an oriented path, then* $(e^{-t\overrightarrow{\mathcal{N}}})_{t\geq 0}$ *is uniformly exponentially stable.*

Proof. We begin by writing $\overrightarrow{\mathcal{N}} = \mathcal{D}^{\text{out}} - \mathcal{A}^{\text{in}}$. We are going to estimate $W(\overrightarrow{\mathcal{N}})$ by $W(\mathcal{D}^{\text{out}}) - W(\mathcal{A}^{\text{in}})$. In order to describe $W(\mathcal{A}^{\text{in}})$, we observe that $\mathcal{A}^{\text{in}} + (\mathcal{A}^{\text{in}})^* = \mathcal{A}^{\text{in}} + \mathcal{A}^{\text{out}} = \mathcal{A}$ is a tridiagonal Toeplitz matrix with vanishing diagonal and with all off-diagonal entries equal to 1. Hence its largest eigenvalue is 2. On the other hand, $\mathcal{D}^{\text{out}} = \text{Id}$ and therefore $W(\mathcal{D}^{\text{out}}) = \{1\}$. Accordingly, by Proposition 2.23.(7) the numerical range of $\overrightarrow{\mathcal{N}}$ is contained in the ball $\overline{B_1(1)}$. Accordingly, $\overrightarrow{\mathcal{N}}$ is dissipative and the assertion follows by the Lumer–Phillips' Theorem.

Likewise, a trivial application of the Lie–Trotter product formula yields

$$e^{-z\overrightarrow{\mathcal{N}}} = e^{-z(\mathcal{D}^{\text{out}} - \mathcal{A}^{\text{in}})} = e^{-z\mathcal{D}^{\text{out}}} e^{z\mathcal{A}^{\text{in}}} = e^{-z} e^{z\mathcal{A}^{\text{in}}}, \qquad z \in \mathbb{C}.$$

For a finite path $\mathcal{D}^{\text{out}} = \text{Id}$ and hence $e^{-z\mathcal{D}^{\text{out}}} = e^{-z}\,\text{Id}$. Also, a direct computation shows that

$$e^{z\mathcal{A}^{\text{in}}} = \begin{pmatrix} 1 & z & \frac{z^2}{2!} & \frac{z^3}{3!} & \cdots \\ 0 & 1 & z & \frac{z^2}{2!} & \ddots \\ 0 & 0 & 1 & z & \ddots \\ 0 & 0 & 0 & 1 & \ddots \\ \vdots & \ddots & \ddots & \ddots & \ddots \end{pmatrix}, \qquad z \in \mathbb{C},$$

i.e., $(e^{t\mathcal{A}^{\text{in}}})_{t\geq 0}$ has polynomial growth. The claim thus follows. □

In order to treat these matrices in the case of infinite graphs, we first need to introduce suitable form methods: This will be done in Chap. 6. For this reason, for the time being we refrain from discussing the generator property of any matrix in the case of infinite graphs and only focus on finite graphs. Because in this case it is already clear that all operators from Sect. 2.1 are generators by (4.1), we rather focus on deducing some qualitative properties of solutions by semigroup methods.

If S is the node set of a graph, we can specialize the criteria found in Example 4.44 to the case of difference operators on graphs introduced in Sect. 2.1. We thus find the following.

Lemma 4.57. *Let* $\mathsf{G} = (\mathsf{V}, \mathsf{E}, \mu)$ *be a finite weighted oriented graph. Then the following assertions hold.*

(1) $(e^{-t\mathcal{L}})_{t\geq 0}$ *is sub-Markovian and stochastic—hence contractive with respect to the* $\ell^p(\mathsf{V})$*-norm for all* $p \in [1,\infty]$. $(e^{t\mathcal{L}})_{t\geq 0}$ *is neither positive nor* $\ell^\infty(S)$*-contractive.*

(2) $(e^{-t\mathcal{Q}})_{t\geq 0}$ *is* $\ell^\infty(\mathsf{V})$*- and* $\ell^1(\mathsf{V})$*-contractive–hence contractive with respect to the* $\ell^p(\mathsf{V})$*-norm for all* $p \in [1,\infty]$. *It is not positive, unlike* $(e^{t\mathcal{Q}})_{t\geq 0}$. *It is dominated by* $(e^{-t\mathcal{L}})_{t\geq 0}$.

(3) Both $(e^{-t\mathcal{L}_{\rm norm}})_{t\geq 0}$ *and* $(e^{-t\mathcal{Q}_{\rm norm}})_{t\geq 0}$ *are contractive. Also* $(e^{-t\mathcal{Q}_{\rm norm}})_{t\geq 0}$ *is dominated by* $(e^{-t\mathcal{L}_{\rm norm}})_{t\geq 0}$. *Furthermore,* $(e^{-t\mathcal{L}_{\rm norm}})_{t\geq 0}$ *is positive but in general not* $\ell^\infty(\mathsf{V})$*-contractive.*

(4) $(e^{t(\mathcal{T}^T-\mathrm{Id})})_{t\geq 0}$ *is contractive and stochastic.*

All the above semigroups are irreducible if and only if G *is connected.*

(5) $(e^{-t\overrightarrow{\mathcal{N}}})_{t\geq 0}$ *and* $(e^{-t\overleftarrow{\mathcal{N}}})_{t\geq 0}$ *are stochastic, whereas* $(e^{-t\mathcal{K}^{\rm out}})_{t\geq 0}$ *and* $(e^{-t\mathcal{K}^{\rm in}})_{t\geq 0}$ *are sub-Markovian.*

(6) $(e^{t(\overrightarrow{\mathcal{T}}^T-\mathrm{Id})})_{t\geq 0}$ *and* $(e^{t(\overleftarrow{\mathcal{T}}-\mathrm{Id})})_{t\geq 0}$ *are stochastic.*

All the semigroups in (5)–(6) are irreducible if and only if G *is strongly connected.*

If G *has finite volume (or more generally if the condition in Proposition 3.8.(2) is satisfied) and additionally* G *is strongly connected, then* $(e^{-t\overrightarrow{\mathcal{N}}})_{t\geq 0}$, $(e^{-t\overleftarrow{\mathcal{N}}})_{t\geq 0}$, $(e^{-t\mathcal{K}^{\rm out}})_{t\geq 0}$, $(e^{-t\mathcal{K}^{\rm in}})_{t\geq 0}$ *converge towards a periodic group.*

(Indeed, also $(e^{-t\mathcal{L}})_{t\geq 0}, (e^{-t\mathcal{L}_{\rm norm}})_{t\geq 0}, (e^{t(\mathcal{T}^T-\mathrm{Id})})_{t\geq 0})$ converge towards a periodic group if G is connected, but in their case the limiting periodic group is trivial, as the *stable subspace* $\ell^p_0(\mathsf{V})$ has codimension 1.)

In particular, (6) justifies Definition 2.30, since—like any other pagerank—Chung's heat kernel pagerank needs to be a probability distribution for all parameters t, if so is f.

Proof. (1) (Minus) the discrete Laplace–Beltrami operator $-\mathcal{L}$ is dissipative in view of (2.9), hence the semigroup it generates is contractive by the Lumer–Phillips Theorem. It has real entries that are negative off-diagonal and sum up to 0 both row-wise and column-wise. Furthermore, $-\mathcal{L}$ is dissipative by (2.10).

(2) Similar considerations hold for $\mathcal{Q}$, the signless discrete Laplace–Beltrami operator. Again, $-\mathcal{Q}$ is dissipative by (2.14). Because its off-diagonal entries are positive, $(e^{-t\mathcal{Q}})_{t\geq 0}$ is not positive. By construction of $\mathcal{Q}$, $(e^{-t\mathcal{L}})_{t\geq 0}$ is the modulus semigroup of $(e^{-t\mathcal{Q}})_{t\geq 0}$ (and $(e^{t\mathcal{Q}})_{t\geq 0}$ is the modulus semigroup of $(e^{t\mathcal{L}})_{t\geq 0}$), hence $(e^{-t\mathcal{Q}})_{t\geq 0}$ is dominated by $(e^{-t\mathcal{L}})_{t\geq 0}$. In particular, by (1) $(e^{-t\mathcal{Q}})_{t\geq 0}$ turns out to be both ℓ^∞-contractive and (by duality, as $\mathcal{Q}$ is symmetric) ℓ^1-contractive.

(3) Again, dissipativity of both operators follows directly from their definition. The assertion on domination is proved in the same way. Moreover, $\mathcal{L}_{\rm norm}$ has real entries that are negative off-diagonal—unlike $\mathcal{Q}_{\rm norm}$. That $(e^{-t\mathcal{L}_{\rm norm}})_{t\geq 0}$ is generally not ℓ^∞-contractive can be seen taking G to be an unweighted star

with more than two nodes and checking that condition (4.11) does not hold if one takes i to be the center of the star.

(4) By (2.23) also $-(\mathrm{Id}-\mathcal{T}) = -\mathcal{D}^{\frac{1}{2}}\mathcal{L}_{\mathrm{norm}}\mathcal{D}^{-\frac{1}{2}}$ is dissipative. Since $\mathcal{T}^T$ is column stochastic, the semigroup generated by $\mathcal{T}-\mathrm{Id}$ is stochastic, too.

(5) We can likewise deduce all properties of the semigroups generated by $-\mathcal{K}^{\mathrm{in}}, -\mathcal{K}^{\mathrm{out}}, -\overrightarrow{\mathcal{N}}, -\overleftarrow{\mathcal{N}}$ from (2.26) and (2.27).

(6) Both $\overrightarrow{\mathcal{T}}^T$ and $\overleftarrow{\mathcal{T}}$ are row stochastic.

The assertion about convergence toward a periodic group follows from Proposition 4.38 and the fact that a linear operator has compact resolvent if and only if its domain is compactly embedded in the ambient space. By □

Example 4.58. Let us consider the abstract Cauchy problem associated with the adjacency matrix $\mathcal{A}$: This is usually known as the *master equation* in applied natural sciences. Let us represent $\mathcal{A}$ as

$$\mathcal{A} = \mathcal{L} - \mathcal{D}.$$

We already know from Lemma 4.4 that if (3.2) holds, then $\mathcal{A}, \mathcal{L}, \mathcal{D}$ are bounded on $\ell^p(\mathsf{V})$, hence the master equation is surely well-posed on all $\ell^p(\mathsf{V})$-spaces.

How to show that $(e^{-t\mathcal{A}})_{t\geq 0}$ is positive on all these spaces? Although Lemma 4.57 only holds in Hilbert spaces, it is clear that all extrapolated semigroups inherit positivity of the extrapolating semigroup (like $(e^{-t\mathcal{L}})_{t\geq 0}$) defined on the given Hilbert space. In this case $(e^{-t\mathcal{A}})_{t\geq 0}$ is not sub-Markovian and hence it is not clear whether the master equation is well-posed on all $\ell^p(\mathsf{V})$-spaces. But this follows indeed applying Lemma 4.27 to the bounded operator $\mathcal{D}$, if (3.2) holds.

Furthermore, $(e^{-t\mathcal{L}})_{t\geq 0}$ is positive and so is $\mathcal{D}$: We conclude that from Lemma 4.57 that also $(e^{-t\mathcal{A}})_{t\geq 0}$ is positive on $\ell^p(\mathsf{V})$ for all p. Indeed, the series representation of $(e^{t(A+B)})_{t\geq 0}$ in Lemma 4.27, which goes back to F. Dyson, shows in particular that $(e^{t(A+B)})_{t\geq 0}$ is positive if $(e^{tA})_{t\geq 0}$ and B are positive, too.

Example 4.59. We will briefly discuss the notion of *connectome* in Chap. 5. In [47] the authors express their criticism of this notion advocating the view that the statistics of the degree distributions and metric properties of a network are not quite significant, whereas more insight may be gained modeling both networks of all neurons connected by electrical and chemical synapses by means of parabolic equation associated with the discrete Laplacian $\mathcal{L}_{\mathrm{ele}}$ and the incoming adjacency matrix $\mathcal{A}^{\mathrm{in}}_{\mathrm{chem}}$ associated with the respective subgraph, respectively. After some simplification and *ad-hoc* hypotheses, they thus find a system of two equations

$$\begin{cases} \frac{df}{dt}(t,\mathsf{v}) = -\mathcal{L}_{\mathrm{ele}} f(t,\mathsf{v}), & t\geq 0,\ \mathsf{v}\in\mathsf{V}, \\ \frac{df}{dt}(t,\mathsf{v}) = \mathcal{A}^{\mathrm{in}}_{\mathrm{chem}} f(t,\mathsf{v}), & t\geq 0,\ \mathsf{v}\in\mathsf{V}, \end{cases}$$

whose unknown is the vector of the membrane potentials. They can then be combined to obtain

$$\frac{df}{dt}(t, \mathsf{v}) = \left(-\mathcal{L}_{\text{ele}} + \mathcal{A}^{\text{in}}_{\text{chem}}\right) f(t, \mathsf{v}), \qquad t \geq 0, \, \mathsf{v} \in \mathsf{V}.$$

As a brain contains finitely many neurons, well-posedness follows already by (4.1). Furthermore, $-\mathcal{L}_{\text{ele}} + \mathcal{A}^{\text{in}}_{\text{chem}}$ has positive off-diagonal entries but generally neither its rows nor its columns sum up to 0. Thus, the semigroup $\left(e^{t\left(-\mathcal{L}_{\text{ele}}+\mathcal{A}^{\text{in}}_{\text{chem}}\right)}\right)_{t\geq 0}$ is positive, but in general neither stochastic nor sub-Markovian: This seems in accordance with the triggering of action potentials.

Example 4.60. Given a (possibly oriented) finite graph with node set V, a famous class of problems in the theory of distributed systems consists in finding an appropriate operator $\mathcal{W}$ on $\ell^2(\mathsf{V})$ such that either the continuous or the discrete dynamical system associated with $\mathcal{W}$ converges to a prescribed *consensus*.

In our language, this amounts to finding a suitable generalized Laplacian $\mathcal{W}$ such that for a certain class of functions $f \in \mathbb{R}^{\mathsf{V}}$

$$e^{-t\mathcal{W}} f \qquad \text{or} \qquad \mathcal{W}^k f$$

converge as $t \to \infty$ or $k \to \infty$ to some given vector in the node space—this is typically the constant function $\mathbf{1}$ if the graph is finite (one calls this specific instance *synchronization* or *average consensus*). The survey paper [37] provides a good introduction to this subject.

It was observed in [38] that the discrete Laplacian yields a protocol for reaching *average consensus* on a non-oriented, connected graph, and so does the outgoing Kirchhoff matrix, if the graph is oriented and strongly connected. Indeed, we have seen in Example 4.57 that both $(e^{-t\mathcal{L}})_{t\geq 0}$ and $(e^{-t\mathcal{K}^{\text{out}}})_{t\geq 0}$ are positive and irreducible. Moreover, $s(\mathcal{L}) = 0$ and by Proposition 4.37 $(e^{-t\mathcal{L}})_{t\geq 0}$ converges exponentially towards the projector onto the subspace spanned by $\mathbf{1}$. Likewise, if follows that $-\mathcal{K}^{\text{in}}, -\mathcal{K}^{\text{out}}$ and hence their transposes $-\overrightarrow{\mathcal{N}}, -\overleftarrow{\mathcal{N}}$ have zero spectral radius, thus we deduce by Proposition 4.37 that they generate semigroups that converge *exponentially* toward a projector. In the case of $(e^{-t\mathcal{K}^{\text{in}}})_{t\geq 0}, (e^{-t\mathcal{K}^{\text{out}}})_{t\geq 0}$, this projector corresponds to the average consensus, whereas the system may generally converge to a different consensus if we adopt $-\overrightarrow{\mathcal{N}}$ or $-\overleftarrow{\mathcal{N}}$ as protocols.

We already know that $\overrightarrow{\mathcal{T}} - \text{Id}$ is dissipative, hence so is $\overrightarrow{\mathcal{T}}^T - \text{Id}$. Because $\overrightarrow{\mathcal{T}}$ is column stochastic, $\left(\overrightarrow{\mathcal{T}}^T - \text{Id}\right) \mathbb{1} = \mathbb{0}$, hence the spectral bound of both $\overrightarrow{\mathcal{T}}^T - \text{Id}$ and its adjoint $\overrightarrow{\mathcal{T}} - \text{Id}$ is 0. The higher the speed of convergence towards the rank-1-projectors is, the more efficient the protocol is considered.

If one focuses on the discrete dynamical system, then the so-called *Metropolis algorithm* introduced in [30] shows that for any given probability distribution on V it is always possible to find a suitable weight function $\mu : \mathsf{E} \to [0, \infty)$ and hence an associated normalized Laplacian $\mathcal{L}_{\text{norm}}$ in such a way that the solutions, which are given by $(\mathcal{L}^n_{\text{norm}} f)_{n\in\mathbb{N}}$ of the discrete dynamical system converge to the given probability distribution.

4.5 Advection on Metric Graphs

Throughout this section

$\mathfrak{G} = (\mathsf{V}, \mathfrak{E})$ is the metric graph associated with a locally finite, weighted oriented graph $\mathsf{G} = (\mathsf{V}, \mathsf{E}, \mu)$.

We are going to study advection problems on $\mathfrak{G}$, i.e., abstract Cauchy problems associated with the first derivative operator introduced in Sect. 2.2.2. More precisely, we study an advective process

$$\frac{\partial u_{\mathsf{e}}}{\partial t}(t,x) = \frac{\partial u_{\mathsf{e}}}{\partial x}(t,x), \qquad t \geq 0,\ x \in \big(0, \mu(\mathsf{e})\big),\ \mathsf{e} \in \mathsf{E}, \tag{4.23}$$

taking place on each edge e of a metric graph $\mathfrak{G}$. We already know that upon considering the isometric isomorphism Ψ defined in (2.37) we can equivalently assume all edges to have unit length, i.e. $\mu(\mathsf{e}) \equiv 1$, and weight the state space by μ instead.

This leads to considering the operator $M_\gamma \overleftarrow{A}$, where $\overleftarrow{A}$ is the first derivative operator

$$(\overleftarrow{A} u)_{\mathsf{e}}(x) := \frac{du_{\mathsf{e}}}{dx}(x), \qquad x \in \big(0, 1\big),\ \mathsf{e} \in \mathsf{E}, \tag{4.24}$$

with standard boundary conditions as in Definition 2.42 and M_γ is the multiplication operator associated with the edgewise constant function γ introduced in Example (2.37).

Actually, in this section we are going to discuss the abstract Cauchy problem (ACP) associated with the more general operator

$$M_c \overleftarrow{A} + M_p,$$

complemented by standard boundary conditions

$$u(1) = \overleftarrow{\mathcal{B}} u(0),$$

where $\overleftarrow{\mathcal{B}}$ is the matrix defined as in (2.31) and $c, p : (0,1) \to \mathbb{C}^{\mathsf{E}}$ are general functions such that $c_{\mathsf{e}}(x), p_{\mathsf{e}}(x)$ are strictly positive for all $x \in (0,1)$ and all $\mathsf{e} \in \mathsf{E}$. (Again, M_c, M_p are the multiplication operators introduced in Example (2.37).)

It was observed by B. Dorn that, whenever $c \equiv 1$ and $p \equiv 0$, the formula in Example 4.25 can be extended to yield the following, where row stochasticity of $\overleftarrow{\mathcal{B}}$ plays an important role.

Proposition 4.61. *Let* (3.2) *hold. Then the first derivative* $\overleftarrow{A}$ *with standard boundary conditions generates on* $L^1\big((0,1);\ell^1_\mu(\mathsf{E})\big)$ *a positive, contractive* C_0*-semigroup given by*

$$e^{t\overleftarrow{A}}u(x) := \overleftarrow{\mathcal{B}}^k u(t+x-k) \qquad \text{if } t+x \in [k,k+1),\ k \in \mathbb{N},\ x \in (0,1),\ t \geq 0. \tag{4.25}$$

In view of Lemma 4.4.(7) condition (3.2) ensures that $\overleftarrow{\mathcal{B}}$ is bounded on $\ell^1_\mu(\mathsf{E})$—thus in particular that the standard boundary condition $u(1) = \overleftarrow{\mathcal{B}}u(0)$ is well defined—and hence that each operator $e^{t\overleftarrow{A}}$ is bounded on $L^1\big((0,1);\ell^1_\mu(\mathsf{E})\big)$.

Formula (4.25) explicitly shows the influence of the graph's connectivity on the evolution of the system. The general case of $c \not\equiv 1$, $p \not\equiv 0$ can however also be treated, by abstract semigroup methods. That is, we want to prove that $M_c\overleftarrow{A} + M_p$ is a generator, where

$$\left(M_c\overleftarrow{A} + M_p\right)u(x) := \operatorname{diag}\left(c_{\mathsf{e}}(x)\frac{du_{\mathsf{e}}}{dx}(x) - p_{\mathsf{e}}(x)u_{\mathsf{e}}(x)\right)_{\mathsf{e}\in\mathsf{E}}, \qquad x \in (0,1),$$

with same domain as $D(\overleftarrow{A})$, i.e.,

$$\left\{u \in W^{1,1}\big((0,1);\ell^1_\mu(\mathsf{E})\big) : u(1) = \overleftarrow{\mathcal{B}}u(0)\right\}.$$

Proposition 4.62. *Let* (3.2) *hold and let*

$$0 < k_0 \leq c_{\mathsf{e}}(x) \leq K_0 \qquad \text{for some } k_0, K_0 > 0 \text{ and all } x \in (0,1).$$

Then $M_c\overleftarrow{A} + M_p$ *generates a* C_0*-semigroup on* $L^1\big((0,1);\ell^1_\mu(\mathsf{E})\big)$*. If in particular* $0 \leq p_{\mathsf{e}}$ *for all* $\mathsf{e} \in \mathsf{E}$*, then* $(e^{t\overleftarrow{A}})_{t\geq 0}$ *is contractive and real. It is positive if* G *is finite.*

Proof. It is easy to see that $\overleftarrow{A}$ is closed and densely defined, and so is $M_c\overleftarrow{A} + M_p$.

Let us first discuss the case of $0 \leq p_{\mathsf{e}}$ for all $\mathsf{e} \in \mathsf{E}$, endowing $L^1(\mathfrak{G})$ with the equivalent norm

$$|||u||| := \sum_{\mathsf{e}\in\mathsf{E}} \int_0^1 |u_{\mathsf{e}}(x)|\frac{\mu(\mathsf{e})}{c_{\mathsf{e}}(x)}dx, \qquad u \in L^1(\mathfrak{G}).$$

If $u \in D(M_c\overleftarrow{A} - M_p)$ and $v \in L^1(\mathfrak{G})$ with $R(\lambda, A)v = u$, then whenever $\operatorname{Re}\lambda > 0$ one has

$$\lambda|u_{\mathsf{e}}(x)| - c_{\mathsf{e}}(x)\frac{d}{dx}|u_{\mathsf{e}}(x)| + p_{\mathsf{e}}(x)|u_{\mathsf{e}}(x)| = \operatorname{sgn} u_{\mathsf{e}}(x)\cdot v_{\mathsf{e}}(x), \quad x \in (0,1),\ \mathsf{e} \in \mathsf{E}.$$

Multiplying by $\mu(\mathsf{e})/c_\mathsf{e}$, integrating over $(0,1)$, and summing over e one obtains

$$\begin{aligned}\lambda |||u||| &= \sum_{\mathsf{e}\in\mathsf{E}} \int_0^1 \frac{d}{dx}|u_\mathsf{e}(x)|\mu(\mathsf{e})dx - \frac{p_\mathsf{e}(x)}{c_\mathsf{e}(x)}|u_\mathsf{e}(x)|\mu(\mathsf{e})dx \\ &+ \sum_{\mathsf{e}\in\mathsf{E}} \int_0^1 \frac{\operatorname{sgn} u_\mathsf{e}(x)}{c_\mathsf{e}(x)} v_\mathsf{e}(x)\mu(\mathsf{e})dx \\ &\le \|u(1)\|_{\ell^1_\mu} - \|u(0)\|_{\ell^1_\mu} + \sum_{\mathsf{e}\in\mathsf{E}} \int_0^1 \frac{\operatorname{sgn} u_\mathsf{e}(x)}{c_\mathsf{e}(x)} v_\mathsf{e}(x)\mu(\mathsf{e})dx,\end{aligned}$$

since the functions p_e are positive. Owing to the node conditions satisfied by u one concludes that

$$\begin{aligned}\lambda |||u||| &\le \|\overleftarrow{\mathcal{B}} u(0)\|_{\ell^1_\mu} - \|u(0)\|_{\ell^1_\mu} + \sum_{\mathsf{e}\in\mathsf{E}} \int_0^1 \frac{\operatorname{sgn} u_\mathsf{e}(x)}{c_\mathsf{e}(x)} v_\mathsf{e}(x)\mu(\mathsf{e})dx \\ &\le (\|\overleftarrow{\mathcal{B}}\| - 1)\|u(0)\|_{\ell^1_\mu} + \sum_{\mathsf{e}\in\mathsf{E}} \int_0^1 \frac{|v_\mathsf{e}(x)|}{c_\mathsf{e}(x)} \mu(\mathsf{e})dx \\ &= (\|\overleftarrow{\mathcal{B}}\| - 1)\|u_\mathsf{e}(0)\|_{\ell^1_\mu} + |||v||| = |||v|||,\end{aligned}$$

where the last equality holds because the matrix $\overleftarrow{\mathcal{B}}$ is row stochastic and thus has norm 1 as a linear operator on ℓ^1_μ. The claim now follows from the theorem of Hille–Yosida.

The case of general p follows from Lemma 4.27, as P defined a bounded operator on $L^1(\mathfrak{G})$.

In order to prove positivity, by Phillips' Theorem 4.46 it suffices to check dispersivity of $\overleftarrow{A}$. We have to face the problem that, by Remark 3.18.(3), $L^1(\mathfrak{G})' \neq L^\infty(\mathfrak{G})$ unless G is finite. However, in the finite case we may take for all $u \in D(\overleftarrow{A})$ a $v := J(u^+) \in \mathcal{J}(u^+)$ defined by

$$v_\mathsf{e} := \frac{1}{c_\mathsf{e}} \mathbf{1}_{\{u_\mathsf{e} \ge 0\}}, \qquad \mathsf{e} \in \mathsf{E}.$$

Now, reasoning as above we find

$$\begin{aligned}\left\langle (M_c \overleftarrow{A} - M_p)u, v \right\rangle &= \sum_{\mathsf{e}\in\mathsf{E}} \int_0^1 \left(c_\mathsf{e} u'_\mathsf{e} \mathbf{1}_{\{u_\mathsf{e}\ge 0\}} - p_\mathsf{e} u_\mathsf{e} \mathbf{1}_{\{u_\mathsf{e}\ge 0\}} \right) \frac{\mu(\mathsf{e})}{c_\mathsf{e}} dx \\ &\le \sum_{\mathsf{e}\in\mathsf{E}} \int_0^1 \left(u'_\mathsf{e} \mathbf{1}_{\{u_\mathsf{e}\ge 0\}} \right) \mu(\mathsf{e}) dx\end{aligned}$$

$$\leq \| \left(\overleftarrow{\mathcal{B}} u(0)\right)^+ \|_{\ell^1_\mu} - \| (u(0))^+ \|_{\ell^1_\mu}$$
$$\leq \| \overleftarrow{\mathcal{B}} (u(0))^+ \|_{\ell^1_\mu} - \| (u(0))^+ \|_{\ell^1_\mu} = 0,$$

due to row stochasticity of $\overleftarrow{\mathcal{B}}$. This concludes the proof. □

Similar computations show that also $(e^{t\overrightarrow{A}})_{t\geq 0}$, $(e^{t\overleftarrow{A}^*})_{t\geq 0}$, $(e^{t\overrightarrow{A}^*})_{t\geq 0}$ are contractive.

Remarks 4.63. (1) If G is finite, then by Remark 3.18.(3) $L^1\big((0,1);\ell^1_\mu(\mathsf{V})\big)$ is the dual space of $L^\infty\big((0,1);\ell^\infty_\mu(\mathsf{V})\big)$, and we conclude that both $(e^{t\overrightarrow{A}})_{t\geq 0}, (e^{t\overleftarrow{A}})_{t\geq 0}$ extrapolate to a C_0-semigroup on $L^p\big((0,1);\ell^p_\mu(\mathsf{V})\big) \equiv L^p(\mathfrak{G})$ for all $p \in [1,\infty)$. The case of $p = 2$ has been considered in particular in [25]: It has been shown therein that the growth bound of the semigroup on $L^2(\mathfrak{G})$ is at most

$$\min_{\mathsf{e}\in\mathsf{E}} c_{\mathsf{e}} \log \| \overleftarrow{\mathcal{B}} \|_2$$

if all c_{e} are constant, where $\|\cdot\|_2$ denotes the spectral norm; and that this estimate is sharp in the sense that the semigroup is contractive if and only if $\| \overleftarrow{\mathcal{B}} \|_2 = 1$. Furthermore, even when a formula like (4.25) is not available, one can prove that $\overleftarrow{A}$ generates a C_0-group (resp., a unitary C_0-group) if and only if $\overleftarrow{\mathcal{B}}$ is invertible (resp., $\overleftarrow{\mathcal{B}}$ is unitary).

(2) This last result shows that the abstract Cauchy problem (ACP) associated with $\overleftarrow{A}$ is in general *not* backward well-posed. This is intuitively due to the fact that (Kr) only prescribes the behavior at each outgoing edge, but not at each incoming one. If the flow would be naively reversed, the backward evolving system would then in general lack proper node conditions. The correct way of reversing time is to consider $\overrightarrow{A}$ instead, but this does in general not yield standard boundary conditions (unless $\overrightarrow{\mathcal{B}} = \overleftarrow{\mathcal{B}}^{-1}$ of course, like in the case of a cycle).

(3) If we want this backward evolving system to coincide with the system associated with $\overleftarrow{A}^*$, and if we simultaneously require the node conditions to be local (i.e., only relating the boundary values of functions that converge in the same node of the network), then $\overleftarrow{\mathcal{B}}$ has to be a block-matrix with each block corresponding to a certain node. The dimension of the block associated with the node v is the number of edges, incoming in v, on whose *terminal* endpoint a condition is imposed in the forward evolving system; and also the number of edges, outgoing from v, on whose *initial* endpoint a condition is imposed in the backward evolving system. In other words, each node has the same number of incident outgoing and incoming edges. Hence, a self-adjoint realization in

$L^2(\mathfrak{G})$ of the momentum operator with local conditions can exist if and only if the graph is orientedly Eulerian, cf. Theorem A.10.

If G is finite, then by Lemma 3.27 $M_c \overleftarrow{A}$ has compact resolvent. In some special cases it is even possible to write down its resolvent operators explicitly.

For $\operatorname{Re}\lambda > 0$, let us denote by $\mathcal{E}_\lambda(s)$ the $\mathsf{E} \times \mathsf{E}$ matrix

$$\mathcal{E}_\lambda(s) := \operatorname{diag}\left(e^{\frac{\lambda(s-1)}{c_\mathsf{e}}}\right)_{\mathsf{e}\in\mathsf{E}}, \qquad s \in [0,1],$$

and by $D_\lambda : \mathbb{C}^\mathsf{V} \to D(\overleftarrow{A})$ the operator defined by

$$(D_\lambda d)(s) := \mathcal{E}_\lambda(s)\mathcal{M}\mathcal{I}^{-T}(\mathcal{D}^{\text{out}})^{-1}d, \qquad d \in \mathbb{C}^\mathsf{V},\ s \in (0,1).$$

Finally, $C := \operatorname{diag}(c_\mathsf{e})_{\mathsf{e}\in\mathsf{E}}$, while $M : D(\overleftarrow{A}) \to \mathbb{C}^\mathsf{V}$ is the operator defined by

$$Mf := \mathcal{I}^+ f(0), \qquad f \in D(\overleftarrow{A}).$$

Proposition 4.64. *If G is finite and each c_e is a constant function, $\mathsf{e} \in \mathsf{E}$, then for all $f \in X$ the resolvent operator of $M_c \overleftarrow{A}$ on $L^1\big((0,1);\ell^1(\mathsf{E})\big)$ is given by*

$$R(\lambda, M_c\overleftarrow{A})f(s) = \left(\mathrm{Id} + D_\lambda(\mathrm{Id} - MD_\lambda)^{-1}M\right)\int_s^1 \mathcal{E}_\lambda(s-\tau+1)C^{-1}f(\tau)d\tau, \quad s \in [0,1]. \tag{4.26}$$

The key point is that the significant vector-valued equation

$$\lambda u - M_c \overleftarrow{A} u = v$$

is equivalent to a family of equations

$$\lambda u_e - c_\mathsf{e}\frac{du_\mathsf{e}}{dx}u_\mathsf{e} = v_\mathsf{e}, \qquad \mathsf{e} \in \mathsf{E},$$

whose fundamental solutions are particularly simple if each c_e is a constant (indeed, in this case

$$u(s) = \frac{1}{c_\mathsf{e}}\int_s^1 e^{\frac{\lambda(s-\tau+1)}{c_\mathsf{e}}} f(\tau)\, d\tau, \qquad s \in [0,1]),$$

and to a family of compatibility conditions that have to be satisfied by the values of the solutions u_e in the endpoints of each interval/metric edge. Checking that such compatibility conditions force the resolvent operator to take the form (4.26) is just a matter of lengthy computations: Details can be found in [26]. (Clearly, the same idea allows to easily find an expression for the resolvent operators of the first derivative with different node conditions.)

Indeed, the integral on the right hand side of (4.26) is simply the resolvent operator of the first derivative with Dirichlet boundary conditions on the right endpoint: In this way the eigenvalue problem is effectively decomposed in a term that agrees with the solution in the classical case of individual, decoupled intervals; and another term—i.e., the matrix $\big(\mathrm{Id} + D_\lambda(\mathrm{Id} - MD_\lambda)^{-1}M\big)$—that encodes the connectivity of the graph. Using this decomposition it is not difficult to prove the following, cf. [26, Lemma 4.4].

Proposition 4.65. *Let* G *be finite and strongly connected. Then* $(e^{tM_c\overleftarrow{A}})_{t\geq 0}$ *is irreducible. It converges towards a periodic group.*

The second assertion follows from Proposition 4.38.

Another possible application of the explicit formula (4.26) is presented in the following.

Example 4.66. If all coefficients $c_\mathsf{e} \equiv 1$, then $\mathcal{E}_\lambda(s - \tau + 1)C^{-1}f(\tau)$ is a diagonal matrix for all $s \in [0, 1]$ and all $\tau \in (s, 1)$, hence the integral term in (4.26) leaves any subspace Y of $L^1\big((0, 1); \ell^1(\mathsf{E})\big)$ invariant. Let e.g. P be an orthogonal projector of $\mathbb{C}^\mathsf{E}$ and consider the subspace

$$Y := \{f \in L^1\big((0, 1); \ell^1(\mathsf{E})\big) : f(x) \in \operatorname{Rg} P \text{ for a.e. } x \in (0, 1)\}.$$

Then, by Corollary 4.29 one sees that Y is invariant under $(e^{t\overleftarrow{A}})_{t\geq 0}$ if and only if

$$D_\lambda(1 - MD_\lambda)^{-1}MY \subset Y \qquad \text{for some } \lambda \in \mathbb{R}.$$

By definition of M, D_λ such a condition only depends on the topology of the graph. Similar results will be obtained with less effort in Chap. 8.

4.6 Notes and References

Section 4.1. Lemma 4.3 was established in [4] and extended to the weighted case in [33]. In the unweighted case it was showed in [31] that the adjacency matrix is bounded on $\ell^p(\mathsf{V})$ for some/all p if and only if G is uniformly locally finite; and compact if and only if G is finite. Remarkably, in [34] V. Müller has disproved the conjecture, due to B. Mohar, that the adjacency operator of an (unweighted) locally finite graph is always self-adjoint—in fact, it is in general not even essentially self-adjoint as shown in [16].

On the other hand, by Lemma 4.3 the discrete Laplacian $\mathcal{L}$ is bounded and symmetric, hence self-adjoint on each uniformly locally finite graph. The question whether $\mathcal{L}$ is generally essentially self-adjoint has been open for a long time. It has been finally answered in the positive in [50] in the unweighted case, and then independently in [21, 23, 45] in the weighted case, provided the node weight ν is bounded away from 0. If the node weight is allowed to degenerate, or if the graph is not locally finite, then $\mathcal{L}$ is in general not essentially self-adjoint.

Since $\mathcal{L} = (\mathcal{I}\mathcal{C}^{\frac{1}{2}})(\mathcal{I}\mathcal{C}^{\frac{1}{2}})^T$, if for $p = 2$ $\mathcal{L}$ is bounded on $\ell^p(\mathsf{V})$ and hence

$$\|\mathcal{I}\mathcal{C}^{\frac{1}{2}}\|^2_{\mathcal{L}(\ell^p_\gamma(\mathsf{E}),\ell^p(\mathsf{V}))} = \|\mathcal{L}\|_{\mathcal{L}(\ell^p(\mathsf{V}))} < \infty,$$

then $\mathcal{I}$ is bounded from $\ell^p_\gamma(\mathsf{E})$ to $\ell^p(\mathsf{V})$, hence by Lemma 4.3.(3) G is uniformly locally finite. This observation has been generalized to all $p \in [1,\infty]$ in [19, Theorem 9.3].

On a graph without sinks (resp., sources) one may also define normalized versions of the outgoing (resp., incoming) Kirchhoff or advection matrices, like $\overrightarrow{\mathcal{L}}_{\text{norm}} := (\mathcal{D}^{\text{out}})^{-\frac{1}{2}}\overrightarrow{\mathcal{N}}(\mathcal{D}^{\text{out}})^{-\frac{1}{2}}$. Because a pendant of Lemma 4.2 holds replacing deg by deg^{out}, and because one sees that $\mathcal{K}^{\text{out}}$ and hence its transpose $\overrightarrow{\mathcal{N}}$ are bounded on $\ell^2_{\text{deg}^{\text{out}}}(\mathsf{V})$, one concludes that $\overrightarrow{\mathcal{L}}_{\text{norm}}$ is bounded on $\ell^2(\mathsf{V})$ regardless of the connectivity of G.

Section 4.2. A modern, convenient survey of the theory of C_0-semigroups can be found in [11] or, in more brief form, in many other monographs including [1,3,20]. The celebrated Theorem 4.20 was obtained simultaneously but independently in 1948 by E. Hille and K. Yosida and marked the birth of this theory. Over the last 60 years, semigroup theory has become very rich in its own right, beyond the elementary interplay with first order Cauchy problems. In particular, dwelling on the classical Perron–Frobenius theory a comprehensive collection of results on long-time behavior of positive C_0-semigroups has been obtained among others in [1, 9, 35, 46]. Proposition 4.31 is a classical invariance criterion obtained by H. Brezis in [2]. The slightly generalized version presented here is a special case of [51, Theorem 2.4]. Theorems 4.23 and 4.46 appeared in [28] and [40], respectively.

In general, in view of Proposition 4.12 there is no point in looking for semigroups that are better than merely strongly continuous: it can be easily proved (cf. [11, Theorem I.3.7]) that any family of bounded linear operators that satisfy the semigroup law and are continuous with respect to the operator norm is necessarily generated by a bounded operator—unlike the operators that appear in partial differential equations, which are typically unbounded. (This is also the reason for us to focus on functional settings based on Lebesgue L^p-spaces for $1 \leq p < \infty$, in the following, thus excluding L^∞ from our considerations. Indeed, by a result due to H.P. Lotz, a C_0-semigroup on some L^∞-space is automatically norm continuous in 0, and hence it is necessarily generated by a bounded operator.)

Theorem 4.26 was first obtained by A.M. Gomilko in [17], cf. also [41]. When it was published, Gomilko's Theorem answered a long-standing question. It has been variously generalized ever since, among others in [8, Theorem 2.4] as follows.

Proposition 4.67. *Let A be a densely defined closed linear operator on a Banach space X and $\sigma(A) \subset \{z \in \mathbb{C} : \operatorname{Re} z \leq 0\}$. If*

$$\int_{\delta-i\infty}^{\delta+i\infty} \left| \left\langle \frac{d}{d\lambda} R(\lambda, A)x, y \right\rangle \right| |d\lambda| < \frac{M}{\delta}(1 + \delta^{-d})\|x\|\|y\| \quad \textit{for all } x \in X,\ y \in X'$$

holds for all $\delta > 0$ *and some* $d \geq 0$*, then* A *generates a* C_0*-semigroup such that*

$$\|e^{tA}\| \leq K(1 + t^d)$$

for some $K > 0$ *and all* $t > 0$.

In both this and Gomilko's original version it is remarkable that A is not a priori assumed to be a generator.

The idea of studying domination as an invariance property was developed by E.M. Ouhabaz in [39, § 3]. The problem of finding the orthogonal projector of $L^2(U, \tilde{\mu})$ onto the L^p-unit ball has been open for a long time, until R. Nittka could determine in [36] a semi-explicit formula for it; this is in turn enough to apply some variants of Proposition 4.31. Lemma 4.41 has been taken from [11, § II.2.3].

As we already mentioned in Chap. 2, in this book we mostly focus on time-continuous evolution equations, but in view of specific applications it is sometimes more appropriate to consider discrete time steps instead. One is then led to *time discrete abstract Cauchy problems* like

$$f(n+1) = \mathcal{Z} f(n), \qquad n \in \mathbb{N},$$

so that in general $f(n) = \mathcal{Z}^n f(0)$ for some operator $\mathcal{Z}$ (typically, a matrix) and the relevant asymptotic issue is whether the discrete operator semigroup $(\mathcal{Z}^n)_{n\in\mathbb{N}}$ converges in some sense as $n \to \infty$. A necessary condition is clearly that the operator be power-bounded. The critical case is now that of eigenvalues that lie on the unit circle $U(1)$ of $\mathcal{Z}$, rather than generator's eigenvalues on $i\mathbb{R}$. In comparison with the analysis of long-time behavior of C_0-semigroups the discrete theory is more involved, essentially due to the fact that—unlike $i\mathbb{R}$—$U(1)$ contains non-trivial finite subgroups. Let us only mention the following pendant of Proposition 4.38, which follows from [22, Theorem 5.6] remembering that Lebesgue L^p-spaces have order continuous norm for all $p \in [1, \infty)$.

Proposition 4.68. *Let* $p \in [1, \infty)$ *and let* $\mathcal{Z}$ *be a positive irreducible power-bounded operator on* $L^p(U, \tilde{\mu})$*. Then* $(\mathcal{Z}^n)_{n\in\mathbb{N}}$ *converges almost weakly to a periodic group* $(\mathcal{U}^n)_{n\in\mathbb{Z}}$*; this means that there exist spaces* L^p_0, L^p_{ap} *such that*

- $L^p(\mathsf{V}) = L^p_0 \oplus L^p_{\mathrm{ap}}$ *and both* L^p_0 *and* L^p_{ap} *are invariant under* $\mathcal{Z}$*;*
- *there exists a periodic group* $(\mathcal{U}^n)_{n\in\mathbb{Z}}$ *on* L^p_{ap} *such that* $\mathcal{Z} \equiv \mathcal{U}$ *on* L^p_{ap}*;*
- *for every* $x \in L^p_0$ *there exists a sequence* $(n_j)_{j\in\mathbb{N}} \subset \mathbb{N}$ *such that* $\mathcal{Z}^{n_j}x$ *converges weakly to* 0.

We stress that (linear) continuous dynamical systems and discrete ones are tightly related. Indeed, let $-\mathcal{W} = (\omega_{\mathsf{vw}})$ be a $\mathsf{V} \times \mathsf{V}$-matrix that generates a Markovian semigroup, which is then by Example 4.44 of the form

$$\mathcal{W} = \begin{pmatrix} \sum_{\mathsf{w}\neq\mathsf{v}} \omega_{\mathsf{vw}} & & -\omega_{\mathsf{vz}} \\ & \ddots & \\ -\omega_{\mathsf{zv}} & & \sum_{\mathsf{w}\neq\mathsf{z}} \omega_{\mathsf{zw}} \end{pmatrix}.$$

Now, let $\lambda := \max_{\mathsf{v}\in\mathsf{V}} \sum_{\mathsf{w}\neq\mathsf{v}} \omega_{\mathsf{vw}}$, so that

$$\mathcal{Z} := \mathrm{Id} - \frac{1}{\lambda}\mathcal{W}$$

is a row sub-stochastic matrix. Then

$$e^{t\mathcal{W}^T} = e^{t\lambda(\mathcal{Z}^T - \mathrm{Id})} = \sum_{k=0}^{\infty} \frac{e^{-\lambda t}(\lambda t)^k}{k!} \mathcal{Z}^{Tk}, \qquad t \geq 0.$$

Observe that terms corresponding to a homogeneous Poisson distribution and the transition-like matrix $\mathcal{Z}$ coexist, yielding some kind of *jump sub-Markovian process*, cf. [5, Chap. 12].

Conversely, if $\mathcal{Z}$ is a stochastic matrix, then $\mathrm{Id} - \mathcal{Z}$ is negative off-diagonal and satisfies (4.11), so that $(e^{t(\mathcal{Z}-\mathrm{Id})})_{t\geq 0}$ generates a Markovian semigroup—and hence, $(e^{t(\mathcal{Z}^T-\mathrm{Id})})_{t\geq 0}$ is a stochastic semigroup: A similar idea is behind the introduction of the heat kernel pagerank in Definition 2.30. In both cases, the properties of the random walk described by $\mathcal{W}$ is enhanced by perturbing it by a lazy walk (at each step the walker has a certain probability of not moving), rather than by random surfing like in Google's algorithm.

Section 4.3. Cosine operator functions were first studied by M. Sova in [44]. We have summarized elementary results, but more detailed overviews can be found in [13, 48] and [1, Sect. 3.14]. Observe that the operator B discussed in Example 4.50 is the square of the operator A introduced in Example 4.16.

Conversely, by a classical theorem of H.O. Fattorini each C_0-cosine operator function generator on L^p-spaces in the reflexive range is, up to a scalar perturbation, the square of a C_0-group generator. Proposition 4.53 is taken from [29], where a previous partially erroneous assertion in [15] was corrected.

Unlike in the case of semigroups, the stability theory of cosine operator functions is rather poor. This is mainly due to the fact, discussed e.g. in [1], that if for a C_0-cosine operator function $(C(t))_{t\geq 0}$ $\lim_{t\to\infty} C(t)x = 0$, then necessarily $x = 0$. Seemingly, the only reasonable asymptotic notion is that of boundedness (or perhaps of *polynomial* boundedness, in the sense of Proposition 4.67 above. A cosine operator function counterpart of Theorem 4.26 has been proved in [27].

Section 4.5. The generator property of $C\overleftarrow{\mathcal{A}} + M_p$ has been proved in [26], while (4.25) has been found by B. Dorn in [6, Proposition 1.2.1 and Corollary 3.2.5].

Proposition 4.64 is [26, Proposition 3.3]. Several subsequent papers of the Tübingen school have been devoted to more and more general advection processes on metric graphs, including infinite ones in [7]. If G is an oriented cycle, then it is clear that $(e^{t\overleftarrow{\mathcal{A}}})_{t\geq 0}$ embeds in a periodic C_0-group; if it is not, one may expect that the transported matter either disappears from the system, or it stabilizes on a set of closed orbits. Actually, the following refinement of Proposition 4.65 holds by [26, Theorem 4.5] and [42, Theorem 2.4.11], where the notions introduced in Proposition 4.68 are extended in a natural way.

Proposition 4.69. *If* G *is strongly connected, then* $(e^{t\overleftarrow{\mathcal{A}}})_{t\geq 0}$ *converges exponentially towards a direct sum of periodic groups, one for each strongly connected component, whose periods are the greatest common divisors of the cycle lengths inside the components.*

Thus, in some sense the metric graph asymptotically splits into the direct sum of its strongly connected components. More refined results are presented in [42, Chap. 2].

If $\overleftarrow{\mathcal{B}}$ is replaced by a general—especially, not necessarily row-stochastic—matrix, we are allowing for absorption and/or generation phenomena in the nodes. This setting can be discussed in a similar way using the idea presented in [6, Remark at page 45]. Also time-dependent-conditions have been considered in [43]. The generation results and growth estimates in [25] have been recently generalized and refined in [10]. The observation in Remark 4.63.(3) appears in [12].

References

1. W. Arendt, C.J.K. Batty, M. Hieber, F. Neubrander, *Vector-Valued Laplace Transforms and Cauchy Problems*, volume 96 of *Monographs in Mathematics* (Birkhäuser, Basel, 2001)
2. H. Brézis, *Operateurs Maximaux Monotones et Semi-Groupes de Contractions dans les Espaces de Hilbert* (North-Holland, Amsterdam, 1973)
3. H. Brezis, *Functional Analysis, Sobolev Spaces and Partial Differential Equations* (Universitext. Springer, Berlin, 2010)
4. S. Cardanobile, The L^2-strong maximum principle on arbitrary countable networks. Lin. Algebra Appl. **435**, 1315–1325 (2011)
5. E.B. Davies, *Linear Operators And Their Spectra* (Cambridge University Press, Cambridge, 2007)
6. B. Dorn, Semigroups for flows on infinite networks. Master's thesis, Eberhard-Karls-Universität, Tübingen, 2005
7. B. Dorn, Flows in Infinite Networks – A Semigroup Approach. PhD thesis, Eberhard-Karls-Universität, Tübingen, 2008
8. T. Eisner, Polynomially bounded C_0-semigroups. Sem. Forum **70**, 118–126 (2005)
9. T. Eisner, *Stability of Operators and Operator Semigroups*. Operator Theory: Advances and Applications, vol. 209 (Birkhäuser, Basel, 2010)
10. K.-J. Engel, Generator property and stability for generalized difference operators. J. Evol. Equ. **13**, 311–334 (2013)
11. K.-J. Engel, R. Nagel, *One-Parameter Semigroups for Linear Evolution Equations*. Graduate Texts in Mathematics, vol. 194 (Springer, New York, 2000)

12. P. Exner, Momentum operators on graphs, in *Spectral Analysis, Differential Equations and Mathematical Physics: A Festschrift in Honor of Fritz Gesztesy's 60th Birthday*, ed. by H. Holden, B. Simon, G. Teschl. Proceedings of Symposia in Pure Mathematics, vol. 87 (American Mathematical Society, Providence, 2013), pp. 105–118
13. H.O. Fattorini, *Second Order Linear Differential Equations in Banach Spaces*. Mathematical Studies, vol. 108 (North Holland, Amsterdam, 1985)
14. M. Fiedler, Algebraic connectivity of graphs. Czech. Math. J. **23**, 298–305 (1973)
15. E. Giustim, Funzioni coseno periodiche. Boll. UMI **22**, 478–485 (1967)
16. S. Golénia, C. Schumacher, Comment on "the problem of deficiency indices for discrete schrodinger operators on locally finite graphs". J. Math. Phys. **54**, 0641010 (2013)
17. A.M. Gomilko, Conditions on the generator of a uniformly bounded C_0-semigroup. Funct. Anal. Appl. **33**, 294–296 (1999)
18. L.J. Grady, J.R. Polimeni, *Discrete Calculus: Applied Analysis on Graphs for Computational Science* (Springer, New York, 2010)
19. S. Haeseler, M. Keller, D. Lenz, R. Wojciechowski, Laplacians on infinite graphs: Dirichlet and Neumann boundary conditions. J. Spectral Theory **2**, 397–432 (2012)
20. B. Jacob, H. Zwart, *Linear Port-Hamiltonian Systems on Infinite-dimensional Spaces*, Operator Theory: Advances and Applications, vol. 223 (Birkhäuser, Basel, 2012)
21. P.E.T. Jorgensen, Essential self-adjointness of the graph-Laplacian. J. Math. Phys. **49**, 073510 (2008)
22. V. Keicher, R. Nagel, Positive semigroups behave asymptotically as rotation groups. Positivity **12**, 93–103 (2008)
23. M. Keller, D. Lenz, Dirichlet forms and stochastic completeness of graphs and subgraphs. J. Reine Angew. Math. **666**, 189–223 (2012)
24. S.V. Kislyakov, Sobolev imbedding operators and the nonisomorphism of certain Banach spaces. Funct. Anal. Appl. **9**, 290–294 (1975)
25. B. Klöss, Difference operators as semigroup generators. Semigroup Forum **81**, 461–482 (2010)
26. M. Kramar, E. Sikolya, Spectral properties and asymptotic periodicity of flows in networks. Math. Z. **249**, 139–162 (2005)
27. S. Król, Resolvent characterisation of generators of cosine functions and C_0-groups. J. Evol. Equ. **13**, 281–309 (2013)
28. G. Lumer, R.S. Phillips, Dissipative operators in a Banach space. Pac. J. Math. **11**, 679–698 (1961)
29. D. Lutz, Periodische operatorwertige Cosinusfunktionen. Results Math. **4**, 75–83 (1981)
30. N. Metropolis, A.W. Rosenbluth, M.N. Rosenbluth, A.H. Teller, E. Teller, Equation of state calculations by fast computing machines. J. Chem. Phys. **21**, 1087 (1953)
31. B. Mohar, The spectrum of an infinite graph. Linear Alg. Appl. **48**, 245–256 (1982)
32. C. Moler, C. Van Loan, Nineteen dubious ways to compute the exponential of a matrix, twenty-five years later. SIAM Rev. **45**, 3–49 (2003)
33. D. Mugnolo, Parabolic theory of the discrete p-Laplace operator. Nonlinear Anal. Theory Methods Appl. **87**, 33–60 (2013)
34. V. Müller, On the spectrum of an infinite graph. Lin. Algebra Appl. **93**, 187–189 (1987)
35. R. Nagel (ed.), *One-Parameter Semigroups of Positive Operators*. Lectures Notes on Mathematics, vol. 1184 (Springer, Berlin, 1986)
36. R. Nittka, Projections onto convex sets and L^p-quasi-contractivity of semigroups. Arch. Math. **98**, 341–353 (2012)
37. R. Olfati-Saber, J.A. Fax, R.M. Murray, Consensus and cooperation in networked multi-agent systems. Proc. IEEE **95**, 215–233 (2007)
38. R. Olfati-Saber, R.M. Murray, Consensus problems in networks of agents with switching topology and time-delays. IEEE Trans. Autom. Control **49**, 1520–1533 (2004)
39. E.M. Ouhabaz, Invariance of closed convex sets and domination criteria for semigroups. Potential Anal. **5**, 611–625 (1996)
40. R.S. Phillips, Semi-groups of positive contraction operators. Czech. Math. J. **12**, 294–313 (1962)

41. D.H. Shi, D.X. Feng, Characteristic conditions of the generation of C_0 semigroups in a Hilbert space. J. Math. Anal. Appl. **247**, 356–376 (2000)
42. E. Sikolya, Semigroups for flows in networks. PhD thesis, Eberhard-Karls-Universität, Tübingen, 2004
43. E. Sikolya, Flows in networks with dynamic ramification nodes. J. Evol. Equ. **5**, 441–463 (2005)
44. M. Sova, Cosine operator functions. Dissertationes Math. **49**, 1–47 (1966)
45. N. Torki-Hamza, Essential self-adjointness for combinatorial Schrödinger operators I - Metrically complete graphs. Confluentes Math. **2**, 333–350 (2010)
46. J. Van Neerven, *The Asymptotic Behaviour of Semigroups of Linear Operators*. Operator Theory: Advances and Applications, vol. 88 (Birkhäuser, Basel, 1996)
47. L.R. Varshney, B.L. Chen, E. Paniagua, D.H. Hall, D.B. Chklovskii, Structural properties of the *Caenorhabditis elegans* neuronal network. PLoS Comput. Biol. **7**, e1001066 (2011)
48. V.V. Vasil'ev, S.I. Piskarev, Differential equations in banach spaces ii. theory of cosine operator functions. J. Math. Sci. **122**, 3055–3174 (2004)
49. J. Voigt, One-parameter semigroups acting simultaneously on different L_p-spaces. Bull. Soc. Royale Sci. Liège. **61**, 465–470 (1992)
50. R.K. Wojciechowski, Stochastic Completeness of Graphs. PhD thesis, City University of New York, 2007
51. T. Yokota, Invariance of closed convex sets under semigroups of nonlinear operators in complex Hilbert spaces. SUT J. Math. **37**, 91–104 (2001)

Chapter 5
And Now Something Completely Different: A Crash Course in Cortical Modeling

In the previous chapters we have introduced a convenient functional analytical framework for studying evolution equations on network-like structures, along with some first examples. In the next chapters we will devote much attention to the investigation of properties of partial differential equations that are motivated by applications, and we will do so by applying the methods developed in the first half of the book. The reader will be perhaps puzzled to see that virtually all evolution equations treated in the next chapters are of diffusion and wave type. Why are wave and, above all, diffusion phenomena so ubiquitous in nature? An answer is not known—or perhaps this question is ill-posed all-together. Still, we feel that we owe the reader some convincing examples of real-life emergence of diffusion-type equations as well as of differential equations in networks.

The emergence of actual evolution equations in applications is a good source of inspiration for the development of the theory, and it is mostly in consideration of applications that we are going to present both the continuous and the discrete theory jointly throughout the book. In several fields of applied sciences, mathematical models that involve the former or the latter coexist, perhaps to capture different scales of a system. This becomes strikingly clear if one looks at how evolution equations on networks usually arise in the context of theoretical neuroscience. Conversely, it seems that several of the mathematical objects we will encounter in this book can be paradigmatically introduced through neuronal models, which will be therefore described in some detail in this chapter.

Indeed, modern neuroscience is a research field largely devoted to the investigation of dynamic behaviors *of* networks as well as *in* networks. While most theoretical neuroscientists are well trained in mathematics, only seldom do mathematicians have more than only a rough idea of brain theory. Therefore, it is often surprising and exciting for the latter to discover how pervasively the mathematics of networks (graph theory, network analysis and discrete dynamical system, above all) influences brain research. In this chapter we have tried to give a brief overview of this influence, focusing on the interactions that are likely to prove

D. Mugnolo, *Semigroup Methods for Evolution Equations on Networks*, Understanding Complex Systems, DOI 10.1007/978-3-319-04621-1_5,

more fascinating for the professional mathematician and neurobiological layman. Diffusion phenomena will indeed appear throughout.

After a dispute that lasted for decades at the end of nineteenth century, neurons were eventually accepted by the scientific community as the building blocks of the nervous system: This was finally certified when the Nobel Prize in Physiology or Medicine was awarded to S. Ramón y Cajal in 1906. Even if his theories had to be variously extended and updated in the light of new experimental findings, it is nowadays still generally maintained that neurons are the fundamental computational units of animal brains. These can in turn be fairly accurately described as incredibly complex and greatly efficient ramified structures based on geometric and functional juxtaposition of neurons.

Each neuron is a cell, whose dimension can be considerably different. It essentially consists of a collection of *dendrites*, a *soma* (the cell's body), and an *axon* (Fig. 5.1).

A dendrite is a thin fiber on whose surface thousands of appendages (*spines*) are found: they collect electrical *synaptic* impulses from other neurons and transmit them toward the dendrite's ending and then, via branching points, to further dendrites. In each neuron dendrites form a ramified structure: If we think of it as a tree—in the graph-theoretical meaning—then the soma is its root. Signal is conducted by dendrites towards the soma, where it is elaborated and eventually propagated along the *axon*. In its final tract, each axon ramifies into many terminals that collectively constitute the *axonal tree*. At their ends a *synapse* is responsible for passing signal to other neurons.

In most common models, dendrites passively transmit electrical potential without any form of self-excitation (this simplifying assumption has been variously disproved though, cf. [16]); while in the soma electric charge is accumulated until it surpasses a certain electrophysiological threshold. Both processes are jointly described by the *lumped soma model* that was first proposed by W. Rall in [15] and several subsequent papers: it consists of a *linear diffusive partial differential equation*

$$\frac{\partial u}{\partial t}(t,x) = \frac{\partial^2 u}{\partial x^2}(t,x) - u(t,x) \tag{5.1}$$

along a single so-called *equivalent cylinder*, which represents the transport on the whole collection of dendrites. This is complemented by a *dynamic boundary condition*

$$\frac{\partial u}{\partial t}(t,\mathsf{v}_s) = -u'(t,\mathsf{v}_s) - u(t,\mathsf{v}_s) \tag{5.2}$$

imposed on the endpoint v_s of the equivalent cylinder corresponding to the soma, which is for simplicity assumed to be isopotential, cf. also [18] for a more detailed derivation. Rall formulated his hypothesis under strong, possibly unrealistic symmetry assumptions on the network structure: His work has been later generalized

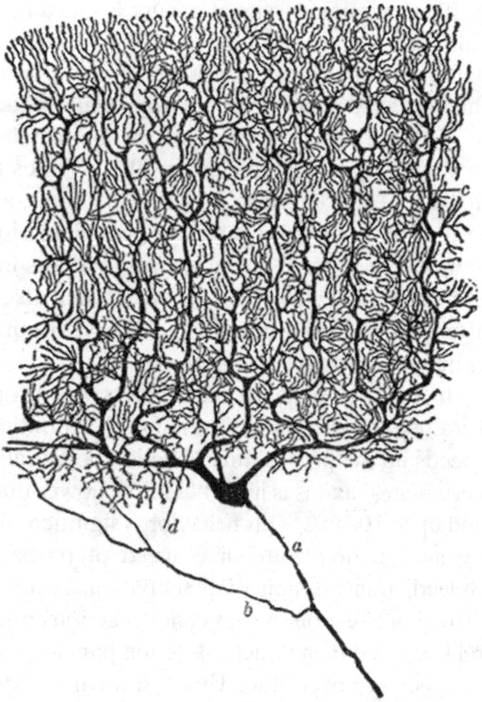

Fig. 5.1 Drawing of a neuron by S. Ramón y Cajal

to arbitrary dendritic networks in a series of papers beginning with [13], where the hypothesis of the equivalent cylinder was dropped and the actual geometric configuration was studied imposing in each branching point v_b a condition of the form

$$\sum u'_{\mathsf{e}}(t, \mathsf{v}_b) = 0 \tag{5.3}$$

along with (5.2). Here the sum of the incoming and outgoing currents u'_{e} is taken over all incident dendrites. The electric potential is assumed not to make jumps at branching points. Of course, this does *not* imply that an analogous continuity condition is also satisfied by the incoming and outgoing currents.

Electric charge accumulates in the soma after being transmitted along the dendrites: this is the mathematical meaning of the above dynamic boundary condition (5.2). Thus, it may seem natural to think of the soma as a capacitor, but this analogy is partially misleading: Since the soma is not perfectly isolated, ions tend to diffuse over time and cause the potential to decrease naturally. This "leak" is modeled by the second term in the right hand side of (5.2). Whenever the potential reaches a phenomenological threshold of approx. −55 mV, though, the soma can finally discharge releasing ions in the axon. This process of charge accumulation, decrease, and transmission is commonly considered as a form of signal elaboration: neural models that rely on this approach are referred to as *leaky integrate-and-fire* (shortly: *LIF*).

Transmission of signal occurs by means of a short, intense wave of potential called *action potential* or *spike*: This can be approximately thought of as a Dirac delta traveling along axons and probably also along dendrites. The hypothesis of this so-called *all-or-none law* was finally underpinned by the experimental findings of C. Sherrington and E. Adrian, for which they were awarded the Nobel Prize in Physiology or Medicine in 1932. Spikes are usually initiated in the so-called *hillock*, at the interface between the soma and the axon, whenever the potential in the soma surpasses a certain phenomenological threshold ξ. The precise way a neuron's membrane potential rises and falls may depend on various factors, usually of biochemical nature. On the other hand, also the threshold may vary in time: Right after a spike has been initiated, the value of ξ typically increases for some time, called *refractory period*.

In order to enhance the conductivity properties of biological fibers, axons of vertebrates are usually covered with a continuous *myelin sheath* that significantly speeds up the propagation of action potentials. Still, speed of signal propagation in vertebrates' axons is high, but fairly slower than that of light: usually approx. 30 m/s, and up to 100 m/s. This behavior is significantly different to what would be predicted by an equation with finite speed of propagation, like the linear heat equation. Indeed, transmission of potential in axons is usually modeled by a semilinear diffusion equation. Axons conduct actions potentials at a speed of approx. 30 m/s by relaxing and then re-activating ion pumps whose work allows to modify membrane conductance over time: This cost them energy. The conduction of action potentials would soon stop, had the neurons not developed a regenerative self-excitation mechanism. The most widely accepted model of signal propagation in the axon was suggested in 1952 by A.L. Hodgkin and A.F. Huxley in [11, Sect. 3], who based on their studies of the "giant axon" that controls the propulsion in squid. As this specific one, discovered in 1936, is much thicker than any other common axon, Hodgkin and Huxley were able to perform pioneering in vitro experiments that led them to describe the transmission of potential by means of a system of differential equations. Its mathematical structure is the following:

$$\begin{cases} \frac{\partial u}{\partial t}(t,x) = \frac{\partial^2 u}{\partial x^2}(t,x) - (u(t,x) - u_{\mathrm{L}}) - (n(t,x))^4\,(u(t,x) - u_{\mathrm{K}}) \\ \qquad\qquad - (m(t,x))^3\,h(t,x)\,(u(t,x) - u_{\mathrm{Na}}) \\ \frac{\partial m}{\partial t}(t,x) = \alpha_m(u(t,x))(1 - m(t,x)) - \beta_m(u(t,x))(m(t,x)), \\ \frac{\partial h}{\partial t}(t,x) = \alpha_h(u(t,x))(1 - h(t,x)) - \beta_h(u(t,x))(h(t,x)), \\ \frac{\partial n}{\partial t}(t,x) = \alpha_n(u(t,x))(1 - n(t,x)) - \beta_n(u(t,x))(n(t,x)). \end{cases} \tag{HH}$$

For simplicity, no boundary conditions were imposed, thus the axon is sketched as a straight line of infinite length, whose membrane potential at time t and at point x is $u(t,x)$ along the axon. So why does propagation of an action potential not eventually stop? The explanation suggested in [11] is ingenious: Chains of ion channels of three different kinds are active in the membrane and may enable or inhibit transmission, depending on whether they are closed or opened. The probability that these channels are open at time t and at point x of the axon is given by $m(t,x), h(t,x), n(t,x)$, respectively, so that the three additional terms in the first equation represent a *l*eak current along with the current carried by K^+ and Na^+ ions, respectively, whereas u_L, u_{Na}, u_K represent some reference potentials. (A lot of phenomenological parameters have been omitted in (HH): In particular, they account for the well-known fact that conduction velocity in thick nerve fibers is greater than in thin ones.)

The first equation in (HH) is an example of *nonlinear diffusion equation*, a ubiquitous object in applied mathematics that is both rich in structure and mathematically well understood and which we have already met in Example 4.14 and Sect. 4.4. This model has been a breakthrough in modern neuroscience and has earned Hodgkin and Huxley a Nobel Prize in Physiology or Medicine in 1963. Nevertheless, as it stands it does not account for several experimentally observed features of neurons (cf. [6, Sect. 1.4]) and has thus been variously patched over the last decades. Let us mention only two of the many proposed improvements.

Apparently, the Hodgkin–Huxley model says nothing about how different neurons communicate. We are going to discuss synaptic coupling of neurons soon, but since the 1940s there has also been experimental evidence of *ephaptic* coupling, i.e., of mutual electric interactions among axons that form bundles. While not strong enough in their own right to trigger action potentials in neighboring neurons, ephaptic couplings can still effectively induce depolarization or hyperpolarization and hence in turn influence the timing of spikes, as shown in [1]. A possible explanation for ephaptic communication is based on ionic diffusion in the extracellular medium [3], but it is more common to see it as an effect of mutual electric excitability of neighboring neurons—possibly of pathological origin. Models for this mode of interaction are based on involved electric circuits that are not series and boil down to coupled systems of diffusion equations of the form

$$\frac{\partial u_{\mathsf{e}}}{\partial t}(t,x) = \sum c_{\mathsf{ef}} \frac{\partial^2 u_{\mathsf{f}}}{\partial x^2}(t,x),$$

cf. [12], or to nonlinear versions thereof. Here (c_{ef}) is usually a (doubly) stochastic, positive definite matrix whose entries encode the influence exercised on the potential $u_{\mathsf{e}}(t,x)$ at point x of axon e by the potential $u_{\mathsf{f}}(t,x)$ at same point x of neighboring axon f. The sum is taken over all such axons f.

It appears that ephaptic coupling, a comparatively usual mode of communications in the invertebrates' brain, is rather rare in vertebrates, and so are the corresponding models.

Another, much more problematic aspect of the original Hodgkin–Huxley system is related to one of the main properties of action potentials: As observed in vitro, spikes are transmitted along the axon without any change in either their profile or their amplitude. However, it is not obvious whether a traveling spike may in fact be a solution of (HH), if one fits all parameters of (HH) in a biologically realistic way, cf. [8, 17]. One might alternatively take a chance on the *advection equation*

$$\frac{\partial u}{\partial t}(t,x) = c\frac{\partial u}{\partial x}(t,x),$$

for some $c \neq 0$. We have seen in Example 4.16 that its solutions are of the form

$$f(x + c^{-1}t)$$

for suitably smooth waveforms f (traveling to the left or to the right, depending on whether c is positive or rather negative) that depend on the initial data. But this is not really an option, as the advection equation has no biophysical interpretation and does not account for many other neuronal phenomena. (For example, it is known that neurons are not directional structures as would be asserted by a model based on a first order partial differential equation, and indeed spikes may actually be propagated also against the usual direction of transmission, both along axons and dendrites, if an action potential happens to be triggered beyond the hillock).

Instead, in order to address these points of criticism several corrections to the Hodgkin–Huxley model have been proposed since [11]: Most of them consist in adding nonlinear terms to the first equation of (HH) in order to guarantee wave-like propagation of action potentials.

Both (HH) and all its modifications depend on certain assumptions on the behavior of ion channels and describe the variations in a cell's *membrane potential*. In very recent years some new models have been proposed that dismiss this approach and instead rely upon different derivations. Most notably, T. Heimburg and A.D. Jackson have argued that a cell's *density change* (with respect to some reference density) should instead be modeled: By a thermodynamical derivation, and arguing that a rather specific kind of traveling waves (so-called "solitons") are the natural transmission mode, they have proposed in [10] to replace the whole system (HH) by the equation

$$\frac{\partial^2 \phi}{\partial t^2}(t,x) = \frac{\partial}{\partial x}\left(\left(1 - \phi + \phi^2\right)\frac{\partial \phi}{\partial x}\right)(t,x) - \frac{\partial^4 \phi}{\partial x^4}(t,x). \tag{5.4}$$

This is a *nonlinear beam equation*, a representative of the class of so-called *dispersive* equations, for whose solutions different wavelengths propagate at different phase velocities. Properties of dispersive wave equations are quite different from those of nonlinear diffusion-type equations and their mathematical theory is much harder. This model seems to be in interesting accordance with some experimental data, but such a paradigm shift from a conductance-based description to a pressure-based one is still far from being part of mainstream neuroscience. (Another classical derivation of a neural model—this time for the activity of the excitatory synapses—that leads to a wave-type equation comparable to (5.4) is due to P.L. Nunez [14].)

So far we have sketched the transmission of signal from the dendrites through the soma and finally along the axon. While axonal endings are ramified structures on their own right, their limited size makes difficult to derive experimentally a description of how signal splits in axonal branching points. Indeed some experimental data seem to suggest that the spikes' amplitude is preserved as they approach the axonal endings, cf. [4], so we will neglect this aspect and jump to final phase: the arrival of signal at a synapse.

A human brain contains approx. 500 trillions of synapses, either of *electrical* or *chemical* kind: Each of them is responsible for transmitting signal from the *presynaptic neuron* to the *postsynaptic* one. Electrical synapses recall nodes in electric circuits: They can transmit signal in both directions, preserve the profile of the signal and attenuate its amplitude (owing to the membrane resistance)—and do so almost without delay, thus significantly reducing reaction times. However, it is chemical synapses that carry out most of the task of signal transmission, as we know owing to the experiments of H. Dale and O. Loewi that earned them the Nobel Prize in Physiology or Medicine in 1936. The operation of chemical synapses relies upon release and subsequent reception of amino acids, peptides and other chemicals (collectively called *neurotransmitters*) by the presynaptic and postsynaptic neuron, respectively, upon arrival of an action potential. The release mechanism of neurotransmitters may be modeled similarly to ion channels in (HH) as they, too, act as gates—this means in particular that signal transmission is only allowed in one direction. Unlike transmission in less sophisticated electrical synapses, this biochemical process claims not negligible delays of approx. 1msec. Indeed, such *neuronal latencies* have been observed to be stimulus-dependent.

In spite of their differences in the description of certain specific neuronal features, the models for dendrites, somata, and axons summarized so far share a basic *Ansatz*: They are derived from electrophysiologic principles and aim at describing actual biochemical processes.

However, the equations delivered by the models that take precisely into account neuron the elaborate morphology of neurons include a very large number of phenomenological constants (which we have omitted for mathematical simplicity) that ought to be fitted. Additionally, a human brain contains approx. 100 billions of neurons: Modeling each and every of them with a system of several coupled partial and ordinary differential equations is not computationally feasible at the moment, and will likely not be for many years to come, either. Even the Blue Brain Project—currently the world's most developed and most realistic cortical modeling program, which does aim at full biological plausibility—only deals with discretized versions of the Hodgkin–Huxley equations.

A possible way to simplify the model consists in theoretically assuming and/or experimentally forcing the whole axon, or perhaps even larger neuronal tracts, to be isopotential: This so-called *space-clamp* eliminates longitudinal voltage gradients and in turn allows to replace all partial differential equations by ordinary ones. One may regard this simplification by thinking at a brain as a gigantic network of interconnected points (viz, dimensionless neurons, or somata) that, in view of the all-or-none law, may be active or not and thereby also influence the neurons nearby, after some delay. The whole neural activity is thus stored into an array, or more precisely in the time-dependent entries of a vector of huge, but finite size. This is a massive simplification, and possibly even an oversimplification, but analyzing a 10^{11}-dimensional dynamical system is still rather impractical.

A nervous system can be studied at least three different scales: the microscopic, mesoscopic, and macroscopic levels of individual neurons (approx. 100 billions); *columns* (in the cortex) and other neuron ensembles (approx. 2 millions); and *functional areas* (approx. a dozen), respectively. Typically, as the focus is laid on higher and higher structures, larger and larger regions are considered as clamped and the task steadily shifts from the accurate description of biological structures to the development of computationally efficient algorithms that may be practically implemented in electronic devices. In particular, the dynamics of functional areas is currently studied mostly by imaging tools, like the fMRI, which have their good share of interesting mathematical questions, in particular in connection with the theory of inverse problems. We will avoid to discuss this aspect as the fine network structure described so far has essentially disappeared at this level of definition.

On the other hand, the activity inside cortical columns still offers interesting applications for network analysis and its mathematical tools. In this context all mathematical models rely upon certain minimal assumptions: There is a hierarchy of regions that are responsible for processing a certain signal, each consisting of subregions—the columns; there are connections both from regions active in the early processing stages to later ones, and back—these are called *feed-forward* and *top-down* connections in the jargon of neuroscience; and finally, there are so-called *recurrent* interconnections within each column and, possibly, *lateral* connections to some columns nearby (which we exclude, for the sake of presentation clarity). While these classes of interconnections pursue different modeling purposes and have different theoretical justifications, they all essentially have the same mathematical structure. To begin with, in view of the all-or-none law it is meaningful to shift from a description based on membrane potential to a model in which all information is stored in the frequency of spike activity—the so-called *firing rate*. Luckily, these objects are not unrelated: At least approximately, the current $I(t, \mathsf{v}_0)$ entering at time t the soma of a neuron v_0 and the firing rates f of all its presynaptic neurons—i.e., of all neurons that projects into v_0 through one of their synapses—satisfy the differential equation

$$\frac{dI}{dt}(t, \mathsf{v}_0) = -I(t, \mathsf{v}_0) + \sum \mu(\mathsf{v}, \mathsf{v}_0) f(t, \mathsf{v}),$$

where the sum is taken over all presynaptic neurons v and, for all v, $\mu(\mathsf{v}, \mathsf{v}_0)$ represents the strength of the synaptic connection between v and v_0, cf. [5, Sect. 7.2]. All these weights may be encoded in a square matrix $(\mu(\mathsf{v}, \mathsf{v}_0))$ indexed by all neurons: We have seen in Chap. 2 that this is the *weighted incoming adjacency matrix* $\mathcal{A}^{\text{in}}$ that describes our schematic neuronal network.

The time-dependent behavior of the firing rate f at neuron v_0 inside a certain column is usually modeled by an ordinary differential equation of the form

$$\begin{aligned}\frac{df}{dt}(t, \mathsf{v}_0) = -f(t, \mathsf{v}_0) + F\Big(&\sum_{\mathsf{w}} \sum_i \omega(\mathsf{w}, \mathsf{v}_0) g(t - \tau_i, \mathsf{w}) \\ &+ \sum_{\mathsf{v}} \sum_i \mu(\mathsf{v}, \mathsf{v}_0) f(t - \tau_i, \mathsf{v})\Big) + \phi(t),\end{aligned} \tag{5.5}$$

where the function F is often taken to be a sigmoid or, in a better approximation, a convolution with some integral kernel. Observe that the firing rate at v_0 at time t depends on all spikes that have reached v_0 in the past, at earlier times $t - \tau_i$. Indeed, in view of complicated neuronal latencies one may even allow such delays τ_i do be activity-dependent. The first sum is over all presynaptic neurons w that belong to the column, the second one over all remaining neurons v in columns within the same cortical region, and the inhomogeneous term ϕ accounts for the activity due to external stimuli (and in particular avoids that the constant zero function is a solution).

The neuroscientist may like to distinguish between firing rates f from recurrent and lateral connections with synaptic weight μ and firing rates g from feed-forward ones, weighted by ω. So-called *Wilson–Cowan models* that, like this, tend to identify all neurons belonging to a same ensemble can be regarded as planar dynamical systems and as such they capture several fundamental features of the neural activity, cf. [22].

Though, a mathematician cannot refrain from grouping together—as their abstract role is just the same—f, g and therefore also μ, ω: Again, the latter will then fill together the entries of a weighted incoming adjacency matrix $\mathcal{A}^{\text{in}}$. In this way one can eventually re-write (5.5) as a higher-dimensional dynamical system

$$\frac{df}{dt}(t, \mathsf{v}) = -f(t, \mathsf{v}) + \mathcal{F}\left(\mathcal{A}^{\text{in}} \sum_i f(t - \tau_i, \mathsf{v})\right) + \phi(t), \qquad (5.6)$$

where the definition of $\mathcal{F}$ in dependence of F is obvious.

Such so-called *firing-rate models* can become almost arbitrarily complicated in order to accommodate the demand for more precise description of the brain, possibly splitting the neuronal and/or synaptic populations into relevant subclasses, and including top-down connections as in [5, Sect. 10.3]; one may want to consider the synaptic delays, which may be different neuron-wise; the functions F can be more or less smooth, or perhaps even boil down to simple multiplications by positive numbers—or conversely, $\mathcal{F}$ may be allowed not to respect the local network structure in order to favor the joint activity of an ensemble of neurons, like in so-called *population coding*; in the spirit of a groundbreaking intuition expressed by D.O. Hebb in [9], the weights ω can be assumed to be time-dependent, e.g. to satisfy the *Hebbian rule* expressed by the differential equation

$$\frac{d\omega}{dt}(t, (\mathsf{v}, \mathsf{v}_0)) = Q\omega(t, (\mathsf{v}, \mathsf{v}_0)), \qquad (5.7)$$

where $Q := \big(\text{corr}(\omega(t, (\mathsf{v}, \mathsf{v}_0)), \omega(t, (\mathsf{w}, \mathsf{v}_0))\big)$ is the correlation matrix of the weights, in order to describe learning abilities by so-called *synaptic plasticity*; or $\mathcal{A}^{\text{in}}$ may even be allowed to modify its zero/nonzero pattern over time, in order to describe the so-called *synaptic pruning*, i.e., death or disconnection of neurons. Nonetheless, the underlying mathematical structure remains almost always that of (5.6), as long as firing-rate models are considered.

The above summary of models in neuroscience is by no means complete, nor it is conclusive. The development of neuroscience has been tumultuous over the last 100 years and promises to keep on for a long while. A detailed synopsis of the manifold of hypotheses coexisting or competing in the mathematical neuroscience can be found in [5], where both biophysical and computational approaches are discussed. Of all further monographs devoted to the mathematics of neural systems, [19] is particularly committed to discuss the role of network-based models. It seems likely that a new plethora of both analytical and statistical tools that were first developed by theorists of complex networks will soon find their applications in neuroscience.

Modern neuroscience does not only makes extensive use of networks as metaphor or convenient formalism. It occasionally goes further on saying that the sole network structure can explain—perhaps exclusively, through the connectivity features—the neural behavior, and in fact the animal behavior altogether. Indeed, some argue that just like the genes are responsible for determining all tracts in animal physiology, the animal behavior is completely determined by the connectivity features of the brain. The word *connectome*, first proposed in [7,21], was coined precisely in analogy with *genome*—and the supporters of this theory suggest that the connectome should be completely analyzed just like the human genome has been completely sequenced. The recent monograph [20] is a most representative manifesto of this agenda.

The concept of connectome is in these days both popular and controversial. In a very few cases, a complete connectome has actually been determined: most notably, in the case of the C. elegans, with its mere 302 neurons and approximately 5,000 synapses. Still, even in those cases it is not clear to what extent knowledge of connectivity alone—i.e., of which neurons project into a given one—can possibly give definite answers on behavioral questions. Indeed, there is broad belief among complex network theorists that certain small interconnection patterns, so-called *network motifs*, carry decisive information about the whole system; but just the example of C. elegans seems to show—it is argued e.g. in [2]—that no patterns are quite distinguished, as any two nodes have almost always very small distance.

Moreover, one cannot neglect plasticity phenomena: On one hand we have already mentioned that the brain network is being pruned continually, in particular in adult animals; on the other hand, synaptic plasticity is responsible for constantly modifying the strength of connection. This suggests that a relevant information is carried by ever-changing weights—and we can usually describe these changes only stochastically. This makes for a very involved and currently still rather inaccurate analysis.

References

1. C.A. Anastassiou, R. Perin, H. Markram, C. Koch, Ephaptic coupling of cortical neurons. Nature Neurosci. **14**, 217–223 (2011)
2. C.I. Bargmann, Beyond the connectome: How neuromodulators shape neural circuits. Bioessays **34**, 458–465 (2012)

3. C. Bédard, H. Kröger, A. Destexhe, Modeling extracellular field potentials and the frequency-filtering properties of extracellular space. Biophys. J. **86**, 1829–1842 (2004)
4. C.L. Cox, W. Denk, D.W. Tank, K. Svoboda, Action potentials reliably invade axonal arbors of rat neocortical neurons. Proc. Natl. Acad. Sci. USA **97**, 9724–9728 (2000)
5. P. Dayan, L.F. Abbott, *Theoretical Neuroscience: Computational and Mathematical Modeling of Neural Systems* (MIT Press, Boston, 2001)
6. R. FitzHugh, Mathematical models of excitation and propagation in nerve, in *Biological Engineering*, ed. by H.P. Schwan. Inter-University Electronics Series, vol. 9 (McGraw-Hill, New-York, 1969)
7. P. Hagmann, *From diffusion MRI to brain connectomics*. PhD thesis, Université de Lausanne, 2005
8. S.P. Hastings, Some mathematical problems from neurobiology. Am. Math. Mon. **82**, 881–895 (1975)
9. D.O. Hebb, *The Organization of Behaviour* (Wiley, New York, 1949)
10. T. Heimburg, A.D. Jackson, On soliton propagation in biomembranes and nerves. Proc. Natl. Acad. Sci. USA **102**, 9790–9795 (2005)
11. A.L. Hodgkin, A.F. Huxley, A quantitative description of ion currents and its applications to conduction and excitation in nerve membranes. J. Physiol. **117**, 500–544 (1952)
12. G.R. Holt, C. Koch, Electrical interaction via the extracellular potential near cell bodies. J. Comp. Neurosci. **2**, 169–184 (1999)
13. G. Major, J.D. Evans, and J.J. Jack, Solutions for transients in arbitrarily branching cables: I. Voltage recording with a somatic shunt. Biophys. J. **65**, 423–449 (1993)
14. P.L. Nunez, The brain wave equation: a model for the EEG. Math. Biosci. **21**, 279–297 (1974)
15. W. Rall, Branching dendritic trees and motoneurone membrane resistivity. Exp. Neurol. **1**, 491–527 (1959)
16. W. Rall, G.M. Shepherd, Theoretical reconstruction of field potentials and dendrodendritic synaptic interactions in olfactory bulb. J. Neurophysiol. **31**, 884–915 (1968)
17. J. Rinzel, J.B. Keller, Traveling wave solutions of a nerve conduction equation. Biophys. J. **13**, 1313–1337 (1973)
18. H. Schlitt, Lösung einer Wärmeleitungsaufgabe durch Analogiebetrachtungen. Arch. Elektrotechnik **43**, 51–58 (1957)
19. O. Sporns, *Networks of the Brain* (MIT Press, Boston, MA, 2010)
20. O. Sporns, *Discovering the Human Connectome* (MIT Press, Boston, MA, 2012)
21. O. Sporns, G. Tononi, R. Kötter, The human connectome: A structural description of the human brain. PLoS Comput. Biol. **1**, e42 (2005)
22. H.R. Wilson, J.D. Cowan, Excitatory and inhibitory interactions in localized populations of model neurons. Biophys. J. **12**, 1–24 (1972)

Chapter 6
Sesquilinear Forms and Analytic Semigroups

Applying the theorems of Hille–Yosida or Lumer–Phillips is sometimes unsatisfactory, as they make no claim about possible regularity gain (either in space or time) of solutions. In this chapter we specialize our previous investigations to parabolic equations. These are evolution equations, typically associated with diffusive processes, whose foremost property is enhanced smoothness of initial data.

6.1 Analytic Semigroups

A favorite setting that captures this feature is that of analytic semigroups, which we briefly review in this section.

In the following we are going to adopt the notation

$$\Sigma_\theta := \{z \in \mathbb{C} : |\arg z| < \theta\} \setminus \{0\}$$

for $\theta \in (0, \frac{\pi}{2}]$: i.e., Σ_θ is the sector in $\mathbb{C}$ of all numbers whose argument is smaller than θ.

Definition 6.1. A C_0-semigroup $(T(t))_{t\geq 0}$ on a Banach space X is said to be *analytic* of *angle* $\theta \in (0, \frac{\pi}{2}]$ if it admits an analytic extension $(T(t))_{t\in\Sigma_\theta\cup\{0\}}$ that satisfies

$$T(t)T(s) = T(t+s) \quad \text{and} \quad T(t) \in \mathcal{L}(X) \qquad \text{for all } t, s \in \Sigma_\theta \cup \{0\}.$$

It is said to be *bounded analytic* if it is analytic and moreover for all $\theta_0 \in (0, \theta)$ there exists $M_{\theta_0} > 0$ such that $\|T(t)\|_{\mathcal{L}(X)} \leq M_{\theta_0}$ for all $t \in \Sigma_{\theta_0}$.

The notion of generator remains unchanged. The pendant of the Hille–Yosida Theorem for analytic semigroups is the following.

D. Mugnolo, *Semigroup Methods for Evolution Equations on Networks*, Understanding Complex Systems, DOI 10.1007/978-3-319-04621-1_6,

Theorem 6.2. *Let A be a closed, densely defined operator on a Banach space X and $\theta \in (0, \frac{\pi}{2}]$. The following are equivalent.*

(a) A generates a bounded analytic C_0-semigroup of angle θ.
(b) A generates a bounded C_0-semigroup, $\operatorname{Rg} e^{tA} \subset D(A)$ for all $t > 0$, and

$$\sup_{t>0} \|tAe^{tA}\|_{\mathcal{L}(X)} < \infty. \tag{6.1}$$

(c) $\Sigma_{\theta+\frac{\pi}{2}} \subset \rho(A)$ and for all $\theta_0 \in (0, \theta)$ there exists $M_{\theta_0} \geq 1$ such that

$$|\lambda| \|R(\lambda, A)\|_{\mathcal{L}(X)} \leq M_{\theta_0} \qquad \text{for all } \lambda \in \overline{\Sigma_{\frac{\pi}{2}+\theta-\theta_0}} \setminus \{0\}. \tag{6.2}$$

If the above conditions hold, then A is called sectorial of angle θ *and the semigroup is given by*

$$e^{tA} := \begin{cases} \frac{1}{2\pi i} \int_\gamma e^{\mu z} R(\mu, A)\, d\mu, & z \in \Sigma_\theta, \\ \mathrm{Id}, & z = 0, \end{cases} \tag{6.3}$$

for any piecewise smooth curve in $\Sigma_{\theta+\frac{\pi}{2}}$ that connects $\infty - i(\frac{\pi}{2} + \theta')$ to $\infty + i(\frac{\pi}{2} + \theta')$ for some $\theta' \in (|\arg z|, \theta)$. The integral in (6.3) converges absolutely in $\mathcal{L}(X)$, uniformly in $z \in \Sigma_{\theta'}$.

Lemma 6.3. *Let A be a closed, densely defined operator on a Banach space X. Then A generates an analytic C_0-semigroup on X if and only if there exists $\omega \geq 0$ such that $A - \omega\,\mathrm{Id}$ generates a* bounded *holomorphic semigroup.*

Lemma 6.3 shows that it is possible to modify Theorem 6.2 in order to study operators that are sectorial only upon a scalar perturbation. The following thus holds.

Theorem 6.4. *Let A be a closed, densely defined operator on a Banach space X. Then the following assertions hold.*

(1) If A generates a bounded C_0-semigroup on X, then $(e^{tA})_{t\geq 0}$ is bounded analytic if and only if $\operatorname{Rg} e^{tA} \subset D(A)$ for all $t > 0$ and

$$\sup_{t>0} \|tAe^{tA}\|_{\mathcal{L}(X)} < \infty.$$

(2) There exist $\omega_0 \in \mathbb{R}$ and $\rho > 0$ such that

$$S_{\omega_0,\rho} := \{\lambda \in \mathbb{C} : \operatorname{Re} \lambda > \omega_0 \text{ and } |\lambda| > \rho\} \subset \rho(A)$$

and

$$\sup_{\lambda \in S_{\omega_0,\rho}} |\lambda| \|R(\lambda, A)\|_{\mathcal{L}(X)} < \infty.$$

Lemma 6.5. *Let A generate an analytic (but not necessarily* bounded *analytic) C_0-semigroup on a Banach space X. If B is a bounded operator on X, or a compact operator from $[D(A)]$ to X, then also $A + B$ with domain $D(A)$ generates an analytic C_0-semigroup on X with same analyticity angle.*

A slight modification of the proof of Lemma 4.27 covers the case of bounded B, but the assumption of boundedness usually excludes the case of differential operators that would be useful to treat lower order terms of—say—elliptic differential operators. On the other hand, the assertion concerning compact operators from $[D(A)]$ to X—which was proved by Desch and Schappacher in [35] and has a nice symbiosis with compactness results like the theorems of Ascoli–Arzelà or Rellich–Kondrachov—has only limited usefulness whenever we treat infinite graphs.

In this case one may apply another result, also proved by Desch and Schappacher in [34], that relies upon the notion of interpolation space. We will instead formulate a related but different (and more elementary) perturbation result that is tailored for our needs in Lemma 6.22 below.

The most relevant property of analytic semigroups is arguably the following, which makes them attractive even if one is not interested in considering a complex time variable t. It follows from condition (b) in Theorem 6.2 and the semigroup law.

Proposition 6.6. *Let A generate a bounded analytic C_0-semigroup on a Banach space X. Then*

$$\operatorname{Rg} e^{tA} \subset D(A^k) \qquad \text{for all } t > 0,\ k \in \mathbb{N}$$

and

$$\sup_{t>0} \|t^k A^k e^{tA}\|_{\mathcal{L}(X)} < \infty \qquad \text{for all } k \in \mathbb{N}.$$

In particular, for each $t > 0$ and each $k \in \mathbb{N}$ e^{tA} is a bounded linear operator from X to $[D(A^k)]$.

(In accordance with the notation introduced in Remark 4.9, $[D(A^k)]$ is here the domain of A^k endowed with its graph norm.)

Thus, an analytic C_0-semigroup yields for all $x \in X$ a solution of (ACP) that is of class $C^k((0,\infty), X) \cap C((0,\infty), D(A^k))$ for any arbitrarily large $k \in \mathbb{N}$.

Among many further properties of analytic semigroups let us mention three that prove rather useful in applications to evolution equations. They are the analogs of Theorem 4.35 and Proposition 4.37.

Proposition 6.7. *If A generates an analytic C_0-semigroup, then the spectrum of e^{tA} (possibly up to 0) coincides with*

$$e^{t\sigma(A)} := \{e^{t\lambda} \in \mathbb{C} : \lambda \in \sigma(A)\}.$$

In particular, the geometric multiplicities of the eigenvalues of e^{tA} *and* A *satisfy*

$$m_g(e^{t\lambda}) \geq m_g(\lambda),$$

with equality if A *is self-adjoint.*

Proposition 6.8. *If* A *generates an analytic* C_0*-semigroup, then* $s(A) = \omega_0(A)$.

Hence, an analytic semigroup $(e^{tA})_{t\geq 0}$ is uniformly exponentially stable if and only if $s(A) < 0$.

Proposition 6.9. *Let* $(U, \tilde{\mu})$ *be a* σ*-finite measure space,* $p \in (1, \infty)$*, and let* A *generate a positive, analytic, bounded* C_0*-semigroup on* $L^p(U, \tilde{\mu})$*. Then* $(e^{tA})_{t\geq 0}$ *converges strongly, i.e.,*

$$Pu := \lim_{t\to\infty} e^{tA}u \qquad \textit{exists for all } u \in L^p(U, \tilde{\mu}),$$

and P *is the projector of* $L^p(U, \tilde{\mu})$ *onto* $\operatorname{Ker} A$.

Remark 6.10. If (ACP2) is well-posed in the sense introduced in Proposition 4.11, then so is (ACP): Indeed, if A generates a C_0-cosine operator function $(C(t, A))_{t\geq 0}$ on a Banach space X, then it also generates a C_0-semigroup that is given by the *Weierstraß formula*

$$e^{tA}x := \int_0^\infty \frac{e^{\frac{-s^2}{4t}}}{\sqrt{\pi t}} C(s, A)x \, ds, \qquad t > 0,\ x \in X.$$

This semigroup is analytic of angle $\frac{\pi}{2}$ and the spectrum of its generator A lies inside a parabola (and not only inside a sector, as usual for general analytic semigroup generators).

Hence, even if one is actually interested in (ACP), it is sometimes convenient to study (ACP2) instead, and hence to check whether A generates a cosine operator function.

We conclude with a result that is rather useful whenever discussing non-self-adjoint Hamiltonians—cf. Stone's theorem below.

Theorem 6.11. *Let* A *generate an analytic* C_0*-semigroup of angle* $\frac{\pi}{2}$ *on a Banach space* X*. Then* $\pm iA$ *generates a* C_0*-semigroup on* X *if and only if*

$$\sup_{z\in D_\pm} \|e^{zA}\|_{\mathcal{L}(X)} < \infty,$$

where $D_\pm := \{z \in \mathbb{C} : |z| \leq 1,\ \operatorname{Re} z > 0$ *and* $\operatorname{Im} z \gtrless 0\}$.

If actually both iA and $-iA$ generate a C_0-semigroup on X, then we have seen in Example 4.16 that iA generates a C_0-group. Such a group is called in this context the *boundary group* of the analytic semigroup $(e^{zA})_{z\in\Sigma_{\frac{\pi}{2}}}$; and by Proposition 4.51, $-A^2$ generates a C_0-cosine operator function.

6.2 General Theory of Elliptic Forms

In Hilbert spaces it is usually possible to show generation of analytic semigroups in an easier way than applying Theorem 6.2. Throughout the remainder of this chapter

V, H are separable, complex Hilbert spaces with $V \overset{d}{\hookrightarrow} H$.

We denote by $(\cdot|\cdot)_V, (\cdot|\cdot)_H$ their inner products.

It turns out that forms provide an elegant and mighty tool to investigate linear parabolic equations. In this section we will present a concise invitation to this theory.

Definition 6.12. A mapping $a : V \times V \to \mathbb{C}$ is called a *sesquilinear form* if

$$a(\alpha f + \beta g, \gamma h) = \alpha\overline{\gamma}\, a(f,h) + \beta\overline{\gamma}\, a(g,h) \quad \text{for all } f, g, h \in V \text{ and all } \alpha, \beta, \gamma \in \mathbb{C}.$$

The space V is its *form domain*. Then a is said to be

- *coercive* if there exists $\mu > 0$ such that

$$\operatorname{Re} a(f,f) \geq \mu \|f\|_V^2 \qquad \text{for all } f \in V;$$

- H*-elliptic* if there exist $\mu > 0$ and $\omega \in \mathbb{R}$ such that

$$\operatorname{Re} a(f,f) + \omega \|f\|_H^2 \geq \mu \|f\|_V^2 \qquad \text{for all } f \in V;$$

- *accretive* if

$$\operatorname{Re} a(f,f) \geq 0 \qquad \text{for all } f \in V;$$

- *continuous* if there exists $M > 0$ such that

$$|a(f,g)| \leq M \|f\|_V \|g\|_V \qquad \text{for all } f, g \in V; \tag{6.4}$$

- *of Lions type* if there exists $M > 0$ such that

$$|\operatorname{Im} a(f,f)| \leq M \|f\|_V \|f\|_H \qquad \text{for all } f \in V. \tag{6.5}$$

- *symmetric* if

$$a(f,g) = \overline{a(g,f)} \qquad \text{for all } f, g \in V;$$

In particular in the contexts of linear algebra and mathematical physics, where one is mostly interested in hermitian matrices and self-adjoint operators, continuous, H-elliptic sesquilinear forms that are symmetric are usually referred to as *quadratic forms*.

Observe that H-ellipticity is equivalent to the condition that

$$V \ni f \mapsto \|f\|_a := \sqrt{\operatorname{Re} a(f,f) + \omega \|f\|_H^2} \in [0,\infty) \tag{6.6}$$

defines an equivalent norm on V: $\|\cdot\|_a$ is then called the *form norm*. It is also clear that symmetry implies the Lions condition.

Definition 6.13. Let $a : V \times V \to \mathbb{C}$ be a continuous, H-elliptic sesquilinear form. The *operator associated with a* is

$$\begin{aligned} D(A) &:= \{f \in V : \exists h \in H \text{ s.t. } a(f,g) = (h \mid g)_H \ \forall g \in V\}, \\ Af &:= -h. \end{aligned}$$

Conversely, an operator A on H is said to *come from a form* if there exists a Hilbert space W densely embedded in H and a continuous, H-elliptic sesquilinear form $b : W \times W \to \mathbb{C}$ such that A is the operator associated with b.

The Lax–Milgram Lemma ensures that h in the definition of $D(A)$ exists and is unique.

Remark 6.14. The sign convention in the definition of the operator associated with a form is tailored for the case of differential operators of even order and their Gauß–Green-formulae, as we will see e.g. in Example 6.21; but has the drawback that if $A = (a_{ij})$ is an $n \times n$-matrix and a is given by

$$a(x,y) := \sum_{i,j=1}^{n} a_{ij} x_j \overline{y_i}, \qquad x, y \in \mathbb{C}^n,$$

then the operator associated with a is $-A$, in contrast with the definition of quadratic form associated with a matrix that is usual in linear algebra.

We assume throughout that

$a : V \times V \to \mathbb{C}$ is sesquilinear, continuous and H-elliptic and A is its associated operator

with constants as in Definition 6.12.

Theorem 6.15. *The operator A generates an ω-quasi-contractive, analytic C_0-semigroup on H. Its part in V generates an analytic C_0-semigroup on V. If in particular a is coercive with constant μ, then $(e^{tA})_{t\ge 0}$ is uniformly exponentially stable with*

$$\|e^{tA}\|_{\mathcal{L}(H)} \le e^{-\mu t}, \qquad t \ge 0,$$

Indeed, one can prove that

$$\Sigma_{\pi-\arctan\frac{M}{\mu}} \subset \rho(A-\omega\,\mathrm{Id}) \tag{6.7}$$

(here M is the constant in (6.4)) and for all $\theta \in [0, \pi - \arctan\frac{M}{\mu})$ there holds

$$|\lambda|\|R(\lambda, A-\omega)\|_{\mathcal{L}(H)} \le \frac{1}{\sin\left(\theta - \arctan\frac{M}{\mu}\right)} \qquad \text{for all } \lambda \in \Sigma_\theta. \tag{6.8}$$

The semigroup generated by A is called the *semigroup associated with* a.

While it is in many cases very difficult to determine the operator domain $D(A)$, the form domain V is known from the very beginning: This is a most appealing property of forms. Even whenever $D(A)$ is unknown and Theorem 6.6 is therefore not immediately useful, one obtains the following smoothing enhancement.

Proposition 6.16. *For each $t > 0$ e^{tA} is a bounded linear operator from H to V.*

Proof. It suffices to apply Theorem 6.6 and the fact that, by definition, the domain of A endowed with the graph norm is densely and continuously embedded in V. □

Remark 6.17. The *adjoint form* $a^* : V \times V \to \mathbb{C}$ is defined by

$$a^*(f, g) := \overline{a(g, f)}, \qquad u, v \in V.$$

It is apparent that a^* is coercive (resp., H-elliptic, accretive, continuous, of Lions type) if and only if so is a. The operator associated with a^* is A^* and by Theorem 6.15 also A^* generates an analytic C_0-semigroup.

Theorem 6.18. *If a is additionally of Lions type, then A generates on H a C_0-cosine operator function with Kisyński space V.*

Remark 6.19. If a is a *coercive* bounded sesquilinear form of Lions type, then

$$f \mapsto \mathrm{Re}\, a(f, f), \qquad f \in V,$$

defines an equivalent norm on V. One then takes the reduction matrix $\mathbf{A}$ defined in (4.12) and checks that—with respect to said equivalent norm on V!—the Lumer–Phillips Theorem applies: More precisely, both $\mathbf{A}$ and $-\mathbf{A}$ are dissipative because for all $\mathbf{x} \in D(\mathbf{A})$

$$\begin{aligned}\mathrm{Re}(\mathbf{Ax}|\mathbf{x})_{V\times X} &= \mathrm{Re}(x_2|x_1)_V + \mathrm{Re}(Ax_1|x_2)_H \\ &= \mathrm{Re}(x_2|x_1)_V - \mathrm{Re}\, a(x_1, x_2) \\ &= \mathrm{Re}(x_2|x_1)_V - \mathrm{Re}(x_1|x_2)_V = 0.\end{aligned}$$

Hence, both $\mathbf{A}$ and $-\mathbf{A}$ generate a contraction C_0-semigroup—i.e., $\mathbf{A}$ generates a C_0-group of operators that are all unitary with respect to the norm of the Hilbert space $V \times H$. One therefore interprets the quantity

$$\|x_0\|_V^2 + \|x_1\|_H^2$$

as the square of the *energy* of the initial data x_0, x_1: By unitarity of $(e^{t\mathbf{A}})_{t\in\mathbb{R}}$, this energy is conserved over time under evolution of the second order problem (ACP2). These considerations apply e.g. to the wave equation with Dirichlet boundary conditions, as we will see in Example 6.21 below.

A C_0-semigroup $(T(t))_{t>0}$ is said to be *compact* (resp., *of trace class*) if so is $T(t)$ for all $t > 0$.

Lemma 6.20. *The C_0-semigroup $(e^{tA})_{t\geq 0}$ is*

- *compact if and only if the embedding $V \hookrightarrow H$ is compact;*
- *of trace class if the embedding $V \hookrightarrow H$ is Hilbert–Schmidt;*
- *of trace class if $H = L^2(U, \tilde{\mu})$ for a finite measure space $(U, \tilde{\mu})$ and e^{tA} maps $L^2(U, \tilde{\mu})$ into $L^\infty(U, \tilde{\mu})$ for all $t > 0$.*

Observe that by the ideal property of Schatten class operators a semigroup cannot be of any class $\mathcal{S}_p$, $1 < p < \infty$, without being already of trace class.

Proof. The claims are proved using the semigroup property, Proposition 6.6, the ideal property of compact and Schatten class operators, and the fact that each operator mapping $L^2(U, \tilde{\mu})$ into $L^\infty(U, \tilde{\mu})$ is Hilbert–Schmidt provided $\tilde{\mu}(X) < \infty$, cf. [4, Thm. 1.6.2]. □

Example 6.21. (1) Let us consider again the setting in Example 4.14. We restrict to the case of $p = 2$ and hence take $H = L^2(0, 1)$. Integrating by parts one finds for all $u \in D(\Delta^D)$ and all test functions $v \in C_c^\infty(0, 1)$

$$(\Delta^D u \mid v)_H = \int_0^1 u''(x)\overline{v(x)}dx = -\int_0^1 u'(x)\overline{v'(x)}dx,$$

due to the Dirichlet boundary conditions satisfied by v (this would not work taking $v \in C^\infty(0, 1)$!). This suggests to define a sesquilinear form by

$$a(u, v) := \int_0^1 u'(x)\overline{v'(x)}dx.$$

This expression is only well-defined for $(u, v) \in W^{1,2}(0, 1) \times W^{1,2}(0, 1)$, hence the form domain V of a has to be a subspace of $W^{1,2}(0, 1)$. Indeed, we want the form norm to define an equivalent norm on V: hence, we close $C_c^\infty(0, 1)$ up in the form norm, thus obtaining $V = \mathring{W}^{1,2}(0, 1)$.

By Lemma B.8.(4) we can endow V with the equivalent norm $u \mapsto \|u'\|_{L^2}$: this establishes coercivity of a. In order to show that the associated operator is precisely Δ^D, take $u \in V$ such that there exists $w \in H$ with

$$a(u,v) = \int_0^1 u'\overline{v}' dx \stackrel{!}{=} \int_0^1 w\overline{v} dx \qquad \text{for all } v \in V.$$

By definition of weak derivative $u' \in W^{1,2}$, i.e., $w = -u'' \in L^2(0,1)$ and $u \in W^{2,2}(0,1)$. Thus, Δ^D generates an analytic, contractive, uniformly exponentially stable C_0-semigroup. Finally, by Lemma B.3 and 6.20 $(e^{t\Delta^D})_{t\geq 0}$ is of trace class.

(2) One can see likewise that taking instead the largest possible domain such that a is still elliptic with respect to $H = L^2(0,1)$, i.e., $V = W^{1,2}(0,1)$, the operator associated with a is the second derivative with Neumann boundary conditions

$$D(\Delta^N) := \{u \in W^{2,2}(0,1) : u'(0) = u'(1) = 0\}$$
$$\Delta^N u := u''.$$

Thus, also Δ^N generates an analytic, contractive C_0-semigroup that is of trace class. However, a is not coercive any more and in fact $(e^{t\Delta^N})_{t\geq 0}$ is not uniformly exponentially stable, since $s(\Delta^N) = 0$.

Once it has been proved that an operator is a generator, one may want to consider its lower-order and/or boundary perturbations. Then, the following perturbation lemmata prove useful.

Lemma 6.22. *Let $\alpha \in (0,1)$ and let H_α be some normed space such that $V \hookrightarrow H_\alpha \hookrightarrow H$ and such that additionally*

$$\|f\|_{H_\alpha} \leq M_\alpha \|f\|_V^\alpha \|f\|_H^{1-\alpha} \qquad f \in V, \tag{6.9}$$

holds. If $b : V \times H_\alpha \to \mathbb{C}$ and $c : H_\alpha \times V \to \mathbb{C}$ are continuous sesquilinear mappings, then $a + b + c : V \times V \to \mathbb{C}$ is H-elliptic and continuous.

If additionally a is of Lions type, b, c are continuous on $V \times H$ and $H \times V$, respectively, and (for $\alpha = \frac{1}{2}$) $d : H_{\frac{1}{2}} \times H_{\frac{1}{2}}$ is continuous for some $H_{\frac{1}{2}}$ that satisfies (6.9), then $a + b + c + d : V \times V \to \mathbb{C}$ is H-elliptic, continuous, and also of Lions type.

Proof. We apply Young's inequality $xy \leq \epsilon x^p + c_{\epsilon,p} y^{p/(p-1)}$, which is valid for every $p \in (1,\infty)$, every $x, y \geq 0$, and every $\epsilon > 0$ with some constant $c_{\epsilon,p} \geq 0$. For $p := \frac{2}{1+\alpha}$ we obtain that

$$\begin{aligned}\operatorname{Re} b(f,f) &\geq -\|b\|\|f\|_V\|f\|_{H_\alpha} \geq -\|b\|M_\alpha\|f\|_V^{1+\alpha}\|f\|_H^{1-\alpha} \\ &\geq -\|b\|M_\alpha\epsilon\|f\|_V^2 - \|b\|M_\alpha c_{\epsilon,p}\|f\|_H^2\end{aligned}$$

for all $f \in V$. For $\epsilon := \frac{\mu}{2\|b\|M_\alpha}$ we thus obtain that

$$\operatorname{Re} a(f,f) + \operatorname{Re} b(f,f) - (\omega - \|b\|M_\alpha c_{\epsilon,p})\|f\|_H^2 \geq \frac{\mu}{2}\|f\|_V^2$$

for all $f \in V$. The other terms can be treated likewise. □

Lemma 6.23. *Let W, Z be Banach spaces that are continuously embedded in H and such that V is continuously embedded in them. Let S, T be compact operators from V into W, Z, respectively. Let $b_0 : V \times W \to \mathbb{C}$ and $c_0 : Z \times V \to \mathbb{C}$ be continuous sesquilinear forms. Define further sesquilinear forms b, c by*

$$b(f,g) := b_0(f, Sg) \quad \text{and} \quad c(f,g) := c_0(Tf, g), \qquad f, g \in V.$$

Then $a + b + c$ is continuous and H-elliptic.

Proof. We deduce from compactness of S that for all $\epsilon > 0$ there exists $c_\epsilon > 0$ with

$$\|Sf\|_Z \leq \epsilon\|f\|_V + c_\epsilon\|f\|_H \qquad \text{for all } f \in V,$$

and thus for all $f \in V$

$$\begin{aligned}
|b(f,f)| &= |b_0(f, Sf)| \\
&\leq \|b_0\|\|f\|_V\|Sf\|_Z \\
&\leq \epsilon\|b_0\|\|f\|_V^2 + c_\epsilon\|b_0\|\|f\|_V\|f\|_H \\
&\leq \epsilon\|b_0\|\|f\|_V^2 + \delta c_\epsilon\|b_0\|\|f\|_V^2 + \frac{c_\epsilon\|b_0\|}{4\delta}\|f\|_H^2
\end{aligned}$$

for all $\delta > 0$ by Young's inequality. An analogous estimate holds for c. Picking first $\epsilon > 0$ small enough, and then $\delta > 0$ small enough to compensate c_ϵ, we can deduce H-ellipticity of $a+b+c$ from H-ellipticity of a as in the proof of Lemma 6.22. □

We have emphasized in Remark 6.47 that forms can be considered as a special instance of the nonlinear theory presented in the Appendix 6.2.2. Applying abstracts results on subdifferentials of suitable energy functionals we can thus complement Corollary 4.18 as follows.

Corollary 6.24. *Let $T > 0$, $x_0 \in H$ and $f \in L^2\big((0,T); H\big)$. Then there exists a unique $\xi \in C\big([0,T]; H\big)$ which is differentiable for a.e. $t \in [0,T]$, satisfies $\xi(t) \in D(A)$ for a.e. $t \in [0,T]$ and such that*

$$\begin{cases} \frac{d\xi}{dt}(t) = A\xi(t) + f(t), & \text{for a.e.} t \in [0,T], \\ \xi(0) = x_0. \end{cases} \tag{iACP}$$

If additionally $x_0 \in V$, *then* $\xi \in H^1\big((0,T);H\big) \cap L^\infty\big((0,T);V\big)$. *If* $x_0 \in D(A)$, *then even* $\xi \in W^{1,\infty}\big((0,T);H\big)$ *and* ξ *is right differentiable for all* $t \geq 0$.

Remark 6.25. For $T > 0$ and $f \in L^2(0,T;V')$ one can also study non-autonomous abstract Cauchy problems like

$$\begin{cases} \frac{d\xi}{dt}(t) = A(t)u(t) + f(t), & t \in [0,T], \\ \xi(0) = x_0, \end{cases} \tag{nACP}$$

by studying time-dependent forms. More precisely, by a classical result in [71] assume that for all $t \in [0,T]$ $A(t)$ comes from a form $a(t,\cdot,\cdot) : V \times V \to \mathbb{C}$. Assume furthermore that $a(t,\cdot,\cdot) \equiv a_1(t,\cdot,\cdot) + a_2(t,\cdot,\cdot)$ such that (with obvious modifications of the notions of Lemma 6.22)

- a_1 and a_2 are equi-continuous on $V \times V$ and $V \times H_\alpha$, respectively;
- a_1 is equi-coercive; and
- $t \mapsto a_1(t,\cdot,\cdot)$ and $t \mapsto a_2(t,u,v)$ are measurable from $[0,T]$ to $\mathbb{C}$ for all $(u,v) \in V \times V$ and all $(u,v) \in V \times H$, respectively.

Then for all $u_0 \in H$ there exists a unique $u \in H^1(0,T;V') \cap L^2\big((0,T);V\big)$ such that the differential equation in (nACP) is satisfied for a.e. $t \in [0,T]$, and moreover $u \in C([0,T];H)$.

A comparable result has been recently proved in [6]: Assume additionally that

- $a_1(t,\cdot,\cdot)$ is symmetric for all t; and
- $t \mapsto a_1(t,\cdot,\cdot)$ is locally Lipschitz continuous from $[0,T]$ to $(V \times V)'$.

Then for all $u_0 \in V$ there exists a unique $u \in H^1\big((0,T);H\big) \cap L^2\big((0,T);V\big)$ such that the differential equation in (nACP) is satisfied for a.e. $t \in [0,T]$, and in fact $u \in C([0,T];V)$.

Under the assumptions of Theorem 6.18, it follows by a result due to McIntosh [79] that V is the domain of the square root of (a scalar perturbation of) A and a complex interpolation space between $D(A)$ and H, see also [3, § 5.6.6]. Let us mention that this allows to study some class of nonlinear evolution equations, by the theory in [76, Chap. 7].

Corollary 6.26. *Let* $a : V \times V \to \mathbb{C}$ *be continuous,* H*-elliptic, and of Lions type with associated operator* A. *Let* $B : [0,T] \times V \to H$ *be a continuous nonlinear operator that is globally Lipschitz continuous in the second variable. Then the nonlinear abstract Cauchy problem*

$$\begin{cases} \dot{u}(t) = Au(t) + B(t,u(t)), & t \geq 0, \\ u(0) = u_0, \end{cases}$$

is globally well-posed, i.e., there exists a function $u \in C([0,\infty);H)$ *that satisfies the variation of constants formula*

$$u(t) = e^{tA}u_0 + \int_0^t e^{(t-s)A} B(s, u(s))ds, \qquad t \geq 0.$$

This assertion has a counterpart that ensures local well-posedness whenever the Lipschitz condition is satisfied by F only locally. It is critical for applications, cf. Example 6.27 below, that unlike in Lemma 6.44 B is here allowed to satisfy a Lipschitz condition only with respect to the norm of V.

Example 6.27. Let us discuss the initial value problem associated with the Hodgkin–Huxley model (HH) presented in Chap. 5, which describes the propagation of an action potential on an axon of length μ. To this aim one introduces the state space $H := L^2(0, \mu)^4$ and the sesquilinear form

$$a((f, \xi), (g, \zeta)) := \int_0^\mu \left(f'(x)\overline{g'(x)} + f(x)\overline{g(x)} + \left(\left(\alpha_m \ \alpha_h \ \alpha_n\right) f(x)\right) \cdot \overline{\zeta(x)} \right) dx$$

with domain $V := W^{1,2}(0, \mu) \times L^2(0, \mu)^3$. The associated operator A governs a linearized version of (HH). Continuity and H-ellipticity of a can be easily checked—observe that the second and third integrands are associated with bounded operators on H. In order to study the original semilinear problem one observes that each of the nonlinear terms, which are polynomials of odd degree, can be split into a maximal monotone operator and an error term that is Lipschitz continuous with from $C([0, \mu])$ to $L^2(0, \mu)$, and hence also from $H^1(0, \mu)$ to $L^2(0, \mu)$, cf. [26, § 6] for details. This latter term can be dealt with by means of Corollary 6.26, whereas well-posedness of the complete problem follows from the classical theory of m-accretive operators as in Sect. 6.2.2.

The following invariance criterion has been obtained by Ouhabaz in [96] in the accretive case and extended in [78] to the case of general elliptic forms.

Theorem 6.28. *Let C be a closed convex subset of H and denote by P_C the orthogonal projector of H onto C. Then the following assertions are equivalent.*

(a) C is invariant under $(e^{tA})_{t \geq 0}$.
(b) $P_C u \in V$ and $\operatorname{Re} a(P_C u, u - P_C u) \geq 0$ for all $u \in V$.

If a is additionally symmetric, then also the following condition is equivalent.

(c) $P_C u \in V$ and $a(P_C u, P_C u) \leq a(u, u)$ for all $u \in V$.

If u belongs to a Sobolev space $W^{1,2}$, then for many relevant subsets K (including, by Lemma B.12.(2), all order intervals) $(P_K u)'$ agrees with u' on $\operatorname{supp}(P_K u) \subset \operatorname{supp} u$. This suggests that whenever a is local (cf. the precise Definition 7.9 below), condition (c) in Theorem 6.28 is often easy to check. This is especially the case for the forms introduced in Example 6.21. We omit the straightforward details and focus on a related negative result.

Example 6.29. Let A be (minus) the fourth derivative with Dirichlet *and* Neumann boundary conditions. Considered as an operator on $H = L^2(0,1)$, A is associated with the form a defined by

$$a(u,v) := \int_0^1 u''(x)\overline{v''(x)}dx,$$
$$u, v \in V := \{w \in W^{2,2}(0,1) : w(0) = w(1) = w'(0) = w'(1) = 0\}.$$

Since V is dense in H and a is continuous, coercive and of Lions type, A generates on H a cosine operator function that governs the *linear beam equation*

$$\frac{\partial^2 u}{\partial t^2}(t,x) = -\frac{\partial^4 u}{\partial x^4}(t,x), \qquad t \geq 0,\ x \in (0,1),$$

with *clamped boundary conditions*

$$u(0) = u(1) = u'(0) = u'(1) = 0.$$

Hence in particular A generates an analytic C_0-semigroup of angle $\frac{\pi}{2}$. Take now $C := \{w \in L^2(0,1) : w(x) \geq 0 \text{ for a.e. } x \in (0,1)\}$, so that $P_C u = u^+$. We know from Remark B.13 that V is not invariant under P_C. Hence, Theorem 6.28 and Proposition 4.29 show that there exist $t > 0$ and $\lambda > 0$ such that e^{tA} and $R(\lambda, A)$ are not positive.

In the remainder of this section we assume that

$$\boxed{(U,\tilde{\mu}) \text{ is a } \sigma\text{-finite measure space and } H := L^2(U,\tilde{\mu}).}$$

Combining Theorem 6.28 and Lemma 4.33 one can prove the following.

Proposition 6.30. *The following hold.*

(1) $(e^{tA})_{t\geq 0}$ *is real if and only if for all* $f \in V$ *one has* $\operatorname{Re} f \in V$ *and* $a(\operatorname{Re} f, \operatorname{Im} f) \in \mathbb{R}$.

(2) $(e^{tA})_{t\geq 0}$ *is positive if and only if it is real and additionally for all* $f \in V$ *one has* $\operatorname{Re} f^+ \in V$ *and* $a(\operatorname{Re} f^+, \operatorname{Re} f^-) \leq 0$.

(3) $(e^{tA})_{t\geq 0}$ *is stochastic if and only if it is positive and additionally* $\mathbf{1} \in V$ *and* $\operatorname{Re} a(f, \mathbf{1}) = 0$.

(4) $(e^{tA})_{t\geq 0}$ *is irreducible if and only if for any measurable subset* U_0 *of* U *one has* $\tilde{\mu}(U_0) = 0$ *or* $\tilde{\mu}(U_0^C) = 0$ *whenever* $\mathbf{1}_{U_0} \in V$ *and* $\operatorname{Re} a(f\mathbf{1}_{U_0}, f\mathbf{1}_{U_0^C}) = 0$ *for all* $f \in V$.

(5) *If a is accretive, then* $(e^{tA})_{t\geq 0}$ *is contractive with respect to* $\|\cdot\|_{L^\infty}$ *if and only if for all* $f \in V$ *one has* $(1 \wedge |f|) \operatorname{sgn} f \in V$ *and* $\operatorname{Re} a((1 \wedge |f|) \operatorname{sgn} f, (|f|-1)^+ \operatorname{sgn} f) \geq 0$.

(6) *Let W be a Hilbert space that is densely and continuously embedded in H and $b : W \times W \to \mathbb{C}$ be another H-elliptic, continuous sesquilinear form with associated operator B. Let both a and b be accretive, and $(e^{tB})_{t\geq 0}$ be positive. Then $(e^{tB})_{t\geq 0}$ dominates $(e^{tA})_{t\geq 0}$ in the sense of positive operators if and only if*

- *V is a lattice ideal of W in the sense of Definition B.5 and additionally*
- *for all $f, g \in V$ such that $f\overline{g} \geq 0$ one has $\operatorname{Re} a(f, g) \geq b(|f|, |g|)$.*

(7) *Under the assumptions of (7), let additionally $(e^{tA})_{t\geq 0}$ be positive, too. Then $(e^{tB})_{t\geq 0}$ dominates $(e^{tA})_{t\geq 0}$ in the sense of positive operators if and only if*

- *V is a lattice ideal of W and additionally*
- *for all positive-valued $f, g \in V$ one has $a(f, g) \geq b(f, g)$.*

An interesting property of C_0-semigroups $(T(t))_{t\geq 0}$ on $L^p(U, \tilde{\mu})$ that are analytic, positive, and irreducible is that if $f \geq 0$ μ-a.e. but $f \neq 0$, then $T(t)f > 0$ μ-a.e., cf. [97, Thm. 2.9 and Def. 2.8]. This shows that whenever some heat is injected in a system at time $t = 0$, this can be immediately sensed at any point. In other words, diffusion equations typically have infinite speed of propagation, cf. Definition 4.54.

Let us recall the *Kantorovich–Vulikh Theorem.*

Theorem 6.31. *Let for $K \in L^\infty(U \times U)$*

$$(T_K u) := \int_U K(\omega, \cdot)u(\omega)\, d\tilde{\mu}(\omega), \qquad u \in L^1(U). \tag{6.10}$$

Then T_K defines a bounded linear operator from $L^1(U)$ to $L^\infty(U)$. Conversely, every bounded linear operator from $L^1(U)$ to $L^\infty(U)$ can be represented as T_K for some $K \in L^\infty(U \times U)$. Furthermore, $\|T_K\|_{\mathcal{L}(L^1, L^\infty)} = \|K\|_{L^\infty}$.

Example 6.32. We mention the following as a typical application of the Kantorovich–Vulikh Theorem. Under our standing assumptions, by Proposition 6.16, e^{tA} is bounded from H to V for all $t > 0$. If in particular $H = L^2(I)$, and V is a subspace of $W^{1,2}(I)$ for some interval $I \subset \mathbb{R}$, then in view of Lemma B.3 e^{tA} is bounded from $L^2(I)$ to $L^\infty(I)$ for all $t > 0$, and so is e^{tA^*} by Remark 6.17. By duality, e^{tA} extends to a bounded linear operator from $L^1(I)$ to $L^\infty(I)$. Hence, each e^{tA} is an integral operator with integral kernel K_t. Then, $(K_t)_{t\geq 0}$ is called the *heat kernel* of $(e^{tA})_{t\geq 0}$. It can be estimated by the following.

Proposition 6.33. *Assume both semigroups $(e^{tA})_{t\geq 0}$ and $(e^{tA^*})_{t\geq 0}$ to be $L^\infty(U, \tilde{\mu})$-contractive. If*

$$\|f\|_{L^2}^3 \leq M\|f\|_V\|f\|_{L^1}^2 \tag{6.11}$$

holds for some constant M and all $f \in V$, or else if the embedding

$$V \hookrightarrow L^{\frac{2n}{n-2}}(U) \tag{6.12}$$

holds for some $n \geq 2$*, then* $(e^{tA})_{t\geq 0}$ *is* ultracontractive of dimension d*, i.e., it satisfies the estimate*

$$\|e^{tA} f\|_{L^\infty} \leq ct^{-\frac{d}{4}} \|f\|_{L^1}, \quad t > 0, \ f \in L^2(U), \tag{6.13}$$

for some constant $c > 0$*, where* $d = 1$ *if* (6.11) *holds, or* $d = n$ *if* (6.12) *holds.*

Clearly, smaller or larger d yield in (6.13) better estimates for smaller or larger t, respectively.

Example 6.34. In the setting of Example 6.21, several qualitative properties of the semigroups generated by Δ^N or Δ^D can be proved by means of Proposition 6.30, thus extending Lemma 4.57. Let $u \in V$. Then, also the functions

$$\begin{aligned} \operatorname{Re} u &: x \mapsto \operatorname{Re}(u(x)), \\ \operatorname{Re} u^+ &: x \mapsto \operatorname{Re}(\max\{u(x), 0\}), \\ (1 \wedge |u|) \operatorname{sgn} u &: x \mapsto (\min\{1, |u(x)|\}) \operatorname{sgn} u(x) \end{aligned}$$

are weakly differentiable and there holds

$$(\operatorname{Re} u^+)' = \operatorname{Re} u', \quad (\operatorname{Re} u^+)' = \operatorname{Re} u' \cdot \mathbf{1}_{\{u\geq 0\}}, \quad ((1 \wedge |f|) \operatorname{sgn} f)' = u' \cdot \mathbf{1}_{\{|u|\leq 1\}}.$$

For example,

$$a(\operatorname{Re} u^+, \operatorname{Re} u^-) = \int_U |(\operatorname{Re} u)'|^2 \mathbf{1}_{\{u\geq 0\}} \mathbf{1}_{\{u\leq 0\}} dx = 0,$$

and similarly $\operatorname{Re} a((1 \wedge |u|) \operatorname{sgn} u, (|u| - 1)^+ \operatorname{sgn} u) = 0$. By Proposition 6.30 $(e^{t\Delta^D})_{t\geq 0}$ is real, positive, and L^∞-contractive. The same properties hold for $(e^{t\Delta^N})_{t\geq 0}$. By interpolation and then by duality either C_0-semigroup extends to a family of C_0-semigroups on all spaces $L^p(0,1)$, $1 < p < \infty$. Moreover, by Proposition 6.30.(6) and Lemma B.6 $(e^{t\Delta^N})_{t\geq 0}$ dominates $(e^{t\Delta^D})_{t\geq 0}$. Finally, it follows from Proposition 6.33 and the Nash inequality (B.3) that $(e^{t\Delta^N})_{t\geq 0}$ is ultracontractive of dimension 1, and hence so is the dominated C_0-semigroup $(e^{t\Delta^D})_{t\geq 0}$.

6.2.1 *Generalized Elliptic Forms*

In this section we keep on imposing the standing assumptions of Sect. 6.2 and discuss the notion of *j-elliptic forms*, a rather recent generalization of the usual notion of forms first proposed in [8].

Definition 6.35. Let $j : V \to H$ a bounded linear map with dense range. A sesquilinear form $a : V \times V \to \mathbb{C}$ is called *j-elliptic* on H with form domain V if there exist $\omega \in \mathbb{R}$ and $\mu > 0$ such that

$$\operatorname{Re} a(u,u) - \omega \|j(u)\|_H^2 \ge \mu \|u\|_V^2 \qquad \text{for all } u \in V. \tag{6.14}$$

If a is continuous as a function from $V \times V$ to $\mathbb{C}$ and j-elliptic, then the unique, densely defined operator A on H given by

$$\begin{aligned} D(A) &:= \{f \in H : \exists u \in V \text{ s.t. } j(u) = f,\ \exists g \in H \text{ s.t. } a(u,v) = (g|j(v))_H\ \forall v \in V\}, \\ Af &:= -g, \end{aligned}$$

is called the *operator associated with* (a, j).

If $j = \mathrm{Id}$, then we recover the classical theory of forms. More generally, j-ellipticity essentially agrees with the usual notion of ellipticity if j is injective, since then V can be identified with a dense subspace of H. All properties of a and A thus follow more or less directly from the decomposition

$$V = V(a) \oplus \operatorname{Ker} j,$$

where

$$V(a) := \{u \in V : a(u,v) = 0 \text{ for all } v \in \operatorname{Ker} j\},$$

in view of the obvious fact that $j_{|V(a)}$ is injective.

Example 6.36. A less obvious application of the theory of j-elliptic forms arises if one looks again at the setting of Example 6.21, i.e.,

$$a(u,v) := \int_0^1 u'(x)\overline{v'(x)}\,dx, \qquad u, v \in V := W^{1,2}(0,1),$$

and take

$$H := \mathbb{C}^2 \qquad \text{and} \qquad j(u) := \begin{pmatrix} u(1) \\ u(0) \end{pmatrix}.$$

Then $V(a)$ is the space of weakly harmonic functions on $(0, 1)$. In particular the operator associated with (a, j) must satisfy $-(Aj(u)|j(v))_H = a(u,v)$ for all $u, v \in V(a)$, i.e.,

$$-\left(A\begin{pmatrix} u(1) \\ u(0) \end{pmatrix} \,\Big|\, \begin{pmatrix} v(1) \\ v(0) \end{pmatrix}\right) \stackrel{!}{=} a(u,v) = u'(x)\overline{v(x)}\Big|_{x=0}^{x=1}, \qquad u, v \in V(a).$$

Fig. 6.1 An interval as a trivial unweighted graph

One hence checks that for $x \in \mathbb{C}^2$ $Ax = -y$ if and only if there exists $u \in V(a)$ such that

$$x = \begin{pmatrix} u(1) \\ u(0) \end{pmatrix} =: j(u) \qquad \text{and} \qquad y = \begin{pmatrix} u'(1) \\ -u'(0) \end{pmatrix}.$$

In other words, A is (minus) the Dirichlet-to-Neumann operator introduced in Sect. 2.3.1. Because of course the (weakly) harmonic function with boundary values $j(u)$ is given by $u : x \mapsto u(0) + (u(1) - u(0))x$, one also sees that $-A$ agrees with the matrix

$$\begin{pmatrix} 1 & -1 \\ -1 & 1 \end{pmatrix}.$$

Thus, the Dirichlet-to-Neumann operator $-A$ agrees with the discrete Laplacian $\mathcal{L}$ for the trivial unweighted graph associated with the interval $(0, 1)$ (Fig. 6.1). This observation can be extended to the case of discrete Laplacians on more interesting graphs, as we will see in Sect. 6.6.

(What happens taking $j : u \mapsto \int_0^1 u(x)\,dx$ instead? The form is clearly j-elliptic, as one sees applying a suitable version of the Poincaré inequality.)

Most of the classical results summarized in Sect. 6.2 can be extended to the case of j-elliptic forms. We consider explicitly the following two instances, taken from [90] and [8] respectively.

Proposition 6.37. *Let $a : V \times V \to \mathbb{C}$ be a j-elliptic, continuous form and denote by A the associated operator. Then A generates an analytic C_0-semigroup on H.*

If additionally there exists $M \geq 0$ such that

$$|\operatorname{Im} a(u, u)| \leq M \|u\|_V \|j(u)\|_H \qquad \text{for all } u \in V(a), \tag{6.15}$$

then A generates a C_0-cosine operator function with Kisyński space $j(V)$ and hence a C_0-semigroup with analyticity angle of $\frac{\pi}{2}$.

Proposition 6.38. *Let C be a closed convex subset of H and denote by P_C the orthogonal projector of H onto C. Then the following assertions are equivalent.*

(a) C is invariant under $(e^{tA})_{t\geq 0}$.
(b) for all $u \in V$ there exists $w \in V$ such that $P_C j(u) = j(w)$ and $\operatorname{Re} a(w, u-w) \geq 0$.

In this way one can e.g. show that the C_0-semigroup generated by Dirichlet-to-Neumann operators is sub-Markovian. In view of the connection with the discrete Laplacians observed above, this Proposition 6.38 may also be used to provide an alternative proof of Theorem 6.54 below.

Using the notion of form convergence that goes back to Mosco, cf. [83], the following has been proved in [90].

Theorem 6.39. *Let $(a_n, j_n)_{n\in\mathbb{N}}$ and (a, j) be densely defined, sesquilinear continuous forms on a Hilbert space H with form domains $(V_n)_{n\in\mathbb{N}}$ and V, respectively. We assume that a_n is j_n-elliptic for all $n \in \mathbb{N}$ and a is j-elliptic, and that the j-ellipticity constants can be chosen to be the same for all $n \in \mathbb{N}$. Then the following are equivalent.*

(a) The sequence of operators $(A_n)_{n\in\mathbb{N}}$ associated with $(a_n, j_n)_{n\in\mathbb{N}}$ converges to the operator $-A$ associated with (a, j) in the strong resolvent sense.
(b) The following conditions are satisfied:

  (i) If $u_n \in V_n$, $j_n(u_n) \rightharpoonup x$ for some $x \in H$ and $\liminf_{n\to\infty} a_n(u_n, u_n) < \infty$, then there exists $u \in V$ such that $j(u) = x$ and $\liminf_{n\to\infty} a_n(u_n, u_n) \geq a(u, u)$;
  (ii) For all $u \in V$ there exists a sequence $(u_n)_{n\in\mathbb{N}}$ with $u_n \in V_n$ such that

$$\lim_{n\to\infty} j_n(u_n) = j(u) \qquad \text{and} \qquad \liminf_{n\to\infty} a_n(u_n, u_n) \leq a(u, u).$$

Observe that strong resolvent convergence of generators is a necessary condition for strong convergence of the associated C_0-semigroups.

Proposition 6.40. *Under the assumptions of Theorem 6.39, let additionally (b, j) be a sesquilinear, continuous, j-elliptic form defined on V. Denote the respective associated operators by A_n, $n \in \mathbb{N}$, A, and B. Assume all the forms to be symmetric. If the equivalent conditions in Theorem 6.39 are satisfied, and additionally*

- *(e^{tB}) is of trace class as well as*
- *for all $n \in \mathbb{N}$*

$$j_n(V_n) \subset j(V) \quad \text{and} \quad a_n(u, u) \geq b(v, v) \text{ whenever } j_n(u) = j(v),$$

then e^{tA_n} converges to e^{tA} for all $t > 0$ in operator norm (and even in trace norm).

6.2.2 Subdifferentials of Energy Functionals

After the Hille–Yosida Theorem paved the way for the development of the theory of C_0-semigroups, it seemed obvious to generalize this theorem—and as much as possible of the linear theory—to the nonlinear setting. Unfortunately, all attempts to find a precise nonlinear counterpart of the Hille–Yosida Theorem have been unsuccessful to date. Partial results are known, but are often rather technical.

Nevertheless, in the 1960s an elegant and efficient theory of nonlinear semigroups could indeed be developed: It regards a class of nonlinear operators that bear many similarities to those linear operators that come from forms, and a class of nonlinear semigroups that enjoy many properties typical of analytic semigroups. This section is devoted to recall this theory: We do so for the sake of completeness of our exposition, even if throughout the book we have chosen to devote our attention almost exclusively to linear models and equations.

Assumptions 6.41. Throughout this section we consider the following functional setting.

- V is a reflexive Banach space.
- H is a Hilbert space.
- V is densely and continuously embedded in H.
- $\mathcal{E} : V \to [0, \infty)$ is a convex, Fréchet differentiable functional (with derivative denoted by $\mathcal{E}'$) such that $\mathcal{E}(0) = 0$.

We can and will extend $\mathcal{E}$ to the whole H by $+\infty$: With an abuse of notation we denote this extension again by $\mathcal{E}$. Then the *subdifferential* of $\mathcal{E} : H \to [0, +\infty]$ at $f \in H$ is defined (cf. e.g. [102, Def. at p. 81]) as the set

$$\partial\mathcal{E}(f) := \begin{cases} \{g \in H : (g|\varphi - f)_H \le \mathcal{E}(\varphi) - \mathcal{E}(f)\ \forall \varphi \in H\}, & \text{if } f \in V, \\ \emptyset, & \text{if } f \in H \setminus V, \end{cases}$$

However, under the above assumptions the subdifferential of $\mathcal{E}$ at each $f \in V$ is by [102, Prop. II.7.6] either empty or a singleton. Thus, we can regard $\partial\mathcal{E}$ as a (single-valued) operator from

$$D(\partial\mathcal{E}) := \{f \in V : \partial\mathcal{E}(f) \neq \emptyset\}$$

to H, and we denote with a slight abuse of notation

$$\partial\mathcal{E}(f) \equiv \{\partial\mathcal{E}(f)\}, \qquad f \in D(\partial\mathcal{E}).$$

Determining a subdifferential is in general a tedious task. Due to our assumption of Fréchet differentiability of $\mathcal{E}$, however, a subdifferential can be described more easily by means of the following result. While it seems to be known, the only precise reference we are aware of is [95, Lemma 2.8.9].

Lemma 6.42. *The subdifferential of $\mathcal{E}$ can be equivalently described by*

$$\begin{aligned} D(\partial\mathcal{E}) &= \{f \in V : \exists g \in H \text{ s.t. } \mathcal{E}'(f)h = (g|h)_H \quad \forall h \in V\}, \\ \partial\mathcal{E}(f) &= g. \end{aligned} \tag{6.16}$$

Unlike in the linear case, in the world of nonlinear evolution equations several discording techniques for finding solutions exist. In particular, subdifferentials of proper, convex, lower semicontinuous functionals are (nonlinear) m-accretive operators, cf. [102, Prop. IV.2.2], i.e., they satisfy the estimate

$$\langle Au - Av, u - v\rangle \geq 0 \qquad \text{for all } u, v \in D(A).$$

(This definition can be even extended to multi-valued operators, but we avoid this general setting for the sake of simplicity.) Therefore the celebrated theorem due to M.G. Crandall and T.M. Liggett—the nonlinear pendant of the theorem of Lumer–Phillips, since obviously a linear operator A is dissipative if and only if $-A$ is accretive—can be applied to find solutions of abstract Cauchy problems associated with subdifferentials in terms of a semigroup of nonlinear operators.

Theorem 6.43. *Let A be a densely defined, m-accretive operator on a Banach space X. Then for any $f_0 \in X$ and any $t \geq 0$ the sequence*

$$J^n_{\frac{t}{n}} f_0 := \left(\mathrm{Id} + \frac{t}{n} A\right)^{-n} f_0, \qquad n \in \mathbb{N},$$

converges uniformly on compact subsets of $[0, \infty)$ and its limit, which we denote by $e^{-tA} f_0$, defines a nonlinear contractive C_0-semigroup, i.e., a strongly continuous family of (in general, nonlinear) contractions on X that satisfy the semigroup law.

Let us observe that, among other things, the perturbation theory for accretive operators is not nearly as mighty as in the linear case. As an example we mention the following, cf. [23, Théo. 3.17 and Rem. 3.14].

Lemma 6.44. *Let A be an m-accretive operator on a Hilbert space H. If $B : H \to H$ is a globally Lipschitz mapping, then $A + B + \omega \,\mathrm{Id}$ is m-accretive, where ω is the Lipschitz constant of B.*

At a first glance, this result is tailored for application to differential equations where the maximal monotone leading term is perturbed by polynomial of odd degree, which can be splitted into globally Lipschitz, bounded part—corresponding to the graph of the polynomial between its smallest and its largest zero—and a maximal monotone part—corresponding to the graph of the function outside this interval. (The case of a perturbation by a polynomial of even degree is more delicate, and we refer to [1, Thm. 3.1] for a comparable generation result.) Unfortunately, polynomials do not usually define globally Lipschitz operators on L^p-spaces for $p < \infty$, but only on spaces of continuous functions.

Just like in the linear case, it can be shown that semigroups yield solutions of abstract Cauchy problems. As shown in [19], however, the solutions obtained in this way have to be defined by means of an approximation scheme and have in general very poor regularity properties. More useful information about solutions can be obtained applying more directly the properties of subdifferentials and using Hilbert space methods.

Theorem 6.45. *Let $T > 0$. Then for all $f \in L^2\big((0,T);H\big)$ and all $f_0 \in H$ there exists a unique $\varphi \in C([0,T];H)$ which is differentiable for a.e. $t \in [0,T]$ and such that*

$$\begin{cases} \varphi(t) \in D(\partial\mathcal{E}), & \text{for a.e. } t \in [0,T], \\ \frac{\partial\varphi}{\partial t}(t) + \partial\mathcal{E}\varphi(t) = f(t), & \text{for a.e. } t \in [0,T], \\ \varphi(0) = f_0. \end{cases} \tag{6.17}$$

Furthermore, $\mathcal{E} \circ u \in L^1(0,T)$, and moreover

- *if $f_0 \in V$, then $\mathcal{E} \circ u \in L^\infty(0,T)$ and $u \in H^1\big((0,T);H\big)$;*
- *if $f_0 \in D(\partial\mathcal{E})$, then $u \in W^{1,\infty}\big((0,T);H\big)$ and u is right differentiable for all $t \geq 0$,*
- *if $f \equiv 0$, then the mapping $t \mapsto \varphi(t)$ agrees for all $t \geq 0$ with $t \mapsto e^{-t\partial\mathcal{E}} f_0$.*

An alternative approach to investigate well-posedness of nonlinear evolution equations goes back to [72] and has been substantially enhanced by Chill and his coauthors in recent years, cf. [29] for a comprehensive exposition. The following collects [29, Thm. 6.1 and § 6.4].

Theorem 6.46. *Additionally to our standing assumptions, let*

- *V be separable,*
- *$\mathcal{E}$ be* coercive*, i.e., the sublevel sets*

$$\{f \in V : \mathcal{E}(f) \leq \alpha\}$$

be bounded with respect to the norm of V for all $\alpha \in \mathbb{R}$, and
- *the Fréchet derivative $\mathcal{E}'$ map bounded sets of V into bounded sets of V'.*

Then for all $f \in L^2\big((0,T);H\big)$ and all $f_0 \in V$ there exists a unique $\varphi \in L^\infty\big((0,T);V\big) \cap H^1\big((0,T);H\big)$ such that

$$\begin{cases} \varphi(t) \in D(\partial\mathcal{E}), & \text{for a.e. } t \in [0,T], \\ \frac{\partial\varphi}{\partial t}(t) + \partial\mathcal{E}\varphi(t) = f(t), & \text{for a.e. } t \in [0,T], \\ \varphi(0) = f_0. \end{cases} \tag{6.18}$$

Furthermore, the energy inequality

$$\int_0^t \left\| \frac{\partial\varphi}{\partial t}(s) \right\|_H^2 ds + \mathcal{E}(\varphi(t)) \leq \mathcal{E}(f_0) + \int_0^t \left(f(s) | \frac{\partial\varphi}{\partial t}(s) \right)_H ds, \qquad t \in [0,T], \tag{6.19}$$

is satisfied. If in particular $V = H$ is finite dimensional and $f \equiv 0$, then the solution φ satisfies the further energy inequality

$$\frac{d}{dt}\mathcal{E}(\varphi(t)) \leq -\frac{1}{2}\left\|\frac{\partial\varphi}{\partial t}(t)\right\|_H^2 - \frac{1}{2}\|\partial\mathcal{E}\varphi(t)\|_H^2, \qquad t \geq 0. \tag{6.20}$$

The main idea of the proof is to consider the weak formulation

$$\left(\frac{\partial\varphi}{\partial t}|h\right)_H + \partial\mathcal{E}\varphi(t)h = (f(t)|h)_H\,, \qquad \text{for a.e. } t \in [0,T] \text{ and all } h \in V,$$

or rather

$$\left(\frac{\partial\varphi}{\partial t}(t)|h\right)_H + \mathcal{E}'(\varphi(t))h = (f(t)|h)_H\,, \qquad \text{for a.e. } t \in [0,T] \text{ and all } h \in V,$$

of the differential equation, and then to discretize it by applying the so-called *Galerkin scheme*:

(1) since V is separable, one can take
   - a total sequence $(e_n)_{n\in\mathbb{N}}$ and hence the sequence of finite dimensional spaces $V_n := \text{span}\{e_m : m \leq n\}$ (with the norm induced by H) such that $\bigcup_{n\in\mathbb{N}} V_n$ is dense in V and
   - a sequence $(f_{0n})_{n\in\mathbb{N}}$ such that $f_{0n} \in V_n$ for all $n \in \mathbb{N}$ and $\lim_{n\to\infty} f_{0n} = f_0$ in V;

(2) for all $n \in \mathbb{N}$, consider $\mathcal{E}_{|V_n}$, take its subdifferential as in (6.16), but with respect to test functions in V_n; and use Carathéodory's Theorem to solve

$$\begin{cases} \frac{\partial\varphi}{\partial t}_n(t) + \partial\mathcal{E}_{|V_n}\varphi_n(t) = P_n f(t), & \text{for a.e. } t \in [0,T], \\ \varphi_n(0) = f_{0n}, \end{cases}$$

where we denote by P_n the orthogonal projection of H onto V_n;

(3) for all $n \in \mathbb{N}$, show that all the solutions ϕ_n admit uniform a priori bounds, which in turn show that the sequence $(\phi_n)_{n\in\mathbb{N}}$ is bounded in $L^\infty\big((0,T);V\big) \cap H^1\big((0,T);H\big)$ and also that $(\mathcal{E}(\phi_n))_{n\in\mathbb{N}}$ is bounded in $L^\infty(0,T;V')$;

(4) extract a converging subsequence and show that its limit is a solution of the abstract Cauchy problem (6.18) with the claimed regularity properties and satisfying the energy inequality (6.19);

(5) use accretivity of $\partial\mathcal{E}$ to prove that there cannot be further solutions.

The latter energy inequality (6.20), which in [29, § 6.4] is reported to be due to De Giorgi, shows that either the solution reaches in finite time a ground state, or its energy decreases indefinitely.

Remark 6.47. There is a connection between the linear theory of quadratic forms and the nonlinear theory we are summarizing in this section: If V is a Hilbert space and $a : V \times V \to \mathbb{R}$ is a bounded, coercive, symmetric sesquilinear form (i.e., a quadratic form), then

$$\mathcal{E} : V \ni f \mapsto \frac{1}{2} a(f, f) \in [0, \infty)$$

defines a convex, coercive, Fréchet differentiable functional and all sublevel sets are bounded in V, hence it satisfies the assumptions of all results in this section. Moreover, one sees directly that the Fréchet derivative of $\mathcal{E}$ is given by

$$\mathcal{E}'(f)g = a(f, g), \qquad f, g \in V.$$

Therefore, by definition the subdifferential/Fréchet derivative of $\mathcal{E}$ is precisely $-A$, where A is the linear operator associated with a in the sense of Definition 6.13.

Finally, making use of semigroup theory it is possible to characterize closed convex sets of H that are left invariant over time, by a nonlinear generalization of Ouhabaz' invariance criterion due to Barthélemy.

Lemma 6.1. *Let C be a closed convex subset of H and denote by P_C the orthogonal projection of H onto C. Then the following assertions are equivalent.*

(a) C is left invariant under $J_\lambda(\partial\mathcal{E})$ for all $\lambda > 0$.
(b) C is left invariant under $e^{-t\partial\mathcal{E}}$ for all $t \geq 0$.
(c) $\mathcal{E}(P_C f_0) \leq \mathcal{E}(f_0)$ for all $f_0 \in V$.

In particular:

- *The semigroup $(e^{-t\partial\mathcal{E}})_{t\geq 0}$ is order preserving, i.e.,*

$$f_0 \leq g_0 \quad \Rightarrow \quad e^{-t\partial\mathcal{E}} f_0 \leq e^{-t\partial\mathcal{E}} g_0 \qquad \forall t \geq 0,\ \forall f_0, g_0 \in H, \tag{6.21}$$

  if and only if

$$\mathcal{E}(f_0 \wedge g_0) + \mathcal{E}(f_0 \vee g_0) \leq \mathcal{E}(f_0) + \mathcal{E}(g_0) \quad \text{for all } f_0, g_0 \in V.$$

- *Let $(U, \tilde{\mu})$ be a σ-finite measure space and $H = L^2(U; \tilde{\mu})$. Then $(e^{-t\partial\mathcal{E}})_{t\geq 0}$ is contractive with respect to the norm of $L^\infty(U; \tilde{\mu})$, i.e.,*

$$\|e^{-t\partial\mathcal{E}} f_0 - e^{-t\partial\mathcal{E}} g_0\|_\infty \leq \|f_0 - g_0\|_\infty \qquad \forall t \geq 0,\ \forall f_0, g_0 \in H,$$

  if and only if

$$\begin{aligned}\mathcal{E}&\left(\frac{g_0 + (f_0 - g_0 + 1)_+}{2} + \frac{g_0 - (f_0 - g_0 - 1)_-}{2}\right)\\ &+\mathcal{E}\left(\frac{f_0 - (f_0 - g_0 + 1)_+}{2} + \frac{f_0 + (f_0 - g_0 - 1)_-}{2}\right)\\ &\leq \mathcal{E}(f_0) + \mathcal{E}(g_0) \quad \text{for all } f_0, g_0 \in V.\end{aligned}$$

6.3 Delay Evolution Equations

In many models arising in biomathematics, and especially in neuroscience, delays have to be considered. Let us mention two examples: due to the duration of pregnancy, the size of a population only determines the number of its offspring at a later time; communication between neurons is slowed down by signal processing inside the individual neurons, as discussed in Chap. 5.

In these and several further models it is therefore appropriate to modify the standard version of an evolution equation and rather discuss

$$\dot{u}(t) = Au(t - \tau), \qquad t \geq 0,$$

for some delay $\tau > 0$ that is determined by the model. For a Banach space X and a function $u : [-\tau, +\infty) \to X$ one introduces the *history segment*

$$u_t : [-\tau, 0] \ni \sigma \mapsto u(t + \sigma) \in X.$$

With this notation we can now introduce a class of delayed abstract Cauchy problems of the form

$$\begin{cases} \dot{u}(t) = Au(t) + \Phi u_t, & t \geq 0, \\ u(0) = u_0, & \\ u_0 = f. & \end{cases} \tag{dACP}$$

We refer to [11, Chapter 3] for a justification of this formulation. Indeed, the condition on the history segment in (dACP) is necessary in order to determine a solution, because (dACP) has to be turned into a certain abstract Cauchy problem on $X \times L^1\big((-\tau, 0; X\big)$ with initial conditions (u_0, f). We want to study this kind of problems in the framework of the theory of elliptic forms and adopt the standing assumption of Sect. 6.2, i.e.,

$a : V \times V \to \mathbb{C}$ is sesquilinear, continuous and H-elliptic and A is its associated operator

Definition 6.48. Let $\tau > 0$, $f \in L^1\big((-\tau, 0); V\big)$, and $u_0 \in H$. A *solution of* (dACP) is a function $u \in C\big([-\tau, \infty); X\big) \cap C^1\big([0, \infty); X\big)$ that satisfies (dACP) and with $u(t) \in D(A)$, $u_t \in W^{1,1}\big((-\tau, 0); V\big)$ for all $t \geq 0$.

Proposition 6.49. *Let $a : V \times V \to \mathbb{C}$ be an H-elliptic, continuous sesquilinear form with associated operator A, $\tau > 0$, and Φ be a bounded linear operator from $W^{1,1}\big((-\tau, 0); V\big)$ to H.*

Then for every $u_0 \in D(A)$ and $f \in W^{1,1}\big((-\tau, 0); V\big)$ such that $f(0) = u_0$ there exists a unique solution of (dACP). *Furthermore, for each pair of sequences*

$(u_{0n})_{n\in\mathbb{N}} \subset D(A)$ *and* $(f_n)_{n\in\mathbb{N}} \subset W^{1,1}\big((-\tau,0);V\big)$ *that tend to 0 in* X *and* $L^1\big((-\tau,0);V\big)$*, respectively, also the sequence* $(u_n)_{n\in\mathbb{N}}$ *of solutions to the corresponding* (dACP) *tends to 0 in* X*, uniformly in compact intervals of* $\mathbb{R}_+$.

An elementary but relevant class of bounded linear operators from $W^{1,1}\big((-\tau,0);V\big)$ to H consists of those operators given by

$$\Phi w := \sum_{k=0}^{n} B_k w(-\tau_k), \qquad w \in W^{1,1}\big((-\tau,0);V\big),$$

for some $n \in \mathbb{N}$, where for all $k = 1,\ldots,n$ B_k is a bounded linear operator from V to H and $\tau_k \in [0,\tau]$. In this case (dACP) can be regarded as an abstract, linearized version of the model in (5.6), as it reads

$$\begin{cases} \dot u(t) = Au(t) + \sum_{k=1}^{n} B_k u(t-\tau_k), & t \ge 0,\\ u(0) = u_0, \\ u_0 = f. \end{cases}$$

Proposition 6.50. *Let* B *be a bounded linear operator from* V *to* H*. If* $(e^{t(A+B)})_{t\ge 0}$ *is uniformly exponentially stable, then there is* $\tau > 0$ *such that for any* $\tau_0 \in [0,\tau]$ *the solution* u *of* (dACP) *with*

$$\Phi w := Bw(-\tau_0), \qquad w \in W^{1,1}\big((-\tau,0);V\big),$$

is exponentially stable, i.e., it satisfies $\|u(t)\|_H \le e^{-\varepsilon t}$ *for some* $\varepsilon > 0$ *and all* $t \ge 0$.

Under the assumptions of Proposition 6.50 the semigroup generated by $A + B$ is analytic by Lemma 6.22. Thus, in order to apply the above result and conclude that sufficiently small delays do not affect the stability of (dACP) it suffices by Proposition 6.8 to show that $s(A+B) < 0$.

6.4 Matrix Semigroups on Networks

In this section we see how to study some of the difference operators introduced in Sect. 2.1 in the light of the theory of forms. We assume throughout that

$$\boxed{\mathsf{G} = (\mathsf{V},\mathsf{E},\gamma) \text{ is a locally finite, weighted oriented graph}}$$

and we generally take over all notational conventions of Sect. 2.1

6.4.1 The Discrete Diffusion Equation

This section is devoted to the study of the *discrete diffusion equation*, i.e., of the evolution equation

$$\frac{df}{dt}(t,\mathsf{v}) = -\mathcal{L}f(t,\mathsf{v}), \qquad t \geq 0,\ \mathsf{v} \in \mathsf{V},$$

(beware the sign!) associated with (minus) the discrete Laplace–Beltrami operator $\mathcal{L}$ on an oriented weighted graph G. If G is finite, then it is natural to look for a solution given by (4.1).

If (3.2) holds, then $\mathcal{L}$ is a bounded operator on $\ell^2(\mathsf{V})$ by Lemma 4.4. Hence, $-\mathcal{L}$ generates a C_0-semigroup (in fact, even an analytic C_0-group) that is given by the exponential formula. Then, by Remark 4.13 no strict restriction and no strict extension of $-\mathcal{L}$ can be a generator as well. But if (3.2) fails, a new phenomenon appears—nothing overly unusual if one is used to work with differential operators: One may also define $-\mathcal{L}$ on some different closed subspace of $w^{1,2}_\gamma(\mathsf{V})$—e.g. on $\mathring{w}^{1,2}_\gamma(\mathsf{V})$, cf. Remark 3.9. (The question whether $w^{1,2}_\gamma(\mathsf{V}) \neq \mathring{w}^{1,2}_\gamma(\mathsf{V})$, i.e., whether two different realizations of $\mathcal{L}$ actually exist, is a difficult one and depends on the properties of G as a metric measure space. We refer to [60] for some preliminary answers.)

For general graphs, the Laplace–Beltrami matrix $\mathcal{L}$ acts on test functions by

$$\mathcal{L}f(\mathsf{v}) := \sum_{\substack{\mathsf{w}\in\mathsf{V}\\ \mathsf{w}\sim\mathsf{v}}} \gamma\big((\mathsf{v},\mathsf{w})\big)(f(\mathsf{v}) - f(\mathsf{w})), \qquad \mathsf{v} \in \mathsf{V},\ f \in c_{00}(\mathsf{V}),$$

but then it can have different extensions: We will consider two of them, defined by

$$\begin{aligned} D(\mathcal{L}^N) &:= \Big\{ f \in w^{1,2}_\gamma(\mathsf{V}) : \exists g \in \ell^2(\mathsf{V}) \\ &\qquad \text{s.t. } \sum_{\mathsf{e}\in\mathsf{E}} \gamma(\mathsf{e})(\mathcal{I}^T f)(\mathsf{e})(\mathcal{I}^T h)(\mathsf{e}) = \sum_{\mathsf{v}\in\mathsf{V}} g(\mathsf{v})h(\mathsf{v})\ \forall h \in w^{1,2}_\gamma(\mathsf{V}) \Big\}, \\ \mathcal{L}^N f &:= -g, \end{aligned}$$

and

$$\begin{aligned} D(\mathcal{L}^D) &:= \Big\{ f \in \mathring{w}^{1,2}_\gamma(\mathsf{V}) : \exists g \in \ell^2(\mathsf{V}) \\ &\qquad \text{s.t. } \sum_{\mathsf{e}\in\mathsf{E}} \gamma(\mathsf{e})(\mathcal{I}^T f)(\mathsf{e})(\mathcal{I}^T h)(\mathsf{e}) = \sum_{\mathsf{v}\in\mathsf{V}} g(\mathsf{v})h(\mathsf{v})\ \forall h \in \mathring{w}^{1,2}_\gamma(\mathsf{V}) \Big\}, \\ \mathcal{L}^D f &:= -g, \end{aligned}$$

respectively. They are defined weakly, i.e., we cannot in general determine their values pointwise, but only integrate—or rather sum—them against suitable test functions. Their form domains are subspaces of $\ell^2(\mathsf{v})$—indeed, the largest and the smallest possible ones, respectively. This reminds us of to the situation in Example 6.21 and we thus regard these realizations as a "Neumann" and a "Dirichlet" discrete Laplace–Beltrami operator, respectively.

Remark 6.51. This setting can be further generalized. Let namely a function $\nu : \mathsf{V} \to \mathbb{R}$ define a measure that is equivalent to the standard counting measure, i.e.,

$$0 < n_0 \le \nu(\mathsf{v}) \le n_1 \qquad \text{for some } n_0, n_1 > 0 \text{ and all } \mathsf{v} \in \mathsf{V}. \tag{6.22}$$

Then $\ell^p(\mathsf{V}) = \ell^p_\nu(\mathsf{V})$, and in particular the sesquilinear form a in (6.23) is $\ell^2_\nu(\mathsf{V})$-elliptic as well. Thus, the operators associated with the form a—either with domain $V^N := w^{1,2}_{\mu,\nu}(\mathsf{V})$, cf. Definition 3.4, or with domain V^D given by the closure in V^N of $c_{00}(\mathsf{V})$—generate C_0-cosine operator functions, and hence analytic C_0-semigroups, on $\ell^2_\nu(\mathsf{V})$. How do these operators look like? A direct computation shows that their action on test functions is

$$f \mapsto \frac{1}{\nu} \sum_{\substack{\mathsf{w}\in\mathsf{V}\\ \mathsf{w}\sim\mathsf{v}}} \gamma\big((\mathsf{v},\mathsf{w})\big)(f(\cdot) - f(\mathsf{w})), \qquad f \in c_{00}(\mathsf{V}).$$

Theorem 6.52. *Both $-\mathcal{L}^D$ and $-\mathcal{L}^N$ generate an analytic, contractive C_0-semigroup on $\ell^2(\mathsf{V})$—in fact, even a C_0-cosine operator function with Kisyński space $w^{1,2}_\gamma(\mathsf{V})$ or $\mathring{w}^{1,2}_\gamma(\mathsf{V})$, respectively. Both these semigroups are compact if both conditions in Proposition 3.8.(2) are satisfied.*

Proof. Define a sesquilinear form a defined either on $V^N := w^{1,2}_\gamma(\mathsf{V})$ or on $V^D := \mathring{w}^{1,2}_\gamma(\mathsf{V})$ by

$$a(f,g) := \left(\mathcal{I}^T f | \mathcal{I}^T g\right)_{\ell^2_\gamma(\mathsf{E})}. \tag{6.23}$$

Then a is sesquilinear, continuous, accretive, and $\ell^2(\mathsf{V})$-elliptic by definition of $w^{1,2}_\gamma(\mathsf{V})$. It is even coercive on V^D. Because a is symmetric, it is also of Lions type. If Proposition 3.8.(2) applies, then compactness of the semigroup follows from Lemma (6.20). Finally, $-\mathcal{L}^N$ and $-\mathcal{L}^D$ are by construction the operators associated with (a, V^N) and (a, V^D), respectively. □

Remark 6.53. An analogous result holds if one considers either the space

$$z^{1,p}_{\gamma,\nu}(\mathsf{V}) := \left\{ f \in \ell^p_\nu(\mathsf{V}) : \mathcal{J}^T f \in \ell^p_\gamma(\mathsf{E}) \right\}$$

with the natural norm, or the closure of $c_{00}(\mathsf{V})$ in $z^{1,p}_{\gamma,\nu}(\mathsf{V})$. Here $\mathcal{J}$ is the signless incidence matrix, cf. Remark 2.4. Taking the form

$$b(f,g) := \left(\mathcal{J}^T f | \mathcal{J}^T g\right)_{\ell^2_\gamma(\mathsf{E})} \tag{6.24}$$

one can deduce a generation result for two different realizations of the signless Laplace–Beltrami matrix $\mathcal{Q}$. We omit the details.

Like in the case of finite graphs discussed in Lemma 4.57, the following holds. Due to non-locality of Laplace–Beltrami matrix, the proof requires ideas than are rather different from those that will prove useful in the continuous case.

Theorem 6.54. *Both* $(e^{-t\mathcal{L}^D})_{t\geq 0}$ *and* $(e^{-t\mathcal{L}^N})_{t\geq 0}$ *are sub-Markovian. They are irreducible if and only if* G *is connected.*

We will exploit throughout that for all $k \in \mathbb{R}$ and $p \in [1,\infty)$ the function

$$f_k : [0,\infty) \ni \alpha \mapsto |k+\alpha|^2 + |k-\alpha|^2 \in [0,\infty) \tag{6.25}$$

is strictly monotonically increasing.

Proof. Reality of the semigroup is clear, as all entries of $\mathcal{I}$ are real.

Let us denote by $\mathcal{E}$ the quadratic form given by the values of a (defined in (6.23)) along the diagonal, i.e.,

$$\mathcal{E}(f) := \frac{1}{2} a(f,f), \qquad f \in V. \tag{6.26}$$

By Theorem 6.28, positivity of the C_0-semigroup has to be proved checking that

$$\mathcal{E}(f^+) \leq \mathcal{E}(f) \qquad \text{for all } f \in w^{1,2}_\gamma(\mathsf{V}).$$

This is actually satisfied, because for all $\mathsf{e} \in \mathsf{E}$ dividing the cases

- $f(\mathsf{e}_{\text{init}}), f(\mathsf{e}_{\text{term}}) \geq 0$,
- $f(\mathsf{e}_{\text{init}}), f(\mathsf{e}_{\text{term}}) \leq 0$,
- $f(\mathsf{e}_{\text{init}}) \leq 0 \leq f(\mathsf{e}_{\text{term}})$,
- $f(\mathsf{e}_{\text{init}}) \geq 0 \geq f(\mathsf{e}_{\text{term}})$.

one sees that

$$|f^+(\mathsf{e}_{\text{init}}) - f^+(\mathsf{e}_{\text{term}})| \leq |f(\mathsf{e}_{\text{init}}) - f(\mathsf{e}_{\text{term}})|.$$

Let us now prove that the semigroups are $\|\cdot\|_{\ell^\infty}$-contractive, i.e., that for all $f,g \in w^{1,2}_\gamma(\mathsf{V})$

$$\mathcal{E}\left(\frac{g+(f-g+1)^+}{2}+\frac{g-(f-g-1)^-}{2}\right)$$
$$+\mathcal{E}\left(\frac{f-(f-g+1)^+}{2}+\frac{f+(f-g-1)^-}{2}\right)$$
$$\leq \mathcal{E}(f)+\mathcal{E}(g).$$

As above, it suffices to take one $\mathsf{e}\in\mathsf{E}$ and to check that for all $f,g\in w^{1,2}_\gamma(\mathsf{V})$

$$\left|\mathcal{I}^T\frac{g+(f-g+1)^+}{2}(\mathsf{e})+\mathcal{I}^T\frac{g-(f-g-1)^-}{2}(\mathsf{e})\right|^2$$
$$+\left|\mathcal{I}^T\frac{f-(f-g+1)^+}{2}(\mathsf{e})+\mathcal{I}^T\frac{f+(f-g-1)^-}{2}(\mathsf{e})\right|^2$$
$$\leq |\mathcal{I}^T f(\mathsf{e})|^2+|\mathcal{I}^T g(\mathsf{e})|^2,$$

or equivalently that for all $x,y,w,z\in\mathbb{R}$

$$\left|\frac{y+(x-y+1)^+}{2}-\frac{z+(w-z+1)^+}{2}+\frac{y-(x-y-1)^-}{2}-\frac{z-(w-z-1)^-}{2}\right|^2$$
$$+\left|\frac{x-(x-y+1)^+}{2}-\frac{w-(w-z+1)^+}{2}+\frac{x+(x-y-1)^-}{2}-\frac{w+(w-z-1)^-}{2}\right|^2$$
$$\leq |x-w|^2+|y-z|^2.$$

Proving this inequality in the nine possible cases

$|x-y|\leq 1$ and $|w-z|\leq 1$, $x-y\leq -1$ and $|w-z|\leq 1$, $x-y\geq 1$ and $|w-z|\leq 1$,
$|x-y|\leq 1$ and $w-z\geq 1$, $x-y\leq -1$ and $w-z\geq 1$, $x-y\geq 1$ and $w-z\geq 1$,
$|x-y|\leq 1$ and $w-z\leq -1$, $x-y\leq -1$ and $w-z\leq -1$, $x-y\geq 1$ and $w-z\leq -1$,

is tedious but not difficult. For example, in the second case (the first being trivial) one has to check that

$$\left|x-\frac{w-z+1}{2}\right|^2+\left|y-\frac{w+z-1}{2}\right|^2\leq |x-w|^2+|y-z|^2.$$

This condition can be re-written as

$$|k+\alpha|^2+|k-\alpha|^2\geq |k+\gamma|^2+|k-\gamma|^2,$$

where

$$k := \frac{x+y-w-z}{2}, \qquad \alpha := \frac{-x+y+w-z}{2}, \qquad \gamma := \frac{x-y-1}{2}.$$

Under the assumption that ($|x-y| \le 1$ and) $w - z \ge 1$, one has

$$x - y - w + z \le x - y - 1 \le y - x + w - z,$$

and hence $|\gamma| < |\alpha|$, whence the assertion follows, by monotonicity of f_k defined in (6.25).

Finally, let us show the assertion about irreducibility. All closed lattice ideals of $\ell^2(\mathsf{V})$ are of the form $\ell^2(\mathsf{V}_0) \equiv \{f \in \ell^2(\mathsf{V}) : \operatorname{supp} f \subset \mathsf{V}_0\}$ for some subset $\mathsf{V}_0 \subset \mathsf{V}$, and the associated orthogonal projections are given by the restriction operators $P_{\mathsf{V}_0} := \mathbf{1}_{\mathsf{V}_0}\cdot$.

Assume G to be connected and $\ell^2(\mathsf{V}_0)$ to be invariant under $(e^{-t\mathcal{L}})$, or equivalently that

$$\mathcal{E}(P_{\mathsf{V}_0} f) \le \mathcal{E}(f) \qquad \text{for all } f \in w^{1,2}_\gamma(\mathsf{V}). \tag{6.27}$$

We have to show that V_0 is a trivial subset of V, i.e., $\mathsf{V}_0 = \mathsf{V}$ or $\mathsf{V}_0 = \emptyset$.

In fact, if $\mathsf{V}_0 \ne \mathsf{V} \ne \mathsf{V}_0^C$ there are two adjacent nodes $\mathsf{v}_0 \in \mathsf{V}_0$ and $\mathsf{v}_1 \in \mathsf{V}_0^C$. Set

$$\tilde{\mathsf{V}} := \mathsf{V} \setminus \{\mathsf{v}_0, \mathsf{v}_1\},$$

so that E is partitioned into $(\mathsf{v}_0, \mathsf{v}_1), \mathsf{E}^0, \mathsf{E}^1, \tilde{\mathsf{E}}$. Here for $i \in \{0,1\}$ E^i consist of those edges *other than* $(\mathsf{v}_0, \mathsf{v}_1)$ one of whose endpoints is v_i (regardless of their orientation), and $\tilde{\mathsf{E}} := \mathsf{E} \setminus ((\mathsf{v}_0, \mathsf{v}_1) \cup \mathsf{E}^0 \cup \mathsf{E}^1)$. In other words,

$$\begin{aligned} 2\mathcal{E}(g) = {} & \gamma(\mathsf{v}_0,\mathsf{v}_1)|g(\mathsf{v}_0) - g(\mathsf{v}_1)|^2 + \sum_{\substack{\mathsf{w}\sim\mathsf{v}_0 \\ \mathsf{w}\ne\mathsf{v}_1}} \gamma(\mathsf{v}_0,\mathsf{w})|g(\mathsf{v}_0) - g(\mathsf{w})|^2 \\ & + \sum_{\substack{\mathsf{w}\sim\mathsf{v}_1 \\ \mathsf{w}\ne\mathsf{v}_0}} \gamma(\mathsf{v}_1,\mathsf{w})|g(\mathsf{v}_1) - g(\mathsf{w})|^2 \\ & + \sum_{\mathsf{e}\in\tilde{\mathsf{E}}} \gamma(\mathsf{e})|g(\mathsf{e}_{\mathrm{init}}) - g(\mathsf{e}_{\mathrm{term}})|^2 \quad \text{for all } g \in w^{1,2}_\gamma(\mathsf{V}). \end{aligned}$$

Let now $f \in w^{1,2}_\gamma(\mathsf{V})$ be defined by

$$f(\mathsf{v}) := \begin{cases} x, & \text{if } \mathsf{v} = \mathsf{v}_0, \\ 1, & \text{if } \mathsf{v} = \mathsf{v}_1, \\ 0, & \text{otherwise,} \end{cases}$$

for some $x > 0$ to be determined later. Accordingly,

$$P_{\mathsf{V}_0} f(\mathsf{v}) := \begin{cases} x, & \text{if } \mathsf{v} = \mathsf{v}_0, \\ 0, & \text{otherwise.} \end{cases}$$

Therefore,

$$2\mathcal{E}(f) = \gamma(\mathsf{v}_0, \mathsf{v}_1)|x - 1|^2 + |x|^2 \sum_{\substack{\mathsf{w}\sim\mathsf{v}_0 \\ \mathsf{w}\neq\mathsf{v}_1}} \gamma(\mathsf{v}_0, \mathsf{w}) + \sum_{\substack{\mathsf{w}\sim\mathsf{v}_1 \\ \mathsf{w}\neq\mathsf{v}_0}} \gamma(\mathsf{v}_1, \mathsf{w}).$$

(the sums on the right hand side are finite, because they are less then $\deg(\mathsf{v}_0)$ and $\deg(\mathsf{v}_1)$, respectively—recall that G is assumed to be locally finite) whilst

$$2\mathcal{E}(P_{\mathsf{V}_0} f) = \gamma(\mathsf{v}_0, \mathsf{v}_1)|x|^2 + |x|^2 \sum_{\substack{\mathsf{w}\sim\mathsf{w}_0 \\ \mathsf{v}\neq\mathsf{v}_1}} \gamma(\mathsf{v}_0, \mathsf{w}).$$

Accordingly,

$$2\mathcal{E}(f) - 2\mathcal{E}(P_{\mathsf{V}_0} f) = \gamma(\mathsf{v}_0, \mathsf{v}_1)\left(|x-1|^2 - |x|^2\right) + \sum_{\substack{\mathsf{w}\sim\mathsf{v}_1 \\ \mathsf{w}\neq\mathsf{v}_0}} \gamma(\mathsf{v}_1, \mathsf{w}) < 0,$$

choosing x large enough, thereby contradicting (6.27). This implies that indeed $\mathsf{V}_0 = \emptyset$ or $\mathsf{V}_0 = \mathsf{V}$.

Conversely, each subspace of $\ell^2(\mathsf{V})$ consisting of functions over a connected component is left invariant under $(e^{-t\mathcal{L}^N})_{t\geq 0}$, hence the semigroup would be reducible if G contained more than one connected components. The same proof holds for $(e^{-t\mathcal{L}^D})_{t\geq 0}$. □

An alternative proof of well-posedness of the abstract Cauchy problem associated with $\mathcal{L}^D$, based instead on the Galerkin scheme, yields a convergence assertion for a sequence of solutions of certain discrete diffusion equations on *finite* graphs. More precisely, a family of graphs $(\mathsf{G}_n)_{n\in\mathbb{N}}$ is said to be *growing* if G_n is an induced subgraph of G_m for all $n, m \in \mathbb{N}$ with $n \leq m$; and that it *exhausts* G if

- G_n is an induced subgraph of G for all $n \in \mathbb{N}$ and
- $\bigcup_{n\in\mathbb{N}} \mathsf{V}_n = \mathsf{V}$.

Then the following holds, by a direct application of the Galerkin scheme described in Appendix 6.2.2, applied to the functional $\mathcal{E}$ in (6.26).

Proposition 6.55. *For all $f_0 \in \mathring{w}^{1,2}_\gamma(\mathsf{V})$ there is a growing family of finite graphs $(\mathsf{G}_n)_{n\in\mathbb{N}}$ that exhausts G and such that the sequence of solutions $(\varphi_n)_{n\in\mathbb{N}}$ of the Cauchy problem*

$$\begin{cases} \dot{\varphi}_n(t,\mathsf{v}) = -\mathcal{L}^{(n)}\varphi_n(t,\mathsf{v}), & t \geq 0,\ \mathsf{v} \in \mathsf{V}_n, \\ \varphi_n(t,\mathsf{v}) = 0, & t \geq 0,\ \mathsf{v} \in \mathsf{V} \setminus \mathsf{V}_n, \\ \varphi_n(0,\mathsf{v}) = f_0(\mathsf{v}), & \mathsf{v} \in \mathsf{V}_n, \end{cases}$$

converges to $e^{-\mathcal{L}^D} f_0$*, weakly in* $W^{1,2}\big((0,\infty);\ell^2(\mathsf{V})\big)$ *and weakly* in* $L^\infty\big((0,\infty);\mathring{w}_\gamma^{1,2}(\mathsf{V})\big)$. *Here* V_n *denotes the node set of* G_n *and for all* $h \in c_{00}(\mathsf{V})$

$$\mathcal{L}^{(n)}h(\mathsf{v}) := \begin{cases} \sum\limits_{\substack{\mathsf{w}\in\mathsf{V}_n \\ \mathsf{w}\sim\mathsf{v}}} \gamma(\mathsf{v},\mathsf{w})\,(h(\mathsf{v}) - h(\mathsf{w})) + h(\mathsf{v}) \sum\limits_{\substack{\mathsf{w}\notin\mathsf{V}_n \\ \mathsf{w}\sim\mathsf{v}}} \gamma(\mathsf{v},\mathsf{w}), & \mathsf{v} \in \mathsf{V}_n, \\ -h(\mathsf{v}) \sum\limits_{\substack{\mathsf{w}\in\mathsf{V}_n \\ \mathsf{w}\sim\mathsf{v}}} \gamma(\mathsf{v},\mathsf{w}), & \mathsf{v} \notin \mathsf{V}_n. \end{cases}$$

Corollary 6.56. *Let* $p \in (1,\infty)$. *Either* C_0*-semigroup* $(e^{-t\mathcal{L}^N})_{t\geq 0}, (e^{-t\mathcal{L}^D})_{t\geq 0}$ *extrapolates to a family of semigroups on* $\ell^q(\mathsf{V})$ *for all* $q \in [1,\infty]$ *as well as on*

$$c_0(\mathsf{V}) := \{f : \mathsf{V} \to \mathbb{R} : \forall \varepsilon > 0\ \exists \mathsf{W} \subset \mathsf{V}, |\mathsf{W}| < \infty,\ s.t.\ |f(\mathsf{v})| < \varepsilon\ \forall \mathsf{v} \notin \mathsf{W}\}.$$

They are strongly continuous for $q \in (1,\infty)$.

Proof. Contractivity of $(e^{-t\mathcal{L}^N})_{t\geq 0}$ with respect to the norm of $\ell^\infty(\mathsf{V})$ yields that the semigroup on $\ell^2(\mathsf{V})$ extends to a contractive C_0-semigroup on the closure of $\ell^2(\mathsf{V})$ in the ℓ^∞-norm, i.e., in $c_0(\mathsf{V})$. By duality we obtain a contractive semigroup on $\ell^1(\mathsf{V})$, and finally again by duality a contractive semigroup on $\ell^\infty(\mathsf{V})$. Now, the assertion follows by the Riesz–Thorin Theorem. □

Some elementary spectral properties follow directly from the definitions and Proposition 3.8.

Proposition 6.57. *The following assertions hold for the Laplace–Beltrami matrix* $\mathcal{L}$ *and for the associated form a defined in* (6.23).

(1) If $\mathsf{G} = (\mathsf{V},\mathsf{E},\gamma)$ *is finite, then the spectrum of* $\mathcal{L}$ *consists only of eigenvalues.*
(2) $\mathbf{1} \in w_{\gamma,\nu}^{1,2}(\mathsf{V})$ *if and only if* $\mathsf{G} = (\mathsf{V},\mathsf{E},\gamma)$ *has finite surface with respect to the node weight* ν*, and in this case* $a(\mathbf{1}, f) = 0$ *for all* $f \in w_{\gamma,\nu}^{1,2}(\mathsf{V})$.
(3) Thus, 0 *is an eigenvalue of the realization of* $\mathcal{L}$ *in* $\ell^2_\nu(\mathsf{V})$ *if and only if there exist connected components of* G *of finite surface with respect to the node weight* ν *(in the sense of Definition A.17); and in this case its multiplicity agrees with the number of such connected components.*

We can likewise deduce similar properties for the signless Laplace–Beltrami matrix $\mathcal{Q}$.

Proposition 6.58. *The following assertions hold for* $\mathcal{Q}$ *and for the associated form* $b : (f,g) \mapsto (\mathcal{C}\mathcal{J}^T f \,|\, \mathcal{J}^g)$.

(1) If $\mathsf{G} = (\mathsf{V}, \mathsf{E}, \gamma)$ *is finite, then the spectrum of* $\mathcal{Q}$ *consists only of eigenvalues.*

(2) Let G *be bipartite, say with* $\mathsf{V} = \mathsf{V}_1 \dot{\cup} \mathsf{V}_2$*, and denote* $\tilde{\mathbf{1}} := \mathbf{1}_{\mathsf{V}_1} - \mathbf{1}_{\mathsf{V}_2}$*. Then* $\tilde{\mathbf{1}} \in w^{1,2}_{\gamma,\nu}(\mathsf{V})$ *if and only if* $\mathsf{G} = (\mathsf{V}, \mathsf{E}, \gamma)$ *has finite surface with respect to the node weight* ν*, and in this case* $b(\tilde{\mathbf{1}}, f) = 0$ *for all* $f \in w^{1,2}_{\gamma}(\mathsf{V})$*.*

(3) Thus, 0 *is an eigenvalue of the realization of* $\mathcal{Q}$ *in* $\ell^2_\nu(\mathsf{V})$ *if and only if there exist connected components of* G *that are both bipartite and of finite surface with respect to the node weight* ν *(in the sense of Definition A.17); and in this case the multiplicity of* 0 *agrees with the number of such connected components.*

Proof. We only prove the assertion in (2). First of all, recall that $\mathcal{Q}$ does not depend on the orientation of G. On the other hand, a bipartite graph can be always re-oriented in such a way that one cell of the partition only comprises sinks and the other only sources. Hence $(\mathcal{J}^T \tilde{\mathbf{1}})_{\mathsf{e}} = 0$ for all $\mathsf{e} \in \mathsf{E}$, i.e., $\tilde{\mathbf{1}}$ (which apparently belongs to $w^{1,2}_{\mu,\nu}(\mathsf{V})$ if and only if G has finite surface with respect to ν. The proof is completed by observing that our reasoning extends component-wise to the case of a non-bipartite graph. □

6.4.2 The Discrete Advection Equation

The C_0-semigroup in Example 4.16, which yields the solution of the space-continuous advection equation, is non-analytic. Indeed, it acts shifting the profile of a function, and therefore there cannot be any gain of regularity—in contrast with the situation described in Proposition 6.6.

We have seen in Sect. 2.1.6 that the relevant operators for the *discrete* advection equations are the advection matrices $\overrightarrow{\mathcal{N}}, \overleftarrow{\mathcal{N}}$. If however G is not uniformly locally finite, then $\mathcal{I}$ and hence $-\overrightarrow{\mathcal{N}} = \mathcal{I}\mathcal{C}\mathcal{I}^{-T}$ may be unbounded, hence we cannot use (4.1) to show that it generates a semigroup.

Consider the "Neumann" discrete advection operator

$$
\begin{aligned}
D\left(\overrightarrow{\mathcal{N}}^N\right) := \Big\{ & f \in w^{1,2}_{\gamma}(\mathsf{V}) : \exists g \in \ell^2(\mathsf{V}) \\
& \text{s.t. } -\sum_{\mathsf{e}\in\mathsf{E}} \gamma(\mathsf{e}) \left(\mathcal{I}^{-T} f\right)(\mathsf{e}) \left(\mathcal{I}^T h\right)(\mathsf{e}) \\
& = \sum_{\mathsf{v}\in\mathsf{V}} g(\mathsf{v}) h(\mathsf{v}) \quad \forall h \in w^{1,2}_{\gamma}(\mathsf{V}) \Big\}, \\
\overrightarrow{\mathcal{N}}^N f := -g, &
\end{aligned}
$$

and the "Dirichlet" discrete advection operator

$$D\left(\overrightarrow{\mathcal{N}}^D\right) := \Big\{ f \in \mathring{w}_\gamma^{1,2}(\mathsf{V}) : \exists g \in \ell^2(\mathsf{V})$$
$$\text{s.t. } -\sum_{\mathsf{e}\in\mathsf{E}} \gamma(\mathsf{e}) \left(\mathcal{I}^{-T} f\right)(\mathsf{e}) \left(\mathcal{I}^T h\right)(\mathsf{e})$$
$$= \sum_{\mathsf{v}\in\mathsf{V}} g(\mathsf{v}) h(\mathsf{v}) \quad \forall h \in \mathring{w}_\gamma^{1,2}(\mathsf{V}) \Big\},$$
$$\overrightarrow{\mathcal{N}}^N f := -g,$$

and the operators $\overleftarrow{\mathcal{N}}^D, \overleftarrow{\mathcal{N}}^D, \mathcal{K}^{\mathrm{out}\,N}, \mathcal{K}^{\mathrm{out}\,D}, \mathcal{K}^{\mathrm{in}\,N}, \mathcal{K}^{\mathrm{in}\,D}$ defined likewise.

Proposition 6.59. *Let* G *be inward (resp., outward) uniformly locally finite. Then* $-\overrightarrow{\mathcal{N}}^N, -\overrightarrow{\mathcal{N}}^D, -\mathcal{K}^{\mathrm{out}\,N}, -\mathcal{K}^{\mathrm{out}\,D}$ *as well as* $-\overleftarrow{\mathcal{N}}^N, -\overleftarrow{\mathcal{N}}^D, -\mathcal{K}^{\mathrm{in}\,N}, -\mathcal{K}^{\mathrm{in}\,D}$*) generate* C_0*-cosine operator functions, hence analytic* C_0*-semigroups on* $\ell^2(\mathsf{V})$*. All these* C_0*-semigroups are compact if the conditions in Proposition 3.8.(2) are satisfied.*

Proof. By definition, $\overrightarrow{\mathcal{N}}^N$ is the operator associated with the form $b = a + b_0$ with domain $w_\gamma^{1,2}(\mathsf{E})$, where a is the continuous and $\ell^2(\mathsf{V})$-elliptic (by Theorem 6.52) form introduced in (6.23) and

$$b_0(f, g) := \left(\mathcal{C}\mathcal{I}^{+^T} f | \mathcal{I}^T g\right)_{\ell^2_\gamma(\mathsf{E})}.$$

Now, under our assumptions b_0 is continuous on $\ell^2(\mathsf{V}) \times w_\gamma^{1,2}(\mathsf{V})$—indeed, $\mathcal{I}^{+^T}$ is bounded on $\ell^2(\mathsf{V})$ by Lemma 4.3. Hence, in view of Lemma 6.22 b is a continuous, $\ell^2(\mathsf{V})$-elliptic form and the associated operator $\overrightarrow{\mathcal{N}}^N$ generates an analytic C_0-semigroup on $\ell^2(\mathsf{V})$. The other advection operators can be treated likewise. The assertions on the Kirchhoff matrices follow by duality. All operators with Dirichlet boundary conditions can be treated replacing $w_\gamma^{1,2}(\mathsf{V})$ by $\mathring{w}_\gamma^{1,2}(\mathsf{V})$. □

Applying Theorem 6.28 one can prove the following.

Proposition 6.60. *For both the Dirichlet and Neumann versions the following assertions hold.*

- *The* C_0*-semigroups* $(e^{-t\overrightarrow{\mathcal{N}}})_{t\geq 0}, (e^{-t\overleftarrow{\mathcal{N}}})_{t\geq 0}, (e^{-t\mathcal{K}^{\mathrm{out}}})_{t\geq 0}, (e^{-t\mathcal{K}^{\mathrm{in}}})_{t\geq 0}$ *are positive. They are also irreducible provided* G *is strongly connected.*
- *The semigroups* $(e^{-t\mathcal{K}^{\mathrm{out}}})_{t\geq 0}, (e^{-t\mathcal{K}^{\mathrm{in}}})_{t\geq 0}$ *are* $\ell^\infty(\mathsf{V})$*-contractive and hence their transpose* $(e^{-t\overrightarrow{\mathcal{N}}})_{t\geq 0}$ *and* $(e^{-t\overleftarrow{\mathcal{N}}})_{t\geq 0}$ *are* $\ell^1(\mathsf{V})$*-contractive.*
- *The semigroups* $(e^{-t\mathcal{K}^{\mathrm{out}}})_{t\geq 0}, (e^{-t\mathcal{K}^{\mathrm{in}}})_{t\geq 0}$ *are* $\ell^1(\mathsf{V})$*-contractive and hence their transpose* $(e^{-t\overrightarrow{\mathcal{N}}})_{t\geq 0}$ *and* $(e^{-t\overleftarrow{\mathcal{N}}})_{t\geq 0}$ *are* $\ell^\infty(\mathsf{V})$*-contractive, provided* $\deg^{\mathrm{out}}(\mathsf{v}) \geq \deg^{\mathrm{in}}(\mathsf{v})$ *and* $\deg^{\mathrm{out}}(\mathsf{v}) \leq \deg^{\mathrm{in}}(\mathsf{v})$ *for all* $\mathsf{v} \in \mathsf{V}$*, respectively.*

6.5 Diffusion on Metric Graphs

Throughout this section

$\mathfrak{G} = (\mathsf{V}, \mathfrak{E})$ is the metric graph over a locally finite, weighted oriented graph $\mathsf{G} = (\mathsf{V}, \mathsf{E}, \mu)$.

We consider the diffusive system

$$\frac{\partial u_{\mathsf{e}}}{\partial t}(t,x) = \frac{\partial^2 u_{\mathsf{e}}}{\partial x^2}(t,x), \qquad t > 0,\ x \in (0, \mu(\mathsf{e})),\ \mathsf{e} \in \mathsf{E}, \tag{Di}$$

complemented by the standard node conditions (Cc) – (KRc) introduced in Sect. 2.2. In analogy with the approach in Example 6.21 it seems natural to consider the sesquilinear form

$$(u, v) \mapsto \int_{\mathfrak{E}} u'(x)\overline{v'(x)}dx, \qquad u, v \in W^{1,2}(\mathfrak{G}).$$

It is however convenient to assume all edges to have unit length and consider instead the Laplacian, or more generally

$$\nabla(c^2\nabla) - M_p,$$

acting on $L^2(\mathfrak{G}) \simeq L^2\big((0,1); \ell^2_\mu(\mathsf{E})\big)$. Here $\nabla(c^2\nabla)$ and M_p are the elliptic operator with standard node conditions and the multiplication operator introduced in Definition 2.40 and Example 2.37, respectively, for two functions c, p defined on $[0, 1]$ that take values in the space of diagonal $\mathsf{E} \times \mathsf{E}$ matrices. This is legitimate upon considering the isometric isomorphism Ψ defined in (2.37), similarly to what we have done in Sect. 4.5 for the case of an advective system.

We are going to show well-posedness of the associated Cauchy problem in $L^p(\mathfrak{G})$ for all $p \in [1, \infty)$, or equivalently of the abstract Cauchy problem (ACP) associated with $\nabla(c^2\nabla) - M_p$ on the natural state space

$$H := L^2\big((0,1); \ell^2_\mu(\mathsf{E})\big).$$

By Lemma 3.20 one sees that for all $u \in W^{2,2}\big((0,1); \ell^2_\mu(\mathsf{E})\big)$ and all $v \in W^{1,2}\big((0,1); \ell^2_\mu(\mathsf{E})\big)$

$$
\begin{aligned}
((cu')'|v)_{L^2\left((0,1);\ell^2_\mu(\mathsf{E})\right)} &= \int_0^1 \left((c^2u')'(x)|v(x)\right)_{\ell^2_\mu} dx \\
&= \left(c^2(1)u'(1)|v(1)\right)_{\ell^2_\mu} - \left(c^2(0)u'(0)|v(0)\right)_{\ell^2_\mu} \\
&\quad - \int_0^1 \left(c^2(x)u'(x)|v'(x)\right)_{\ell^2_\mu} dx \\
&= \left(\begin{pmatrix} c^2(1)u'(1) \\ -c^2(0)u'(0) \end{pmatrix} \mid \begin{pmatrix} v(1) \\ v(0) \end{pmatrix}\right)_{\ell^2_\mu \times \ell^2_\mu} \\
&\quad - \int_0^1 \left(c^2(x)u'(x)|v'(x)\right)_{\ell^2_\mu} dx.
\end{aligned}
\tag{6.28}
$$

A direct computation shows that the standard node conditions (Cc) – (KRc) are transformed under the isomorphism Ψ from (2.37) into node conditions for $\nabla(c^2\nabla)$ that can be equivalently written as

$$
\begin{pmatrix} u(1) \\ u(0) \end{pmatrix} \in Y \tag{Cc'}
$$

and

$$
\begin{pmatrix} c^2(1)u'(1) \\ -c^2(0)u'(0) \end{pmatrix} + \mathcal{W}\begin{pmatrix} u(1) \\ u(0) \end{pmatrix} \in Y^\perp, \tag{KRc'}
$$

where Y is the (closed, by Lemma 4.3.(1)) subspace of $\ell^2_\mu(\mathsf{E}) \times \ell^2_\mu(\mathsf{E})$ defined by

$$
Y := \mathrm{Rg}\begin{pmatrix} (\mathcal{I}^+)^T \\ (\mathcal{I}^-)^T \end{pmatrix}. \tag{6.29}
$$

(We stress that the orthogonality relation is considered with respect to the (weighted) inner product of $\ell^2_\mu(\mathsf{E}) \times \ell^2_\mu(\mathsf{E})$.)

Remark 6.61. Even if this does not respect our standing assumption that multiple edges be forbidden, let us for a moment informally consider the case where G is a "rose", in the terminology of [69], i.e., it contains infinitely many edges all of whose endpoints coincide (in other words: there is just one "central" node v).

Then (Cc′) corresponds to continuity if Y is the subspace of $\ell^2_\mu(\mathsf{E}) \times \ell^2_\mu(\mathsf{E}) = \ell^2_\mu(\mathsf{E}_\mathsf{v}) \times \ell^2_\mu(\mathsf{E}_\mathsf{v})$ spanned by the constant vector $\mathbf{1}$: In order to write continuity as above it is therefore necessary to assume that in particular

$$
\infty > \sum_{\mathsf{e}\in\mathsf{E}_\mathsf{v}} 1^2\mu(\mathsf{e}) + \sum_{\mathsf{e}\in\mathsf{E}_\mathsf{v}} 1^2\mu(\mathsf{e}) = \deg_\mu(\mathsf{v})
$$

i.e., v to have finite degree, or equivalently $\mathsf{G} = (\mathsf{V}, \mathsf{E}, \mu)$ to be locally finite. (This restrictive condition is superfluous if one works in $L^\infty\left((0,1);\ell^\infty_\mu(\mathsf{E})\right)$ instead—a tempting feature that has been exploited in [16] and subsequent papers.)

Whenever u and v satisfy (Cc′) – (KRc′) and (Cc′), respectively, we can rewrite (6.28) as

$$\int_0^1 ((c^2u')'(x)|v(x))_{\ell^2_\mu}\,dx = -\left(\mathcal{W}\begin{pmatrix}u(1)\\u(0)\end{pmatrix} \mid \begin{pmatrix}v(1)\\v(0)\end{pmatrix}\right)_{\ell^2_\mu(\mathsf{E})\times\ell^2_\mu(\mathsf{E})} - \int_0^1 (c^2(x)u'(x)|v'(x))_{\ell^2_\mu}\,dx. \tag{6.30}$$

We thus introduce a sesquilinear form $a_\mathcal{W} : V \times V \to \mathbb{C}$ defined by

$$a_\mathcal{W}(u,v) := \int_0^1 \left((c^2(x)u'(x)|v(x))_{\ell^2_\mu} + (p(x)u(x)|v(x))_{\ell^2_\mu}\right) dx + \left(\mathcal{W}\begin{pmatrix}u(1)\\u(0)\end{pmatrix} \mid \begin{pmatrix}v(1)\\v(0)\end{pmatrix}\right)_Y, \tag{6.31}$$

with (dense, by Lemma 3.27.(1)) domain

$$V := W^{1,2}_Y((0,1);\ell^2_\mu(\mathsf{E})),$$

where

$$W^{1,2}_Y((0,1);\ell^2_\mu(\mathsf{E})) := \left\{u \in W^{1,2}((0,1);\ell^2_\mu(\mathsf{E})) : \begin{pmatrix}u(1)\\u(0)\end{pmatrix} \in Y\right\} \simeq W^{1,2}(\mathfrak{G}) \tag{6.32}$$

under the standing assumptions that

$$\boxed{p \in L^1(0,1;\ell^\infty_\mu(\mathsf{E}))},$$

$$\boxed{c \in L^\infty(0,1;\ell^\infty_\mu(\mathsf{E})) \text{ with } c_\mathsf{e}(x) \geq c_0 \text{ for some } c_0 > 0, \text{ all } \mathsf{e} \in \mathsf{E} \text{ and a.e. } x \in (0,1),}$$

and that

$$\boxed{\mathcal{W} \text{ is a bounded linear operator on } Y}.$$

Remark 6.62. By definition of $W^{1,p}_Y((0,1);\ell^2_\mu(\mathsf{E}))$ and in view of Lemma 4.3 one sees that the operator $u \mapsto u_{|\mathsf{V}}$ that evaluates u at the nodes of $\mathfrak{G}$ is surjective from $W^{1,p}_Y((0,1);\ell^2_\mu(\mathsf{E}))$ to $\ell^2_{\deg_\mu}(\mathsf{V})$.

Lemma 6.63. *The sesquilinear form $a_\mathcal{W}$ is well-defined, continuous, H-elliptic, and of Lions type. Additionally, $a_\mathcal{W}$ is symmetric if and only if $\mathcal{W}$ is self-adjoint.*

Proof. The form $a_{\mathcal{W}}$ is well-defined, as by Lemma 4.3 $\mathcal{I}^{\pm}u(x) \in \ell^2_{\deg_\mu}(\mathsf{V})$ for all $x \in [0,1]$ and all $u \in W^{1,2}\big((0,1);\ell^2_\mu(\mathsf{E})\big)$. Moreover, if $p(x) \in \ell^\infty_\mu(\mathsf{E})$ for a.e. $x \in (0,1)$, then $p(x)\cdot u(x) \in \ell^2_\mu(\mathsf{E})$ for a.e. $x \in (0,1)$ and hence $pu \in L^1\big((0,1);\ell^2_\mu(\mathsf{E})\big)$ as by Lemma 3.21.(1) $u \in L^\infty\big((0,1);\ell^2_\mu(\mathsf{E})\big)$ for all $u \in V$.

By (3.6), the leading term $a_0 : (u,v) \mapsto (c^2u'|v')_{L^2((0,1);\ell^2_\mu)}$ is continuous, H-elliptic, symmetric, and hence of Lions type. Let us consider the remaining terms

$$a_1 : (u,v) \mapsto (pu|v)_{L^2\left((0,1);\ell^2_\mu\right)} + \left(\mathcal{W}\begin{pmatrix}u(1)\\u(0)\end{pmatrix} \;\Big|\; \begin{pmatrix}v(1)\\v(0)\end{pmatrix}\right)_Y$$

as lower order perturbations. By the Hölder inequality (first for sequences, then for functions) we can estimate for all $u, v \in V$

$$\begin{aligned}|a_1(u,v)| \le & \int_0^1 \|p(x)\|_{\ell^\infty_\mu}\|u(x)\|_{\ell^2_\mu(\mathsf{E})}\|v(x)\|_{\ell^2_\mu(\mathsf{E})}dx\\ & + \|\mathcal{W}\|_{\mathcal{L}(Y)}\left|\left(\begin{pmatrix}u(1)\\u(0)\end{pmatrix} \;\Big|\; \begin{pmatrix}v(1)\\v(0)\end{pmatrix}\right)_Y\right|\\ \le & \left(\|p\|_{L^1} + 2\|\mathcal{W}\|_{\mathcal{L}(Y)}\right)\|u\|_\infty\|v\|_\infty.\end{aligned}$$

For $u = v$ we can refine this estimated applying Lemma 3.21.(5) for $p = \infty$, $q = r = 2$, and hence $\alpha = \frac{1}{2}$, thus obtaining $|a_1(u,u)| \le \left(\|p\|_{L^1} + 2\|\mathcal{W}\|_{\mathcal{L}(Y)}\right)\|u\|_{W^{1,2}}\|v\|_{L^2}$. We conclude that $a_{\mathcal{W}} := a_0 + a_1$ is H-elliptic, continuous, and of Lions type because we can apply Lemma 6.22 with $H_{\frac{1}{2}} = C\big([0,1];\ell^2_\mu(\mathsf{E})\big)$. □

Remark 6.64. In view of the equivalence between (Cc′) and the continuity condition in the nodes, the space Y is lattice isomorphic to $\ell^2_{\deg_\mu}(\mathsf{V})$ via the identification

$$\begin{pmatrix}u(1)\\u(0)\end{pmatrix} \simeq u_{|\mathsf{V}},$$

cf. (2.43). Our standing assumption on $\mathcal{W}$ can thus be reformulated by requiring that

$$\boxed{\mathcal{W} \text{ is a bounded linear operator on } \ell^2_{\deg_\mu}(\mathsf{V}).}$$

This motivates to write (6.31) equivalently as

$$\begin{aligned}a_{\mathcal{W}}(u,v) := & \int_{\mathfrak{G}}\left(c^2(x)u'(x)\overline{v'(x)} + p(x)u(x)\overline{v(x)}\right)dx\\ & + \left(\mathcal{W}u_{|\mathsf{V}} \mid v_{|\mathsf{V}}\right)_{\ell^2_{\deg_\mu}}, \qquad u,v \in W^{1,2}(\mathfrak{G}),\end{aligned} \tag{6.33}$$

as $W^{1,2}(\mathfrak{G}) \simeq W^{1,2}_Y\big((0,1);\ell^2_\mu(\mathsf{E})\big)$.

It remains to determine the operator associated with $a_{\mathcal{W}}$ in $L^2((0,1);\ell^2_\mu(\mathsf{E}))$.

Lemma 6.65. *The operator A associated with $a_{\mathcal{W}}$ in $L^2((0,1);\ell^2_\mu(\mathsf{E}))$ is*

$$D(\nabla\cdot(c^2\nabla)) := \left\{u \in W^{2,2}((0,1);\ell^2_\mu(\mathsf{E})) : u \textit{ satisfies } (\mathrm{Cc'})\text{-}(\mathrm{KRc'})\right\},$$
$$\nabla(c^2\nabla)u := (c^2u')'.$$

Proof. The arguments following Remark 6.61 show that $\nabla(c^2\nabla)$ is contained in the operator associated with $a_{\mathcal{W}}$. In order to prove the converse inclusion, take $u \in V$ such that there exists $w \in H$ satisfying

$$a_{\mathcal{W}}(u,v) = (w \mid v)_H = \int_0^1 (w(x)|v(x))_{\ell^2_\mu}\,dx \qquad \text{for all } v \in V. \tag{6.34}$$

Hence, (6.34) is satisfied in particular for all functions $v \equiv \tilde{v} \otimes e_{\mathsf{e}}$, where e_{e} is the e-th vector of the canonical basis of $\ell^2_\mu(\mathsf{E})$ and $\tilde{v}_{\mathsf{e}}$ is a scalar-valued $\mathring{W}^{1,2}(0,1)$-function, i.e.,

$$\int_0^1 \left(c^2_{\mathsf{e}}(x)u'_{\mathsf{e}}(x)\overline{\tilde{v}'_{\mathsf{e}}(x)} + p_{\mathsf{e}}(x)u_{\mathsf{e}}(x)\overline{\tilde{v}_{\mathsf{e}}(x)}\right)\mu(\mathsf{e})dx$$
$$= \int_0^1 w_{\mathsf{e}}(x)\overline{\tilde{v}_{\mathsf{e}}(x)}\mu(\mathsf{e})dx \qquad \text{for all } \mathsf{e} \in \mathsf{E}. \tag{6.35}$$

By definition of weak derivative, this means that $c_{\mathsf{e}}u'_{\mathsf{e}} \in W^{1,2}(0,1)$ and hence $u'_{\mathsf{e}} \in W^{1,2}(0,1)$ with weak derivative $w_{\mathsf{e}} - p_{\mathsf{e}}u_{\mathsf{e}}$, for all $\mathsf{e} \in \mathsf{E}$. As moreover (6.35) holds in particular for all functions $v \in V$ such that $v'_{\mathsf{e}} \equiv 1$ for any $\mathsf{e} \in \mathsf{E}$ on arbitrarily large closed subsets of $(0,1)$, one sees that $u' \in W^{1,2}((0,1);\ell^2_\mu(\mathsf{E}))$, i.e., $u \in W^{2,2}((0,1);\ell^2_\mu(\mathsf{E}))$. Moreover, taking into account (6.34)

$$\left(\begin{pmatrix} c^2(1)u'(1) \\ -c^2(0)u'(0)\end{pmatrix} \mid \begin{pmatrix} v(1) \\ v(0)\end{pmatrix}\right)_Y = -\left(\mathcal{W}\begin{pmatrix} u(1) \\ u(0)\end{pmatrix} \mid \begin{pmatrix} v(1) \\ v(0)\end{pmatrix}\right)_Y.$$

By Remark 6.62 the operator of evaluation at the nodes is bounded and surjective from V to Y, hence

$$\begin{pmatrix} c(1)u'(1) \\ -c(0)u'(0)\end{pmatrix} = -\mathcal{W}\begin{pmatrix} u(1) \\ u(0)\end{pmatrix} \quad \text{in } Y^\perp.$$

Thus, u satisfies (KRc$'$) and therefore $u \in D(\nabla(c^2\nabla))$. Furthermore,

$$\int_0^1 \left((c^2(x)u'(x)|v'(x))_{\ell^2_\mu} - (p(x)u(x)|v(x))\right)_{\ell^2_\mu} dx = \int_0^1 (w|v)_{\ell^2_\mu} dx \qquad \text{for all } v \in V$$

and this finally implies that $\nabla(c^2\nabla u) = -w$. □

Corollary 6.66. *If $\mathfrak{G}$ is finite, then the C_0-semigroup $(e^{tA})_{t\geq 0}$ associated with $a_{\mathcal{W}}$ consists of integral operators, i.e.,*

$$e^{tA}u(x) = \int_{\mathfrak{G}} K_t(x,y)u(y)dy \qquad \textit{for a.e. } x \in \mathfrak{G},\ t > 0, \tag{6.36}$$

for a heat kernel $(K_t)_{t\geq 0} \subset L^\infty(\mathfrak{G} \times \mathfrak{G})$, and there holds $\|e^{tA}\|_{\mathcal{L}(L^1,L^\infty)} = \|K_t\|_{L^\infty(\mathfrak{G}\times\mathfrak{G})}$, $t \geq 0$.

Proof. If G is finite, then V is densely and continuously embedded in $L^\infty(\mathfrak{G})$ by assumption. By Lemma 3.21 and Proposition 6.16, and by duality the analytic C_0-semigroup generated by $\nabla(c^2\nabla) - M_p$ maps $L^1(\mathfrak{G})$ to $L^\infty(\mathfrak{G})$, because by the semigroup law

$$L^1(\mathfrak{G}) \overset{e^{\frac{t}{2}(\nabla(c^2\nabla)-M_p)}}{\longrightarrow} L^2(\mathfrak{G}) \overset{e^{\frac{t}{2}(\nabla(c^2\nabla)-M_p)}}{\longrightarrow} L^\infty(\mathfrak{G}),$$

i.e., $e^{t(\nabla(c^2\nabla)-M_p)}$ is a bounded linear operator from $L^1(\mathfrak{G})$ to $L^\infty(\mathfrak{G})$, $t > 0$. In view of Theorem 6.31, the claim follows. □

We can summarize all our findings as follows.

Theorem 6.67. *The operator $\nabla(c^2\nabla) - M_p$ with standard node conditions (with domain as in Lemma 6.65) generates on $L^2\big((0,1);\ell^2_\mu(\mathsf{E})\big)$ a cosine operator function with Kisyński space $W^{1,2}_Y\big((0,1);\ell^2_\mu(\mathsf{E})\big)$, hence also an analytic C_0-semigroup of angle $\frac{\pi}{2}$. This semigroup is given by an integral kernel $(K_t)_{t\geq 0} \subset L^\infty\big((0,1)\times(0,1);\ell^2_\mu(\mathsf{E})\big)$, as in (6.36). Furthermore, $(e^{t\Delta})_{t\geq 0}$ is of trace class if the graph $\mathsf{G} = (\mathsf{V},\mathsf{E},\mu)$ has finite volume.*

If furthermore $-\mathcal{W}$ is dissipative and $p_{\mathsf{e}}(x) \geq 0$ for all $\mathsf{e} \in \mathsf{E}$ and a.e. $x \in (0,1)$, then $(e^{t\Delta})_{t\geq 0}$ is contractive. Assume additionally that $-\mathcal{W}$ is ϵ-quasi-dissipative for some $\epsilon < 0$ or $p_{\mathsf{e}}(x) \geq p_0$ for some constant $p_0 > 0$, all $\mathsf{e} \in \mathsf{E}$ and a.e. $x \in (0,1)$: Then $(e^{t\Delta})_{t\geq 0}$ is also uniformly exponentially stable.

Proof. In view of Lemmata 6.63 and 6.65, the assertion about generation follows from Theorems 6.15 and 6.18. Existence of an integral kernel with said properties is a consequence of Corollary 6.66. Finally, $(e^{tA})_{t\geq 0}$ is contractive if and only if a is accretive, i.e., if and only if for all $u \in V$

$$\begin{aligned} 0 \leq \operatorname{Re} a_{\mathcal{W}}(u,u) = \sum_{\mathsf{e}\in\mathsf{E}} \int_0^1 \left(|c^2_{\mathsf{e}}(x)u'_{\mathsf{e}}(x)|^2 + p_{\mathsf{e}}(x)|u_{\mathsf{e}}(x)|^2\right)\mu(\mathsf{e})\,dx \\ + \sum_{\mathsf{v},\mathsf{w}\in\mathsf{V}} \omega_{\mathsf{vw}} u(\mathsf{w})\overline{u(\mathsf{v})}\deg_\mu(\mathsf{v}). \end{aligned}$$

This holds whenever $-\mathcal{W}$ is dissipative and $p_{\mathsf{e}}(x) \geq 0$ for all $\mathsf{e} \in \mathsf{E}$ and a.e. $x \in (0,1)$. Exponential stability of $(e^{t\Delta})_{t\geq 0}$ can be established in a similar way. □

Example 6.68. Conversely, the matrix $-\mathcal{W}$ need not be dissipative if the operator associated with $a_{\mathcal{W}}$ is dissipative and hence the associated C_0-semigroup is contractive. For $c \equiv 1$ the matrix

$$\mathcal{W} = \begin{pmatrix} -1 & 1 \\ 1 & -1 \end{pmatrix}$$

is negative definite (its eigenvalues are 0 and -2), but nevertheless the corresponding operator Δ on $L^2(0,1)$ is dissipative: To see this, observe that

$$\begin{aligned}\int_0^1 u''(x)\overline{u(x)}\,dx &= u'(1)\overline{\big(u(1)-u(0)\big)} - u'(0)\overline{\big(u(1)-u(0)\big)} \\ &\quad - \int_0^1 |u'(x)|^2 dx \\ &= |u(1)-u(0)|^2 - \int_0^1 |u'(x)|^2 dx \\ &= \left|\int_0^1 u'(x)\,dx\right|^2 - \int_0^1 |u'(x)|^2 dx \le 0,\end{aligned}$$

by Jensen's inequality. Indeed, 0 is an eigenvalue of Δ with multiplicity 2, associated with the eigenfunctions $x \mapsto 1$ and $x \mapsto x$. (As already remarked in Sect. 2.2.1, this is the Krein–von Neumann extension of the second derivative on $L^2(0,1)$, cf. also [99, Example 14.14].)

The solution of the parabolic initial value problem with node conditions (Cc′) – (KRc′) automatically satisfies additional compatibility conditions in the nodes. If u_0 is the initial data, then by Proposition 6.6

$$u(t) := e^{t\nabla(c^2\nabla)}u_0 \in \bigcup_{k\in\mathbb{N}} D(\nabla(c^2\nabla)^k) \qquad \text{for all } t > 0$$

and the derivatives of u of even and odd order satisfy a continuity and Kirchhoff–Robin node condition, respectively. In the special case of the Laplacian of Sect. 2.2.1, e.g., we can hence deduce the following for the Laplacian on $\mathfrak{G}$ with standard node conditions.

Corollary 6.69. *Let $c \equiv 1$, $p \equiv 1$, and $\mathcal{W} = 0$ (i.e., let $A = \Delta$ with continuity and Kirchhoff node conditions). Then for all $N \in \mathbb{N}$*

$$\begin{aligned} u_{\mathsf{e}}^{(2N)}(t,\mathsf{v}) = u_{\mathsf{f}}^{(2N)}(t,\mathsf{v}) &=: u^{(2N)}(t,\mathsf{v}), && \mathsf{e},\mathsf{f} \in \mathsf{E}_{\mathsf{v}},\ \mathsf{v}\in\mathsf{V},\ t>0, \\ \sum_{\mathsf{w}\in\mathsf{V}} \omega_{\mathsf{vw}} u^{(2N)}(t,\mathsf{w}) &= \sum_{\mathsf{e}\in\mathsf{E}} \iota_{\mathsf{ve}} u_{\mathsf{e}}^{(2N+1)}(t,\mathsf{v}), && \mathsf{v}\in\mathsf{V},\ t>0. \end{aligned} \tag{6.37}$$

Proposition 6.70. *The following assertions hold for the form $a_{\mathcal{W}}$ associated with the elliptic operator with standard node conditions* (Cc) – (KRc) *introduced in Sect. 2.2.1.*

(1) If $\mathsf{G} = (\mathsf{V}, \mathsf{E}, \mu)$ has finite volume, then the spectrum of $\nabla(c^2\nabla) - M_p$ consists only of eigenvalues.
(2) $\mathbf{1} \in V$ if and only if $\mathsf{G} = (\mathsf{V}, \mathsf{E}, \mu)$ has finite volume, and in this case $a_{\mathcal{W}}(\mathbf{1}, u) = 0$ for all $u \in V$ if and only if

$$\sum_{\mathsf{w} \in \mathsf{V}} \omega_{\mathsf{vw}} \deg_\mu(\mathsf{v}) = 0 \qquad \text{for all } \mathsf{v} \in \mathsf{V} \tag{6.38}$$

and $p \equiv 0$. Then, 0 is an eigenvalue of $\nabla(c^2\nabla)$ whose multiplicity agrees with the number of connected components of G.

Proof. (1) By Lemma 3.17 $\mathfrak{G}$ is a (finite) measure space and hence V is compactly embedded in H. Therefore, $\nabla(c^2\nabla)$ has compact resolvent and its spectrum consists only of eigenvalues.
(2) The first assertion is clear by definition of volume, as $\mathbf{1}$ satisfies (Cc′). Now, $\mathbf{1} \in D(\nabla(c^2\nabla))$ (and in this case, $\nabla(c^2\nabla\mathbf{1}) = 0$) if and only if $\mathbf{1}$ satisfies (KRc′), i.e., if and only if (6.38) holds. The claim follows by a localization argument applied to test functions.

□

Since $\mathcal{W}$ and the multiplication operator M_p are bounded operators on $\ell^2_{\deg_\mu}(\mathsf{V})$ and $L^2\big((0,1); \ell^2_\mu(\mathsf{E})\big)$, respectively, they generate C_0-semigroups $(e^{-t\mathcal{W}})_{t\ge 0}$ and $(e^{-tM_p})_{t\ge 0}$. Several qualitative properties of $(e^{t\nabla(c^2\nabla)})_{t\ge 0}$ can be characterized by analogous properties of these semigroups, cf. Example 4.44.

Theorem 6.71. *The semigroup $(e^{t(\nabla(c^2\nabla)-M_p)})_{t\ge 0}$ is real if and only if so are $(e^{-t\mathcal{W}})_{t\ge 0}$ and $(e^{-tM_p})_{t\ge 0}$. If $(e^{-tM_p})_{t\ge 0}$ is real, then $(e^{t(\nabla(c^2\nabla)-M_p)})_{t\ge 0}$ is positive if and only if so is $(e^{-t\mathcal{W}})_{t\ge 0}$.*

Proof. By Proposition 6.30 $(e^{t\nabla(c^2\nabla)})_{t\ge 0}$ is real and positive if and only if

- $u \in V \Rightarrow \operatorname{Re} u \in V$ and $a_{\mathcal{W}}(\operatorname{Re} u, \operatorname{Im} u) \in \mathbb{R}$, and
- $u \in V \Rightarrow (\operatorname{Re} u)^+ \in V$, $a_{\mathcal{W}}(\operatorname{Re} u, \operatorname{Im} u) \in \mathbb{R}$, $a_{\mathcal{W}}((\operatorname{Re} u)^+, (\operatorname{Re} u)^-) \le 0$,

respectively.

Let now $u \in V$. One has $\operatorname{Re}(u_\mathsf{e}) = (\operatorname{Re} u)_\mathsf{e}$, $\mathsf{e} \in \mathsf{E}$. It follows from the above arguments that $\operatorname{Re} u \in W^{1,2}\big((0,1); \ell^2_\mu(\mathsf{E})\big)$ and the continuity of the values attained by u in the nodes is preserved upon taking the real part. All in all, $\operatorname{Re} u \in V$ and then one has $a(\operatorname{Re} u, \operatorname{Im} u) \in \mathbb{R}$ if and only if

$$\sum_{\mathsf{v},\mathsf{w} \in \mathsf{V}} \omega_{\mathsf{vw}} \operatorname{Re} u(\mathsf{w}) \operatorname{Im} u(\mathsf{v}) \deg_\mu(\mathsf{v}) \in \mathbb{R}$$

(where $\mathcal{W} = (\omega_{\mathsf{vw}})$) and

$$\sum_{\mathsf{e}\in\mathsf{E}} \int_0^1 p_{\mathsf{e}}(x)\,\mathrm{Re}\, u_{\mathsf{e}}(x)\,\mathrm{Im}\, u_{\mathsf{e}}(x)\mu(\mathsf{e})\, dx \in \mathbb{R}.$$

By Theorem 6.28 this is indeed equivalent to reality of$(e^{-t\mathcal{W}})_{t\geq 0}$ and $(e^{-tM_p})_{t\geq 0}$.

One also sees that $((\mathrm{Re}\, u)^+)_{\mathsf{e}} = (\mathrm{Re}(u_{\mathsf{e}}))^+$ for all $u \in V$ and $\mathsf{e} \in \mathsf{E}$, and hence $(\mathrm{Re}\, u)^+ \in V$. Accordingly,

$$a((\mathrm{Re}\, u)^+, (\mathrm{Re}\, u)^-) = \sum_{\mathsf{v},\mathsf{w}\in\mathsf{V}} \omega_{\mathsf{vw}} (\mathrm{Re}\, u)^+(\mathsf{w})(\mathrm{Re}\, u)^-(\mathsf{v}) \deg_\mu(\mathsf{v}).$$

In particular, for all $\mathsf{v} \in \mathsf{V}$ there holds

$$(\mathrm{Re}\, u)^+(\mathsf{v}) = \begin{cases} 0 & \text{if } (\mathrm{Re}\, u)(\mathsf{v}) \leq 0, \\ (\mathrm{Re}\, u)(\mathsf{v}) & \text{if } (\mathrm{Re}\, u)(\mathsf{v}) \geq 0, \end{cases}$$

and

$$(\mathrm{Re}\, u)^-(\mathsf{v}) = \begin{cases} -(\mathrm{Re}\, u)(\mathsf{v}) & \text{if } (\mathrm{Re}\, u)(\mathsf{v}) \leq 0, \\ 0 & \text{if } (\mathrm{Re}\, u)(\mathsf{v}) \geq 0. \end{cases}$$

Since by Theorem 6.28 $a((\mathrm{Re}\, u)^+, (\mathrm{Re}\, u)^-) \leq 0$ has to hold for all $u_{|\mathsf{V}}$, the claim follows. □

Theorem 6.72. *Let $-\mathcal{W}$ be dissipative. If $(e^{-t\mathcal{W}})_{t\geq 0}$ is $\ell^\infty_{\deg_\mu}(\mathsf{V})$-contractive, then $(e^{t(\nabla(c^2\nabla)-M_p)})_{t\geq 0}$ is $L^\infty\big((0,1);\ell^2_\mu(\mathsf{E})\big)$-contractive. The converse implication holds if additionally $p_{\mathsf{e}} \equiv 0$ for all e.*

This result is in fact simply a special instance of Theorem 6.85 below, but we have chosen to give here a more direct proof.

Proof. By Proposition 6.30 the semigroup $(e^{tA})_{t\geq 0}$ is $L^\infty\big((0,1);\ell^2_\mu(\mathsf{E})\big)$-contractive if and only if

$$u \in V \Rightarrow (1 \wedge |u|)\,\mathrm{sgn}\, u \in V \text{ and } \mathrm{Re}\, a_{\mathcal{W}}((1 \wedge |u|)\,\mathrm{sgn}\, u, (|u| - 1)^+\,\mathrm{sgn}\, u) \geq 0.$$

Take $u \in V$. It can be deduced from Lemma B.12.(2) that the functions defined by

$$((1 \wedge |u|)\,\mathrm{sgn}\, u)(x) = \begin{cases} u(x) & \text{if } |u(x)| \leq 1, \\ \frac{u(x)}{|u(x)|} & \text{if } |u(x)| \geq 1, \end{cases}$$

as well as

$$((|u|-1)^+ \operatorname{sgn} u)(x) = \begin{cases} 0 & \text{if } |u(x)| \le 1, \\ u(x) - \frac{u(x)}{|u(x)|} & \text{if } |u(x)| \ge 1 \end{cases}$$

are in $W^{1,2}((0,1);\ell^2_\mu(\mathsf{E}))$, with $((1 \wedge |u|) \operatorname{sgn} u)' = u' \mathbf{1}_{\{|u| \le 1\}}$ and $((|u| - 1)^+ \operatorname{sgn} u)' = u' \mathbf{1}_{\{|u| \ge 1\}}$: Hence, they have disjoint support.

Accordingly, if $u \in V$, then $(1 \wedge |u|) \operatorname{sgn} f \in V$, as the continuity in the nodes is preserved because

$$((1 \wedge |u|) \operatorname{sgn} u)\,(\mathsf{v}) = \begin{cases} u(\mathsf{v}) & \text{if } |u(\mathsf{v})| \le 1, \\ \operatorname{sgn} u(\mathsf{v}) & \text{if } |u(\mathsf{v})| > 1, \end{cases}$$

as well as

$$\left(|u| - 1)^+ \operatorname{sgn} u\right)(\mathsf{v}) = \begin{cases} 0 & \text{if } |u(\mathsf{v})| \le 1, \\ u(\mathsf{v}) - \operatorname{sgn} u(\mathsf{v}) & \text{if } |u(\mathsf{v})| > 1, \end{cases}$$

for all $\mathsf{v} \in \mathsf{V}$. Now a direct computation yields

$$\begin{aligned} &\operatorname{Re} a_{\mathcal{W}}((1 \wedge |u|) \operatorname{sgn} u, (|u|-1)^+ \operatorname{sgn} u) \\ &\quad = \sum_{\mathsf{e} \in \mathsf{E}} \int_0^1 p_{\mathsf{e}}(x)(|u_{\mathsf{e}}(x)| - 1) \mathbf{1}_{\{|u_{\mathsf{e}}(x)| \ge 1\}} \mu(\mathsf{e}) dx \\ &\quad\quad - \operatorname{Re} \sum_{\mathsf{v},\mathsf{w} \in \mathsf{V}} \omega_{\mathsf{vw}} (1 \wedge |u(\mathsf{w})|) \operatorname{sgn} u(\mathsf{w}) \overline{(|u(\mathsf{v})| - 1)^+ \operatorname{sgn} u(\mathsf{v})} \deg_\mu(\mathsf{v}) \\ &\quad \ge - \operatorname{Re} w((1 \wedge |u_{|\mathsf{V}}|) \operatorname{sgn} u_{|\mathsf{V}}, (|u_{|\mathsf{V}}| - 1)^+ \operatorname{sgn} u_{|\mathsf{V}}), \end{aligned}$$

with equality if $p \equiv 0$. Here w denotes the sesquilinear, accretive form on $\ell^2_{\deg_\mu}(\mathsf{V})$ associated with the matrix $\mathcal{W}$. By Remark 6.62, the claim follows. □

Example 6.73. What does the above theorem say about the semigroup that governs the heat equation on $\mathfrak{G}$ with standard node conditions, i.e., under continuity condition and additionally under generalized Kirchhoff conditions of the form

$$\partial_{\gamma^2} u(\mathsf{V}) + \mathcal{W} u_{|\mathsf{V}} = 0\,?$$

If $\mathcal{W} = 0$ (pure Kirchhoff conditions), then $e^{-t\mathcal{W}} = \mathrm{Id}$ for all $t \ge 0$ and hence the semigroup is sub-Markovian—in fact, Markovian.

More generally, under continuity and Kirchhoff–Robin conditions (i.e., if $\mathcal{W}$ is diagonal), then positivity also holds, and so does L^∞-contractivity provided all (diagonal) entries of $\mathcal{W}$ have non-negative real part.

On the other hand, if $\mathcal{W} = -\mathcal{L}$ (i.e., if we consider the Krein–von Neumann extension of the second derivative), then the semigroup that governs the heat equation on $\mathfrak{G}$ is neither positive nor L^∞-contractive: Indeed, by Lemma 4.57 the semigroup generated by $-\mathcal{W}$, i.e., $(e^{t\mathcal{L}})_{t \ge 0}$, is not positive, nor it is ℓ^∞-contractive.

By the usual arguments based on the Riesz–Thorin Theorem, cf. Proposition 4.40, $(e^{t\nabla(c^2\nabla)})_{t\geq 0}$ extrapolates to an analytic C_0-semigroup on all spaces $L^p\big((0,1);\ell^2_\mu(\mathsf{E})\big)$, $2 \leq p < \infty$, whenever Theorem 6.72 applies and thus yields L^∞-contractvity. By Lemma 3.21.(5) and Corollary 4.42, its generator is the $L^p\big((0,1);\ell^2_\mu(\mathsf{E})\big)$-realization of $\nabla(c^2\nabla)$, with domain

$$D(\nabla(c^2\nabla)) := \left\{u \in W^{2,p}\big((0,1);\ell^2_\mu(\mathsf{E})\big) : u \text{ satisfies (Cc') and (KRc')}\right\}.$$

Remark 6.74. While (KRc′) and hence the operator domain $D(\nabla(c^2\nabla))$ change upon reorienting the graph G, (Cc′) and hence the form domain V do not. This explains why all the properties of the system that depend on the energy methods presented in Sect. 6.2 do not depend on the orientation.

Example 6.75. Further qualitative properties can be deduced from Theorem 6.28. To mention a simple instance, let G be an unweighted inbound star with two edges, $\mathcal{W} = 0$, and consider the closed convex set

$$\mathbf{C} := \{u \in L^2\big((0,1);\mathbb{C}^2\big) : |u_1(x)| \leq u_2(x) \text{ for a.e. } x \in (0,1)\}.$$

We have seen in Lemma 4.43 that the orthogonal projector $P_{\mathbf{C}}$ onto $\mathbf{C}$ is given by

$$P_{\mathbf{C}}u = \frac{1}{2}\Big(\big(|u_1| + \min\{|u_1|, \operatorname{Re} u_2\}\big)^+ \operatorname{sgn} u_1, \big(\max\{|u_1|, \operatorname{Re} u_2\} + \operatorname{Re} u_2\big)^+\Big).$$

Then an edgewise domination result can be deduced from Ouhabaz' criterion: If u is the solution of the diffusion problem (Di) – (Cc′) – (Kc′) and the initial data u_0 verifies $|u_{01}(x)| \leq u_{02}(x)$ for a.e. $x \in (0,1)$, then u satisfies the inequality $|u_1(t,x)| \leq u_2(t,x)$ for a.e. $x \in (0,1)$ and all $t \geq 0$.

Remark 6.76. If $-\mathcal{W}$ is dissipative and

$$\operatorname{Re}\omega_{\mathsf{vv}} \deg_\mu(\mathsf{v}) \geq \sum_{\substack{\mathsf{w}\in\mathsf{V}\\ \mathsf{w}\neq\mathsf{v}}} |\omega_{\mathsf{vw}}| \deg_\mu(\mathsf{v}) \quad\text{and}\quad \operatorname{Re}\omega_{\mathsf{vv}} \geq \sum_{\substack{\mathsf{w}\in\mathsf{V}\\ \mathsf{w}\neq\mathsf{v}}} |\omega_{\mathsf{wv}}| \deg_\mu(\mathsf{v}), \quad \mathsf{v}\in\mathsf{V}, \tag{6.39}$$

then in view of Example 4.44 and Lemma 3.27.(4) Proposition 6.33 applies and yields

$$\|e^{t\nabla(c^2\nabla)}u\|_{C(\mathfrak{G})} \leq ct^{-\frac{1}{4}}\|u\|_{L^2(\mathfrak{G})}, \quad t > 0,\ u \in L^2(\mathfrak{G}), \tag{6.40}$$

for some constant $c > 0$. By duality, the adjoint semigroup maps $L^1(\mathfrak{G})$ (and even $M(\mathfrak{G})$ if G has finite measure) into $L^2(\mathfrak{G})$ and satisfies

$$\|e^{t\nabla(c^2\nabla)}\mu\|_{L^2} \leq ct^{-\frac{1}{4}}\|\mu\|_{M(\mathfrak{G})}, \quad t > 0,\ f \in M(\mathfrak{G}),$$

i.e., $(e^{t\nabla(c^2\nabla)})_{t\geq 0}$ maps $L^1(\mathfrak{G})$ in $C(\mathfrak{G})$ with

$$\|e^{t\nabla(c^2\nabla)}\phi\|_\infty \leq ct^{-\frac{1}{2}}\|\phi\|_{L^1(\mathfrak{G})} \qquad \text{for all } t>0,\ \phi\in M(\mathfrak{G}). \tag{6.41}$$

In particular, $(e^{t\nabla(c^2\nabla)})_{t\geq 0}$ is ultracontractive of dimension 1.

The semigroup on $L^2(\mathfrak{G})$ is strongly continuous if $(e^{-t\mathcal{W}})_{t\geq 0}$ is positive, by Proposition 4.40 and Theorem 6.71.

Let us finally address the issue whether the semigroup that governs the diffusive problem is irreducible, in the sense of Definition 4.30. In view of Theorem 3.28, irreducibility of an operator in $L^2\big((0,1);\ell^2_\mu(\mathsf{E})\big)$ can be discussed studying irreducibility of the corresponding operator on $L^2(\mathfrak{E})$.

Proposition 6.77. *If* G *is connected, then* $(e^{t\nabla(c^2\nabla)})_{t\geq 0}$ *is irreducible. Also the converse implication holds, if additionally* $\mathcal{W}$ *is a diagonal matrix.*

Proof. To begin with, let us observe that if for all $f\in V$ $f\mathbf{1}_{\mathfrak{G}_0}\in V$ for some metric subgraph $\mathfrak{G}_0$ of $\mathfrak{G}$, and if $\mathfrak{G}_0$ contains an interior point x of some edge e_0, then the whole e_0 belongs to $\mathfrak{G}_0$ (otherwise $f\mathbf{1}_{\mathfrak{G}_0}$ would be discontinuous at x, and in particular not of class $W^{1,2}$). Clearly, the same also applies to $\mathfrak{G}_0^C$, since $f\mathbf{1}_{\mathfrak{G}_0^C}=f-f\mathbf{1}_{\mathfrak{G}_0}$.

Let us now prove the first assertion. If $(e^{t\nabla(c^2\nabla)})_{t\geq 0}$ would be reducible, then both $\mathfrak{G}_0$ and $\mathfrak{G}_0^C$ would contain interior points by the initial remark, and they would define a partition of E. Because $f\mathbf{1}_{\mathfrak{G}_0}\in V$ for all $f\in V$, it would be then possible to construct a function $f\in V$ that is identically 1 along any path of finite length that connects an interior point of $\mathfrak{G}_0$ and an interior point of $\mathfrak{G}_0^C$—such a path must exist, if $\mathfrak{G}$ is connected, and moreover there are no summability issues for $f(\mathsf{v})$ at any node, since G is assumed to be locally finite. But because $f\mathbf{1}_{\mathfrak{G}_0}$ is identically 1 on $\mathfrak{G}_0$ and vanishes on $\mathfrak{G}_0^C$, it must be discontinuous at some point in between—a contradiction to $f\mathbf{1}_{\mathfrak{G}_0}\in V$.

In order to check the second assertion, let $\mathcal{W}$ be diagonal. Then $a_{\mathcal{W}}$ is local and the condition in Proposition 6.30.(5) boils down to the following: Let $\tilde{\mathfrak{G}}$ be a subdivision of $\mathfrak{G}$ and $\mathfrak{G}_0$ be an induced (metric) subgraph of $\tilde{\mathfrak{G}}$. If $\mathbf{1}_{\mathfrak{G}_0}\in V$, then $\mathrm{vol}_\mu(\mathfrak{G}_0)=0$ or $\mathrm{vol}_\mu(\mathfrak{G}_0^C)=0$. If this would not be the case, then both $\mathrm{vol}_\mu(\mathfrak{G}_0)=0$ and $\mathrm{vol}_\mu(\mathfrak{G}_0^C)=0$ would contain at least one edge each, and the assertion could be proved essentially as above. □

An alternative, possibly more elegant proof of the former assertion will be presented in Remark 8.34.

Remark 6.78. In view of Remark 6.64, all the results of this section are based on the properties of the form $a_{\mathcal{W}}$ defined on (a subspace isomorphic to) $W^{1,2}(\mathfrak{G})$. What if we restrict it to $\mathring{W}^{1,2}(\mathfrak{G})$ instead? While clearly $\mathring{W}^{1,2}(\mathfrak{G})=W^{1,2}(\mathfrak{G})$ in the finite case, $\mathring{W}^{1,2}(\mathfrak{G})$ and $W^{1,2}(\mathfrak{G})$ may actually differ if the underlying graph is infinite, under certain volume growth conditions. Like in the case of the discrete diffusion equations treated in Sect. 6.4.1, working with such restriction correspond to studying the (weak formulation of) a heat equation on which, additionally to standard node conditions, also a certain growth condition "at infinity" is imposed on solutions. Just like in Theorem 6.52 we may thus obtain two different analytic, sub-Markovian C_0-semigroups. By Proposition 6.30.6, the one associated with the restriction of $a_{\mathcal{W}}$ to $\mathring{W}^{1,2}(\mathfrak{G})$ is dominated by $(e^{t(\nabla(c^2\nabla)-M_p)})_{t\geq 0}$—the semigroup discussed in Theorem 6.67—since $\mathring{W}^{1,2}(\mathfrak{G})$ is a lattice ideal of $W^{1,2}(\mathfrak{G})$.

6.5.1 Generalized Node Conditions

We have so far always considered standard node conditions. While we have explained in Sect. 2.2.1 why we regard them as the "natural ones", other conditions might be relevant as well for specific purposes. As we are going to see, in many cases form methods may be adapted—but not always. In [75], Lumer considered different realizations of elliptic operators on $C(\mathfrak{G})$.

Proposition 6.79. *Consider the Laplacian of Sect. 2.2.1 whose domain is the space of all functions of class $C^2(\mathfrak{G})$ that satisfy the continuity condition* (Cc′) *as well as a Kirchhoff-type condition*

$$\sum_{\mathsf{e}\in\mathsf{E}} c_{\mathsf{ve}} \frac{\partial_{\gamma^2} u_{\mathsf{e}}}{\partial \nu}(\mathsf{v}) = 0, \qquad \textit{for all } \mathsf{v}\in\mathsf{V}. \tag{6.42}$$

If $c_{\mathsf{ve}} > 0$ for each $\mathsf{v}\in\mathsf{V}$, $\mathsf{e}\in\mathsf{E}$, then the associated abstract Cauchy problem is well-posed (in a suitable sense).

An easier extension of the theory presented in Sect. 6.5 can be performed upon a close examination of our methods. Our main technical Lemma 6.63 strongly depends on the representation of the node conditions as in (Cc′) – (KRc′), which leads to the description of V in (6.32) and to Lemma 3.20. One may wonder whether other choices of the space Y may be relevant: Picking a certain suitable space Y may e.g. yield a realization of Δ where Dirichlet conditions (instead of (Cc′) – (KRc′)) are imposed on some nodes; but in fact any closed subspace Y of $\ell^2_\mu(\mathsf{E})\times\ell^2_\mu(\mathsf{E})$ gives rise to a well-posed diffusive system. We are going to briefly discuss this setting, in which usual node conditions are replaced by algebraic relations between node values of functions over a metric graph, which however need not respect the graph structure—i.e., the "node conditions" may be non-local with respect to the actual connectivity of the graph.

Proposition 6.80. *Let Y be a closed subspace of $\ell^2_\mu(\mathsf{E})\times\ell^2_\mu(\mathsf{E})$ and $\mathcal{W}\in\mathcal{L}(Y)$. Then the space*

$$W^{1,2}_Y((0,1);\ell^2_\mu(\mathsf{E})) := \left\{u\in W^{1,2}((0,1);\ell^2_\mu(\mathsf{E})) : \begin{pmatrix} u(1)\\ u(0)\end{pmatrix}\in Y\right\} \tag{6.43}$$

is dense in $L^2\big((0,1);\ell^2_\mu(\mathsf{E})\big)$ *and the form* $a_{\mathcal{W}}$ *defined as in* (6.31) *is* H*-elliptic, continuous, and of Lions type. Furthermore, the form is accretive (resp., coercive) if* $-\mathcal{W}$ *is dissipative (resp.,* ω*-quasi-dissipative for some* $\omega < 0$*).*

The associated operator is $\nabla(c^2\nabla) - M_p$ *with node conditions* (Cc′) – (KRc′). *It is self-adjoint if and only if* $\mathcal{W}$ *is self-adjoint.*

Thus, $\nabla(c^2\nabla) - M_p$ *with node conditions* (Cc′) – (KRc′) *generates on* $L^2\big((0,1);\ell^2_\mu(\mathsf{E})\big)$ *a cosine operator function with Kisyński space* $W^{1,2}_Y\big((0,1);\ell^2_\mu(\mathsf{E})\big)$*, hence also an analytic* C_0*-semigroup of angle* $\frac{\pi}{2}$*. This semigroup is of trace class if the graph* $\mathsf{G} = (\mathsf{V},\mathsf{E},\mu)$ *has finite volume.*

Remark 6.81. For each choice of Y and for $\mathcal{W} = 0$ it is of course possible to consider the operator associated with the form $a_{\mathcal{W}}$ defined on $W^{1,2}_Y\big((0,1);\ell^2_\mu(\mathsf{E})\big)$ as well as the operator associated with the same form but this time defined on $W^{1,2}_{Y^\perp}\big((0,1);\ell^2_\mu(\mathsf{E})\big)$. All these forms are accretive, hence the associated semigroups are contractive. The fact that these semigroups come in pairs is theoretically interesting and a few results are known that show some kind of duality between them, cf. Sect. 7.2.1 and (7.18) below.

If e.g. Y is defined as in (6.29), then the Laplacian obtained switching the roles of Y and $Y^\perp$ is the second derivative with *anti-Kirchhoff* node conditions, i.e., with conditions

$$\frac{\partial u_{\mathsf{e}}}{\partial \nu}(\mathsf{v}) = \frac{\partial u_{\mathsf{f}}}{\partial \nu}(\mathsf{v}) =: \frac{\partial u}{\partial \nu}(\mathsf{v}), \qquad \text{for all } \mathsf{e},\mathsf{f} \in \mathsf{E}_{\mathsf{v}},\ \mathsf{v} \in \mathsf{V},$$

along with

$$\sum_{\mathsf{e}\in\mathsf{E}_{\mathsf{v}}} u_{\mathsf{e}}(\mathsf{v}) = 0, \qquad \text{for all } \mathsf{v} \in \mathsf{V}.$$

By a direct application of Proposition 6.40 we obtain the following.

Corollary 6.82. *Let* $\mathsf{G} = (\mathsf{V},\mathsf{E},\mu)$ *have finite volume. Let* $(Y_n)_{n\in\mathbb{N}}$ *be a sequence of closed subspaces of* $\ell^2_\mu(\mathsf{E}) \times \ell^2_\mu(\mathsf{E})$ *and let* $(\nabla(c^2\nabla)_{Y_n})_{n\in\mathbb{N}}$ *be a sequence of Laplacians on* $L^2\big((0,1);\ell^2_\mu(\mathsf{E})\big)$ *with node conditions*

$$\begin{pmatrix} u(1) \\ u(0) \end{pmatrix} \in Y_n \qquad \text{and} \qquad \begin{pmatrix} c^2(1)u'(1) \\ -c^2(0)u'(0) \end{pmatrix} + \mathcal{W}\begin{pmatrix} u(1) \\ u(0) \end{pmatrix} \in Y_n^\perp, \qquad n \in \mathbb{N}.$$

Assume that there exist a closed subspace of $\ell^2_\mu(\mathsf{E}) \times \ell^2_\mu(\mathsf{E})$ *and a family* $(J^{\downarrow n})_{n\in\mathbb{N}}$ *of unitary operators on* $L^2\big((0,1);\ell^2_\mu(\mathsf{E})\big)$ *converging to the identity* Id *such that* $J^{\downarrow n}Y_n = Y$ *for all* $n \in \mathbb{N}$*. Denote by* $\nabla(c^2\nabla)_Y$ *the elliptic operator with analogous node conditions defined by* Y*. Then*

$$\lim_{n\to\infty} e^{t\nabla(c^2\nabla)_{Y_n}} = e^{t\nabla(c^2\nabla)_Y} \qquad \text{for every } t > 0$$

in trace norm.

Like in the case of first derivatives on metric graphs it is possible to find an explicit formula for the resolvent operator of the second derivative with (Cc′) – (KRc′), cf. [63, § 4]. However, we prefer to present an equivalent representation that is in accordance with our formalism. For the sake of simplicity we focus on the case of $\mu \equiv 1$. Observe that in view of Proposition 4.28.(3) may in principle be used to deduce an explicit formula for the associated C_0-semigroup, too.

Proposition 6.83. *Let G be finite. Let Y be a closed subspace of $\mathbb{C}^{\mathsf{E}} \times \mathbb{C}^{\mathsf{E}}$ and $\mathcal{W} \in \mathcal{L}(Y)$. Let us denote by A the operator associated with $a_{\mathcal{W}}$ with domain $W_Y^{1,2}((0,1);\mathbb{C}^{\mathsf{E}})$. For all $k \neq 0$, $k^2 \in \rho(A)$ if and only if*

$$P_Y R_3(k) + (\mathcal{W} + P_Y^{\perp}) R_4(k)$$

is invertible, and in this case the resolvent operator $R(k^2, A)$ is given by

$$R(k^2, A)u(x) := \int_{\mathfrak{G}} r(x,y,k)u(y)\, dy$$

with integral kernel

$$r(x,y,k) := \frac{i}{2k} \left(r_d(x,y,k) - \Phi(x,k)\Sigma(k)\Psi(y,k) \right). \tag{6.44}$$

Here

$$\Sigma(k) := \left(P_Y R_3(k) + (\mathcal{W} + P_Y^{\perp}) R_4(k) \right)^{-1} \left(P_Y R_1(k) + (\mathcal{W} + P_Y^{\perp}) R_2(k) \right),$$

$r_d(x,y,k)$ is the $\mathsf{E} \times \mathsf{E}$ matrix defined by

$$r_d(x,y,k)_{\mathsf{ef}} := \delta_{\mathsf{ef}} e^{ik|x-y|}, \quad \mathsf{e}, \mathsf{f} \in \mathsf{E},$$

and P_Y is the orthogonal projector onto Y. Furthermore,

$$R_1(k) := \begin{pmatrix} ik\mathbf{1}_{\mathbb{C}^{\mathsf{E}}} & 0 \\ 0 & ike^{ik}\mathbf{1}_{\mathbb{C}^{\mathsf{E}}} \end{pmatrix}, \qquad R_2(k) := \begin{pmatrix} \mathbf{1}_{\mathbb{C}^{\mathsf{E}}} & 0 \\ 0 & e^{ik}\mathbf{1}_{\mathbb{C}^{\mathsf{E}}} \end{pmatrix},$$

$$R_3(k) := \begin{pmatrix} ike^{ik}\mathbf{1}_{\mathbb{C}^{\mathsf{E}}} & -ike^{-ik}\mathbf{1}_{\mathbb{C}^{\mathsf{E}}} \\ -ik\mathbf{1}_{\mathbb{C}^{\mathsf{E}}} & ik\mathbf{1}_{\mathbb{C}^{\mathsf{E}}} \end{pmatrix}, \qquad R_4(k) := \begin{pmatrix} e^{ik}\mathbf{1}_{\mathbb{C}^{\mathsf{E}}} & e^{-ik}\mathbf{1}_{\mathbb{C}^{\mathsf{E}}} \\ \mathbf{1}_{\mathbb{C}^{\mathsf{E}}} & \mathbf{1}_{\mathbb{C}^{\mathsf{E}}} \end{pmatrix}.$$

Finally,

$$\Phi(x,k) := \left(e^{ikx}\mathbf{1}_{\mathbb{C}^{\mathsf{E}}}\ e^{-ikx}\mathbf{1}_{\mathbb{C}^{\mathsf{E}}}\right), \qquad \Psi(x,k) := \begin{pmatrix} e^{iky}\mathbf{1}_{\mathbb{C}^{\mathsf{E}}} \\ e^{-iky}\mathbf{1}_{\mathbb{C}^{\mathsf{E}}} \end{pmatrix}, \qquad x \in (0,1),$$

where the entries are diagonal matrices whose entries are functions with arguments from the corresponding edges.

The node conditions have no influence on the solution of the eigenvalue equation inside each edge, so that the usual formula for the Green function prevails on each individual edge. Deducing a correct expression for the resolvent is then essentially a matter of correctly gluing together such a family of $|\mathsf{E}|$ Green functions, in accordance with the node conditions. A detailed proof in a more general case has been delivered in [54, § 6.3]. As usual, close inspection of the formulae also reveals this: Indeed, on the right hand side of (6.44) the first term is simply the Green function of the elliptic equation $\frac{d^2}{dx^2} - k^2$, whereas the connectivity of the networks is fully encoded in the second term.

This formula for the resolvent may be used to discuss invariance properties by Proposition 4.29. However, it turns out that Ouhabaz' criterion is more convenient to apply. It leads to a generalization of Theorem 6.72 that we present next. The main ingredient in the proof is the following auxiliary result. For a closed convex subset C of $\ell^2_\mu(\mathsf{E})$ we adopt the notation

$$C_H := \left\{u \in L^2\big((0,1);\ell^2_\mu(\mathsf{E})\big) : u(x) \in C \text{ for a.e. } x \in (0,1)\right\},$$

$$C_{\partial H} := \left\{f = (f(0), f(1)) \in \ell^2_\mu(\mathsf{E}) \times \ell^2_\mu(\mathsf{E}) : f(0), f(1) \in C\right\}.$$

Lemma 6.84. *Let C be a closed convex subset of $\ell^2_\mu(\mathsf{E})$ and Y be a closed subspace of $\ell^2_\mu(\mathsf{E}) \times \ell^2_\mu(\mathsf{E})$. Then the inclusion $P_{C_H} V_Y \subset V_Y$ holds if and only if the inclusion $P_Y C \subset C$ holds, where P_{C_H} and P_Y denote the orthogonal projectors of H onto C_H and of $\ell^2_\mu(\mathsf{E}) \times \ell^2_\mu(\mathsf{E})$ onto Y, respectively.*

Proof. By a simple property of general orthogonal projectors, cf. [78, Lemma 2.3],

$$P_Y C \subset C \quad \text{if and only if} \quad P_C Y \subset Y. \tag{6.45}$$

Since any orthogonal projector of a Hilbert space onto a closed convex subset is Lipschitz continuous, by Lemma B.12 one has that pointwise projecting does not affect weak differentiability. That is, $P_{C_H} u \in W^{1,2}\left((0,1);\ell^2_\mu(\mathsf{E})\right)$ for all $u \in V_Y$. Thus, $P_{C_H} V_Y \subset V_Y$ if and only if the node condition $(P_{C_H} u)(z) \in Y$ is satisfied for $z = 0,1$ and for all $u \in V_Y$, i.e., if and only if for all $u \in W^{1,2}\left((0,1);\ell^2_\mu(\mathsf{E})\right)$

$$f(z) \in Y \text{ for } z = 0,1 \quad \text{implies} \quad P_C(u(z)) \in Y \text{ for } z = 0,1, \tag{6.46}$$

or rather, by surjectivity of the trace operator, if and only if $P_C Y \subset Y$. By (6.45), this concludes the proof. □

For $\alpha, \beta \in \mathbb{R}$ let us define order intervals by

$$[\alpha,\beta]_H := \{u \in H : u_{\mathsf{e}}(x) \in [-\alpha,\beta] \text{ for a.e. } x \in (0,1) \text{ and all } \mathsf{e} \in \mathsf{E}\},$$

and

$$[\alpha,\beta]_{\partial H} := \Big\{ f = (f(0), f(1)) \in \ell^2_\mu(\mathsf{E}) \times \ell^2_\mu(\mathsf{E}) : f_{\mathsf{e}}(0), f_{\mathsf{e}}(1) \in [-\alpha,\beta] \text{ for all } \mathsf{e} \in \mathsf{E} \Big\}.$$

(Semi-infinite order intervals are defined likewise.)

Theorem 6.85. *Let Y be a closed subspace of $\ell^2_\mu(\mathsf{E}) \times \ell^2_\mu(\mathsf{E})$ and $\mathcal{W} \in \mathcal{L}(Y)$ and let $\nabla(c^2\nabla)$ be the elliptic operator on $L^2((0,1); \ell^2_\mu(\mathsf{E}))$ with node conditions*

$$\begin{pmatrix} u(1) \\ u(0) \end{pmatrix} \in Y \quad \text{and} \quad \begin{pmatrix} c^2(1)u'(1) \\ -c^2(0)u'(0) \end{pmatrix} + \mathcal{W}\begin{pmatrix} u(1) \\ u(0) \end{pmatrix} \in Y^\perp.$$

Then for all $\alpha, \beta \in \mathbb{R} \cup \{\pm\infty\}$ the C_0-semigroup $(e^{t\nabla(c^2\nabla)})_{t\ge 0}$ leaves invariant $[\alpha,\beta]_H$ if and only if both the C_0-semigroup $(e^{-t\mathcal{W}})_{t\ge 0}$ on Y and the orthogonal projector P_Y of H onto Y leave invariant $[\alpha,\beta]_{\ell^2_\mu(\mathsf{E})} \times [\alpha,\beta]_{\ell^2_\mu(\mathsf{E})}$.

Proof. To begin with, observe that the former condition in Theorem 6.28.(b) can be dealt with using Lemma 6.84. Furthermore, there holds

$$\begin{aligned}
\operatorname{Re} a_{\mathcal{W}}(P_{[\alpha,\beta]_H} u, u - P_{[\alpha,\beta]_H} u) &= \operatorname{Re} \int_U ((P_{[\alpha,\beta]_H} u)'(x) | (u - P_{[\alpha,\beta]_H} u)'(x))_{\ell^2_\mu}\, dx \\
&\quad + \operatorname{Re} \left(\mathcal{W} P_{[\alpha,\beta]_{\partial H}} \begin{pmatrix} u(1) \\ u(0) \end{pmatrix} \,\Big|\, (\operatorname{Id} - P_{[\alpha,\beta]_{\partial H}}) \begin{pmatrix} u(1) \\ u(0) \end{pmatrix} \right)_{\ell^2_\mu \times \ell^2_\mu} \\
&= \operatorname{Re} \int_U ((P_{[\alpha,\beta]_H} u')(x) | (u - P_{[\alpha,\beta]_H} u')(x))_{\ell^2_\mu}\, dx \\
&\quad + \operatorname{Re} \left(\mathcal{W} P_{[\alpha,\beta]_{\partial H}} \begin{pmatrix} u(1) \\ u(0) \end{pmatrix} \,\Big|\, (\operatorname{Id} - P_{[\alpha,\beta]_{\partial H}}) \begin{pmatrix} u(1) \\ u(0) \end{pmatrix} \right)_{\ell^2_\mu \times \ell^2_\mu} \\
&= \operatorname{Re} w \left(P_{[\alpha,\beta]_{\partial H}} \begin{pmatrix} u(1) \\ u(0) \end{pmatrix} \,\Big|\, (\operatorname{Id} - P_{[\alpha,\beta]_{\partial H}}) \begin{pmatrix} u(1) \\ u(0) \end{pmatrix} \right)_{\ell^2_\mu \times \ell^2_\mu},
\end{aligned}$$

where w is the sesquilinear form associated with $\mathcal{W}$. Applying Theorem 6.28—Ouhabaz' invariance criterion—to both $a_{\mathcal{W}}$ and w the claim follows. □

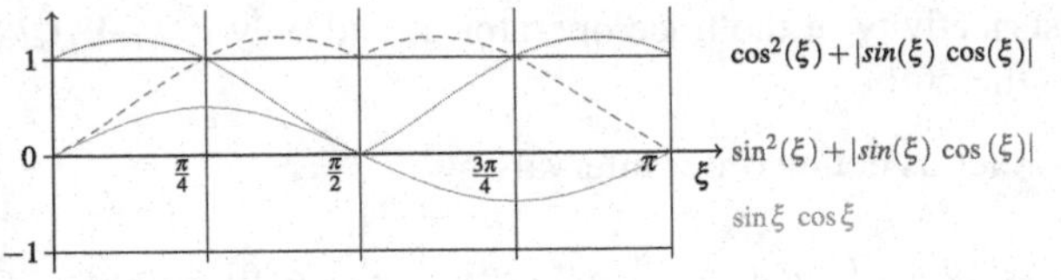

Fig. 6.2 The heat semigroup on a metric graph consisting of a single edge with coupled conditions parametrized by ξ. The semigroup is positive if and only if the corresponding point in the *solid line* is above the 0. It is L^∞-contractive if and only if the corresponding point in the *dashed and dotted lines* both lie within the interval $[-1, 1]$ (colour figure online)

Example 6.86. Theorem 6.85 yields a surprising fact: While for Y as in (6.29) the node conditions (Cc′)–(KRc′) are "natural", they also represent a singularity among all generalized node conditions. This can be seen already in the elementary case of only one (unweighted) edge, i.e., $\ell^2_\mu(\mathsf{E}) \times \ell^2_\mu(\mathsf{E}) \equiv \mathbb{C}^2$. Neglecting the trivial (uncoupled) node conditions defined by $Y = \{0\}$ and $Y = \mathbb{C}^2$ we can consider all one-dimensional subspaces $Y \equiv Y_\xi$ of $\mathbb{C}^2$ by means of the parametrization

$$P_{Y_\xi} := \begin{pmatrix} \cos^2 \xi & \sin \xi \cos \xi \\ \sin \xi \cos \xi & \sin^2 \xi \end{pmatrix}, \qquad \xi \in [0, \pi),$$

where Y_ξ denotes the range of the orthogonal projection P_{Y_ξ}. Observe that $\xi = 0$, $\xi = \frac{\pi}{4}$, $\xi = \frac{\pi}{2}$ and $\xi = \frac{3\pi}{4}$ correspond to uncoupled Dirichlet/Neumann, to Kirchhoff, to uncoupled Neumann/Dirichlet and to anti-Kirchhoff node conditions, respectively, as can be checked directly.

We are going to discuss—in dependence of ξ, and assuming for the sake of simplicity that $R = 0$—the sub-Markovian property of the semigroup associated with these subspaces. A direct computation shows that the semigroup $(e^{t\Delta_{Y_\xi,0}})_{t\geq 0}$ is positive if and only if $\xi \in [0, \frac{\pi}{2}]$; and it is $L^\infty(0,1)$-contractive if and only if P_{Y_ξ} is ℓ^∞-contractive, i.e., if and only if the inequalities

$$\cos^2 \xi + |\sin \xi \cos \xi| \leq 1 \qquad \text{and} \qquad |\sin \xi \cos \xi| + \sin^2 \xi \leq 1$$

hold simultaneously. The former (resp, the latter) inequality holds if and only if $\xi \notin (0, \frac{\pi}{4}) \cup (\frac{3\pi}{4}, \pi)$ (resp., if and only if $\xi \notin (\frac{\pi}{4}, \frac{\pi}{2}), (\frac{\pi}{2}, \frac{3\pi}{4})$). Therefore, L^∞-contractivity of the semigroup associated with node conditions (Cc) – (Kc) represents a singularity (Fig. 6.2). In particular, a sub-Markovian semigroup is generated *exactly* in the following five cases:

- with uncoupled Dirichlet/Dirichlet boundary conditions,
- with uncoupled Neumann/Neumann boundary conditions,
- with uncoupled Dirichlet/Neumann boundary conditions,
- with uncoupled Neumann/Dirichlet boundary conditions and finally
- with continuity and Kirchhoff boundary conditions.

One can prove as in Theorem 6.85 a number of further properties of the semigroups that govern diffusion problems associated with forms $a_{\mathcal{W}}$, which essentially boil down to checking related but much accessible properties of the orthogonal projectors P_Y. We mention the following example. A metric graph $\mathfrak{G}$ over a weighted oriented graph $\mathsf{G} = (\mathsf{V}, \mathsf{E}, \mu)$ is called *equilateral* if $\mu \equiv 1$, i.e., if the graph G is effectively unweighted. (We are going to state in the following a few spectral and symmetry results that hold for diffusion equations on equilateral graphs: By linearity, all of them extend to the case of $\mu \equiv c$ for some positive real number c.)

Definition 6.87. Let $\mathfrak{G}$ be an equilateral metric graph over an oriented star graph, cf. Definition A.3. We call a function $u : \mathfrak{G} \to \mathbb{C}$ *odd* or *even* if

$$\sum_{\mathsf{e}\in\mathsf{E}} u_{\mathsf{e}}(x) = 0 \quad \text{or} \quad u_{\mathsf{e}}(x) = u_{\mathsf{f}}(x) \quad \text{for a.e. } x \in (0,1) \text{ and all edges } \mathsf{e},\mathsf{f} \in \mathsf{E},$$

respectively.

Proposition 6.88. *Let $\mathfrak{G}$ be an equilateral metric graph over a finite oriented star graph. Then the semigroup generated by the elliptic operator $\nabla(c^2\nabla)$ with standard node conditions* (Cc) – (Kc) *leaves invariant both the subspaces of odd and even functions. The same is true for the semigroup generated by the elliptic operator $\nabla(c^2\nabla)$ with anti-Kirchhoff node conditions, cf. Remark 6.81.*

Observe that in the leaves (i.e., in the exterior nodes of the star) node conditions (Cc) – (Kc) are equivalent to Neumann boundary conditions. Likewise, anti-Kirchhoff node conditions impose Dirichlet boundary conditions in the leaves. Furthermore, for *continuous* even functions anti-Kirchhoff node conditions boil down to Dirichlet boundary conditions on a family of decoupled intervals.

Proof. In the case of an inbound star, node conditions (Cc) – (Kc) can be equivalently written as (Cc′) – (KRc′) for $\mathcal{W} = 0$, where Y is the one-dimensional subspace spanned by the vector $(0, 1, 0, 1, \ldots, 0, 1) \in \mathbb{C}^{\mathsf{E}} \times \mathbb{C}^{\mathsf{E}}$. Up to a permutation the orthogonal projector onto Y is a block matrix

$$\begin{pmatrix} 0 & 0 \\ 0 & J \end{pmatrix},$$

where

$$J = \frac{1}{|\mathsf{E}|} \begin{pmatrix} 1 & 1 & \ldots & 1 \\ 1 & \ddots & & \vdots \\ \vdots & & \ddots & \vdots \\ 1 & 1 & \ldots & 1 \end{pmatrix}$$

Now, it suffices to observe that J is precisely the orthogonal projector onto the subspace spanned by the vector $(1, 1, \ldots, 1) \in \mathbb{C}^{\mathsf{E}}$. Similarly, anti-Kirchhoff conditions are obtained replacing Y by $Y^\perp$. The orthogonal projector onto $Y^\perp$ is a block matrix

$$\begin{pmatrix} \mathrm{Id} & 0 \\ 0 & \mathrm{Id} - J \end{pmatrix},$$

and $\mathrm{Id} - J$ is the orthogonal projector onto the subspace of all vectors whose entries sum up to 0. □

6.6 Hybrid Evolution Equations

6.6.1 The Dirichlet-to-Neumann Operator

We have already showed in Sect. 6.5.1 how to use the theory of j-elliptic forms in order to discuss the Dirichlet-to-Neumann operator on one interval. The corresponding operator on a subset V_0 of the node set V of a network can be discussed likewise, taking the form $a_{\mathcal{W}}$ (for arbitrary $\mathcal{W} \in \mathcal{L}(Y)$; the standard operator introduced in Sect. 2.3.1 is recovered letting $\mathcal{W} = 0$) defined on $V := W^{1,2}(\mathfrak{G})$ and letting throughout this section

$$H := \ell^2_{\deg_\mu}(\mathsf{V}_0) \qquad \text{and} \qquad j(u) := u_{|\mathsf{V}_0},$$

where j is surjective from V onto H by Remark 6.62. In view of Lemma 3.27.(6), it is immediate to check that $a_{\mathcal{W}}$ is j-elliptic.

Lemma 6.89. *The operator associated with $(a_{\mathcal{W}}, j)$ is (minus) the Dirichlet-to-Neumann operator $\mathbb{DN}$ with respect to V_0, where $\mathbb{DN}$ is the operator on $\ell^2_\mu(\mathsf{V}_0)$ defined by*

$$\left.\begin{array}{r} f \in D(\mathbb{DN}) \\ \mathbb{DN} f := g \end{array}\right\} \quad \Leftrightarrow \quad \left\{\begin{array}{l} f \in \ell^2_\mu(\mathsf{V}_0) \\ \exists u \text{ solution of } (2.50) \\ g = \partial_{\nu^2} u_{|\mathsf{V}_0} \in \ell^2_\mu(\mathsf{V}_0). \end{array}\right.$$

(Here the shorthand ∂_{ν^2} is defined as usual as in (2.40).*)*

Proof. By definition, the domain $D(A)$ of the operator A associated with $(a_{\mathcal{W}}, j)$ is given by

$$\left\{ \begin{array}{r} x \in \ell^2_{\deg_\mu}(\mathsf{V}_0) : \exists u \in W^{1,2}(\mathfrak{G}) : u_{|\mathsf{V}_0} = x \text{ and } \exists y \in \ell^2_{\deg_\mu}(\mathsf{V}_0) \text{ s.t.} \\ \int_{\mathfrak{G}} u'(x)\overline{w'(x)} = \sum_{\mathsf{v} \in \mathsf{V}_0} y(\mathsf{v})\overline{w(\mathsf{v})} \text{ for all } w \in W^{1,2}(\mathfrak{G}) \end{array} \right\}.$$

But, up to the isomorphism Ψ in (2.37),

$$\int_{\mathfrak{G}} u'\overline{w'} = -\sum_{\mathsf{e}\in\mathsf{E}} \int_0^1 u''(x)\overline{w(x)}\mu_{\mathsf{e}} dx + \sum_{\mathsf{e}\in\mathsf{E}} \left(u'_{\mathsf{e}}(1)\overline{w_{\mathsf{e}}(1)} - u'_{\mathsf{e}}(0)\overline{w_{\mathsf{e}}(0)} \right) \mu_{\mathsf{e}}.$$

Accordingly, $x \in D(A)$ with $Ax = -y$ for a certain $y \in \ell^2_{\deg_\mu}(\mathsf{V}_0)$ if and only if there exists $u \in W^{1,2}(\mathfrak{G})$ (and hence, in particular continuous in all nodes) such that $u'' = 0$ (weakly), $u_{|\mathsf{V}_0} = x$, and finally

$$\sum_{\mathsf{v}\in\mathsf{V}_0} y(\mathsf{v})\overline{w(\mathsf{v})} = \sum_{\mathsf{e}\in\mathsf{E}} \left(u'_{\mathsf{e}}(1)\overline{w_{\mathsf{e}}(1)} - u'_{\mathsf{e}}(0)\overline{w_{\mathsf{e}}(0)} \right) \mu_{\mathsf{e}},$$

i.e., $y \equiv \partial_{\mu^2} u_{|\mathsf{V}_0}$. □

Then by Proposition 6.37 we can deduce the following, of which Theorem 6.52 becomes a special case since the Dirichlet-to-Neumann operator agrees with the Laplacian–Beltrami matrix whenever $\mathsf{V}_0 = \mathsf{V}$.

Theorem 6.90. *Let $\mathsf{V}_0 \subset \mathsf{V}$. Then (minus) the Dirichlet-to-Neumann operator $\mathbb{D}\mathbb{N}$ with respect to V_0 generates on $H = \ell^2_{\deg_\mu(\mathsf{V}_0)}$ a cosine operator function with Kisyński space $j(V)$, hence also an analytic C_0-semigroup of angle $\frac{\pi}{2}$. This semigroup is contractive if $-\mathcal{W}$ is dissipative, and uniformly exponentially stable if $-\mathcal{W}$ is ϵ-quasi-dissipative for some $\epsilon < 0$.*

Finally, for all $\alpha, \beta \in \mathbb{R} \cup \{\pm\infty\}$ the C_0-semigroup $(e^{-t\mathbb{D}\mathbb{N}})_{t\geq 0}$ leaves invariant

$$[\alpha,\beta]_{\ell^2_{\deg_\mu}(\mathsf{V}_0)} := \left\{ f \in \ell^2_{\deg_\mu}(\mathsf{V}_0) : f(\mathsf{v}) \in [-\alpha,\beta] \text{ for a.e. } \mathsf{v} \in \mathsf{V}_0 \right\},$$

if and only if both the C_0-semigroup $(e^{-t\mathcal{W}})_{t\geq 0}$ on Y and the orthogonal projector P_Y of $\ell^2_\mu(\mathsf{E}) \times \ell^2_\mu(\mathsf{E})$ onto Y leave invariant

$$[\alpha,\beta]_{\ell^2_\mu(\mathsf{E})\times\ell^2_\mu(\mathsf{E})} := \left\{ \phi \in \ell^2_\mu(\mathsf{E}) \times \ell^2_\mu(\mathsf{E}) : \phi(\mathsf{e}) \in [-\alpha,\beta] \text{ for all } \mathsf{e} \in \mathsf{E} \right\}.$$

In the last assertion we have used the notation of Sect. 6.5.1. Observe that the Kisyński space agrees with $H = \ell^2_{\deg_\mu(\mathsf{V}_0)}$ whenever G has finite volume, but is in general different.

6.6.2 The Laplacian with Dynamic Node Conditions

Also in view of the relevant role played by dynamic node conditions in applications, we will now turn to properties of the hybrid Laplacian $\mathbb{\Delta}$ introduced in Definition 2.49. A special motivation to study parabolic properties of the hybrid Laplacian

$\mathbb{\Delta}$ with standard/dynamic node conditions comes from the Rall-type lumped soma model discussed in Chap. 5, see also [77]. Indeed, if we neglect some non-essentials lower-order perturbations, the abstract Cauchy problem (ACP) associated with $\mathbb{\Delta}$ is exactly the initial value problem that arises from the (biologically more realistic) generalization of Rall's model suggested in [77].

Another motivation comes from elastic systems: Whenever modeling flexible structures one sometimes has to describe the behavior of those mechanical elements that join different parts of the system. On the one hand they may be massive enough that their modeling has to take into account the force of gravity; on the other hand they may be small enough that it is approximately correct to describe them as point masses. This leads to studying the second order abstract Cauchy problem (ACP2) associated with $\mathbb{\Delta}$, or with related operators.

The well-posedness of both Cauchy problems can be discussed by means of form methods. To this aim we assume throughout this section that

$$\boxed{\begin{array}{c}\mathfrak{G} = (\mathsf{V}, \mathfrak{E}) \text{ is the metric graph associated with} \\ \text{a locally finite, weighted oriented graph } \mathsf{G} = (\mathsf{V}, \mathsf{E}, \mu), \\ \text{and } \mathsf{V}_0 \text{ is a subset of } \mathsf{V}.\end{array}}$$

and

$$\boxed{c \in L^\infty(0,1;\ell^\infty_\mu(\mathsf{E})) \text{ with } c_{\mathsf{e}}(x) \geq c_0 \text{ for some } c_0 > 0, \text{ all } \mathsf{e} \in \mathsf{E} \text{ and a.e. } x \in (0,1),}$$

letting for the sake of simplicity $p \equiv 0$. We then introduce for all $p \in [1, \infty]$ a normed product space $\mathbb{L}^p$ by

$$\|\mathfrak{u}\|^p := \|u\|^p_{L^p_\mu} + \|f\|^p_{\ell^p_{\deg_\mu}} \qquad \text{for } \mathfrak{u} \equiv \begin{pmatrix} u \\ f \end{pmatrix} \in \mathbb{L}^p := L^p_\mu(\mathsf{E}) \times \ell^p_{\deg_\mu}(\mathsf{V}_0).$$

Observe that $\mathbb{L}^p$ is separable for all $p \in [1, \infty)$ and reflexive for all $p \in (1, \infty)$, while $\mathbb{L}^2$ is a Hilbert space.

Lemma 6.91. *The vector space*

$$\mathbb{V} := \left\{ \begin{pmatrix} u \\ f \end{pmatrix} \in W^{1,2}((0,1);\ell^2_\mu(\mathsf{E})) \times \ell^2_{\deg_\mu}(\mathsf{V}_0) : \begin{array}{l} \exists u_{|\mathsf{V}} \in \mathbb{C}^{\mathsf{V}} \text{ with} \\ (\mathcal{I}^-)^\top u_{|\mathsf{V}} = u(0),\ (\mathcal{I}^+)^\top u_{|\mathsf{V}} = u(1), \\ \text{and } u_{|\mathsf{V}_0} = f \end{array} \right\}$$

is densely and continuously embedded in

$$\mathbb{H} := \mathbb{L}^2.$$

If G *is finite, then this embedding is compact—and even of trace class for* $p = 2$.

The proof depends on Lemma 3.21 and the following elementary result.

Lemma 6.2. *Let X_1, X_2, Y_1, Y_2 be Banach spaces, such that $X_1 \hookrightarrow X_2$ and $Y_1 \hookrightarrow Y_2$. Consider a surjective operator $L \in \mathcal{L}(X_1, Y_1)$ such that* $\operatorname{Ker} L$ *is dense in X_2 and* $\operatorname{Rg} L = Y_1$ *is dense in Y_2. Then*

$$\operatorname{Graph} L = \left\{ \begin{pmatrix} x \\ y \end{pmatrix} \in X_1 \times Y_1 : Lx = y \right\}$$

is densely and continuously embedded in $X_2 \times Y_2$. The embedding is compact/of p-th Schatten class if both embeddings $X_1 \hookrightarrow X_2$ and $Y_1 \hookrightarrow Y_2$ are compact/of p-th Schatten class, respectively.

Proof. Let $x \in X_2$, $y \in \operatorname{Rg} L$, $\epsilon > 0$. Take $z \in \operatorname{Rg} L \cap Y_1$ such that $\|y - z\|_Y < \epsilon$. In particular there exists $u \in X_1$ such that $Lu = z$. Take $\tilde{u}, \tilde{x} \in \operatorname{Ker} L$ such that $\|u - \tilde{u}\|_{X_2} < \epsilon$ and $\|x - \tilde{x}\|_{X_2} < \epsilon$. Let $w := \tilde{x} + u - \tilde{u} \in X_1$. Then

$$\begin{aligned} &\left\| \begin{pmatrix} x \\ y \end{pmatrix} - \begin{pmatrix} w \\ z \end{pmatrix} \right\|_{X_2 \times Y} \\ &\le \left\| \begin{pmatrix} x - \tilde{x} \\ 0 \end{pmatrix} \right\|_{X_2 \times Y} + \left\| \begin{pmatrix} u - \tilde{u} \\ 0 \end{pmatrix} \right\|_{X_2 \times Y} + \left\| \begin{pmatrix} 0 \\ y - z \end{pmatrix} \right\|_{X_2 \times Y} < 3\epsilon. \end{aligned}$$

Since $Lw = Lu = z$, density of $\operatorname{Graph} L$ in $X_1 \times Y$ is proved.

The assertion on compactness/Schatten class property of the embedding follows from the obvious observations that

$$\operatorname{Graph} L \hookrightarrow X_1 \times Y_1$$

and that the embedding of $X_1 \times Y_1$ in $X_2 \times Y_2$ is compact/of Schatten class if and only if so are the embeddings $X_1 \hookrightarrow X_2$ and $Y_1 \hookrightarrow Y_2$. □

Again following the ideas from Sect. 2.2.1 we can also write

$$\mathbb{V} \equiv \left\{ \begin{pmatrix} u \\ f \end{pmatrix} \in W^{1,2}((0,1); \ell^2_\mu(\mathsf{E})) \times \ell^2_{\deg_\mu}(\mathsf{V}_0) : \begin{pmatrix} u(1) \\ u(0) \end{pmatrix} \in Y \text{ and } u_{|\mathsf{V}_0} = f \right\}$$

for the space Y defined in (6.29).

Throughout this section we assume that

$$\boxed{\mathcal{W}_1, \mathcal{W}_2 \text{ are bounded linear operators on } \ell^2_{\deg_\mu}(\mathsf{V}_0),\ \ell^2_{\deg_\mu}(\mathsf{V}_0^C), \text{ respectively}}$$

and study the densely defined sesquilinear form $\mathfrak{a}_{\mathcal{W}}$ given by

$$\mathfrak{a}_{\mathcal{W}}\left(\begin{pmatrix} u \\ u_{|\mathsf{V}_0} \end{pmatrix}, \begin{pmatrix} v \\ v_{|\mathsf{V}_0} \end{pmatrix} \right) := \int_0^1 \left(c^2(x) u'(x) | v'(x)\right)_{\ell^2_\mu(\mathsf{E})} dx + \left(\mathcal{W} u_{|\mathsf{V}} \mid v_{|\mathsf{V}}\right)_{\ell^2_{\deg_\mu}(\mathsf{V})},$$

with form domain $\mathbb{V}$. Here $\mathcal{W}$ denotes the diagonal block matrix

$$\mathcal{W} := \begin{pmatrix} \mathcal{W}_1 & 0 \\ 0 & \mathcal{W}_2 \end{pmatrix}.$$

The following assertions can be proved in a way similar to Lemmas 6.63 and 6.65 and Theorem 6.71.

Lemma 6.92. *The densely defined form $\mathfrak{a}_{\mathcal{W}} : \mathbb{V} \times \mathbb{V} \to \mathbb{C}$ is continuous, $\mathbb{H}$-elliptic, and of Lions type.*

Lemma 6.93. *The operator associated with $\mathfrak{a}_{\mathcal{W}}$ is the hybrid elliptic operator $\mathbb{A}$ defined by*

$$D(\mathbb{A}) := \Bigg\{ \mathfrak{u} = \begin{pmatrix} u \\ f \end{pmatrix} \in W^{2,2}\big((0,1); \ell^2_\mu(\mathsf{E})\big) \times \ell^2_{\deg_\mu}(\mathsf{V}) \text{ s.t. } \begin{array}{l} \exists u_{|\mathsf{V}} \in \mathbb{C}^{\mathsf{V}} \text{ with} \\ (\mathcal{I}^-)^\top u_{|\mathsf{V}} = u(0),\ (\mathcal{I}^+)^\top u_{|\mathsf{V}} = u(1), \\ \mathcal{I}^+_{\mathsf{V}_0^C} c^2(1)u'(1) - \mathcal{I}^-_{\mathsf{V}_0^C} c^2(0)u'(0) + \mathcal{W}_2 u_{|\mathsf{V}_0^C} = 0, \\ \text{and } u_{|\mathsf{V}_0} = f \end{array} \Bigg\},$$

$$\mathbb{A}\mathfrak{u} := \begin{pmatrix} \nabla(c^2\nabla) & 0 \\ -\partial_{c^2} & -\mathcal{W}_1 \end{pmatrix} \mathfrak{u} := \begin{pmatrix} \nabla(c^2\nabla u) \\ -\partial_{c^2} u - \mathcal{W}_1 f \end{pmatrix}.$$

The Cauchy problem associated with $\mathbb{A}$ is different from (Di) – (Cc) – (KRc). Indeed, aside (Di) on each edge, (Cc) on each node, and (KRc) on the nodes that belong to V_0^C, we also have the dynamic node condition

$$\frac{\partial u}{\partial t}(t, \mathsf{v}) = -\partial_{c^2} u(t, \mathsf{v}) - \mathcal{W}_1 u(t, \mathsf{v}), \qquad t > 0,\ \mathsf{v} \in \mathsf{V}_0. \tag{Dc}$$

Theorem 6.94. *The operator $\mathbb{A}$ defined in Lemma 6.93 generates on $\mathbb{H}$ a cosine operator function with Kisyński space $\mathbb{V}$, hence also an analytic C_0-semigroup of angle $\frac{\pi}{2}$. This semigroup is contractive and given by an integral kernel. Furthermore, $(e^{t\mathbb{A}})_{t\ge 0}$ is of trace class if G is finite. If additionally $-\mathcal{W}$ is dissipative (resp., ϵ-quasi-dissipative for some $\epsilon < 0$), then $(e^{t\mathbb{A}})_{t\ge 0}$ is contractive (resp., uniformly exponentially stable).*

Remark 6.95. The above results could have been alternatively obtained by the theory of j-elliptic forms. Indeed, $\mathbb{A}$ is easily seen to be also the operator associated with $(a_{\mathcal{W}}, j)$, with domain as in (6.32), where

$$j : u \mapsto \begin{pmatrix} u \\ u_{|\mathsf{V}_0} \end{pmatrix}.$$

One possible advantage of the approach through j-forms is the ease of showing convergence results by means of Theorem 6.39 and Corollary 6.96. For instance, reasoning just as in [90, § 4.7] one can show that

$$\begin{pmatrix} \nabla(c\nabla) & 0 \\ \beta\partial_{c^2} & -\mathcal{W}_1 \end{pmatrix}$$

converges to $\Delta_D \oplus 0$ on $\mathbb{H}$ in the strong resolvent sense as $\beta \to 0$, where Δ_D is Laplacian on $L^2((0,1);\ell^2_\mu(\mathsf{E}))$ with (decoupled) Dirichlet conditions in each node; whereas the same operator does not converge to any closed operator—not even in the weak resolvent sense!—as either $c \to \infty$ or $\beta \to \infty$ (one would heuristically expect convergence towards the Dirichlet-to-Neumann operator and the Laplacian with standard node conditions, respectively).

Applying Lemma 6.5 we can deduce the following from Theorem 6.94.

Corollary 6.96. *Let* G *be finite. If* $\beta \in \mathbb{R}^{\mathsf{V}}$, *then*

$$\begin{pmatrix} \nabla(c\nabla) & 0 \\ \beta\partial_{c^2} & -\mathcal{W}_1 \end{pmatrix}$$

with domain $D(\mathbb{A})$ *generates an analytic* C_0*-semigroup on* $\mathbb{H}$.

Properties of $(e^{t\mathbb{A}})_{t\geq 0}$ can be studied by energy methods like those of $(e^{t\Delta})_{t\geq 0}$ in the previous section. Indeed, the proofs are essentially the same, as one sees that $\mathbb{V}$ is lattice isomorphic to $V = W^{1,2}(\mathfrak{G})$ and—upon identifying these spaces accordingly—the forms $a_{\mathcal{W}}, \mathfrak{a}_{\mathcal{W}}$ agree.

Theorem 6.97. *The semigroup* $(e^{t\mathbb{A}})_{t\geq 0}$ *is real (resp., positive) if and only if so is* $(e^{-t\mathcal{W}})_{t\geq 0}$.

If $-\mathcal{W}$ *is dissipative, then* $(e^{t\mathbb{A}})_{t\geq 0}$ *is* $L^\infty((0,1);\ell^2_\mu(\mathsf{E}))$*-contractive if and only if* $(e^{-t\mathcal{W}})_{t\geq 0}$ *is* $\ell^\infty_{\deg_\mu}(\mathsf{V})$*-contractive.*

If $(e^{-t\mathcal{W}})_{t\geq 0}, (e^{-t\mathcal{W}^*})_{t\geq 0}$ *are* $\ell^\infty_{\deg_\mu}(\mathsf{V})$*-contractive, then* $(e^{t\mathbb{A}})_{t\geq 0}$ *extends to a contractive, analytic* C_0*-semigroup on all spaces* $\mathbb{L}^p$ *whose generator is the* $\mathbb{L}^p$*-realization of* $\mathbb{A}$. *Furthermore,* $(e^{t\mathbb{A}})_{t\geq 0}$ *is ultracontractive of dimension 1.*

Corollary 6.98. *Let* G *be a connected graph of finite volume. If* $\mathcal{W}$ *is a diagonal matrix whose entries are all non-negative, then* $(e^{t\mathbb{A}})_{t\geq 0}$ *is a sub-Markovian semigroup that is strongly-continuous on* $C_0(\mathfrak{G})$*—a so-called* Feller semigroup*—and is hence the transition semigroup of a Feller process. Its generator is the part of* $\mathbb{A}$ *in*

$$\left\{ \begin{pmatrix} u \\ f \end{pmatrix} \in C_0(\mathfrak{G}) \times \mathbb{C}^{\mathsf{V}_0} : u_{|\mathsf{V}_0} = f \right\}.$$

Proof. Under our assumptions we know from Theorem 6.97 that $(e^{t\mathbb{A}})_{t\geq 0}$ is a sub-Markovian (even Markovian, if $\mathcal{W} = 0$) C_0-semigroup . Furthermore, $\mathbb{V}$ is isomorphic to $W^{1,2}(\mathfrak{G})$ by the isomorphism $u \mapsto (u, u_{|\mathsf{V}_0})$. If in particular G is finite, then same operator also acts as an isomorphism between $\mathbb{C}_0(\mathfrak{G})$ and $C(\mathfrak{G})$. Now, $\mathbb{C}(\mathfrak{G})$ embeds in $\mathbb{L}^2$ and $\mathbb{L}^2$ is mapped into $\mathbb{V}$ by the analytic semigroup $(e^{t\mathbb{A}})_{t\geq 0}$. Because by Lemma 3.27 $\mathbb{V}$ embeds into $\mathbb{C}_0(\mathfrak{G})$, we conclude that, up to the above isomorphism, $(e^{t\mathbb{A}})_{t\geq 0}$ map $C_0(\mathfrak{G})$ into itself. By Lemma 3.27.(5) (for $r = q = 2$ and $p = \infty$) $(e^{t\mathbb{A}})_{t\geq 0}$ is strongly continuous also with respect to the ∞-norm, hence on $C_0(\mathfrak{G})$. Finally, Corollary 4.42 yields the claim on the generator of the semigroup on $C_0(\mathfrak{G})$. □

The solution u of the initial value problem associated with $\mathbb{A}$, complemented by node conditions (Cc′) on V, (KRc′) on V_0^C, and (Dc′) on V_0 is by Lemma 6.93 a $W^{2,2}$-function, hence it is a priori not obvious that its second derivative can be evaluated at the nodes. However, it is straightforward to see that

$$D(\mathbb{A}^2) \subset W^{3,2}\big((0,1);\ell^2_\mu(\mathsf{E})\big) \times \ell^2_{\deg_\mu}(\mathsf{V})$$

and hence by analyticity of $(e^{t\mathbb{A}})_{t\geq 0}$ and in view of Proposition 6.6 we conclude that

$$\frac{\partial^2 u}{\partial x^2}(t,\mathsf{v}) = \frac{\partial u}{\partial t}(t,\mathsf{v}) = -\partial_{c^2} u(t,\mathsf{v}) - \mathcal{W}u(t,\mathsf{v}), \qquad t \geq 0,\ \mathsf{v} \in \mathsf{V}_0, \qquad \text{(WRC)}$$

where the first identity follows evaluating (Di) at the nodes in V_0 and the second is (Dc), and all terms are well defined. In other words, (WRC) is satisfied—for all $t > 0$—by all solution of the abstract Cauchy problem associated with the hybrid Laplacian on $\mathbb{H}$. We have hence recovered (2.52).

Example 6.99. The Hodgkin–Huxley model considered in Example 6.27 can be easily extended to the case of a network, possibly to complement a lumped soma model of Rall's type, if different intervals (corresponding to different axons and/or dendrites, or perhaps to simplified models of whole neurons) are coupled by standard and/or dynamic node conditions. We omit the mathematical details—which are not difficult and can be found in [26, § 6]—and focus instead on the modeling issue. If the interface between these elements is an electric synapse, then our current understanding of these biological devices suggests that said class of node conditions actually delivers a convenient description: This is indeed the very simple setting studied in [26].

If however a dendrite and an axon (represented by two intervals $\mathsf{e}_1, \mathsf{e}_2$) are incident in a chemical synapse v, which is terminal endpoint of e_1 and initial endpoint of e_2, then we have already seen in Chap. 5 that the synaptic input coming from e_1 undergoes a delay τ_{del} before reaching e_2 and cannot turn back (Fig. 6.3).

The synaptic input is of course an action potential that lets neurotransmitters be released by synaptic vesicles, but experimental observations seem to suggest that no obvious (linear) algebraic relation exists between the pre- and post-synaptic

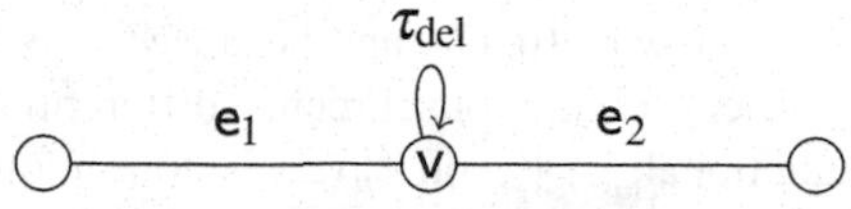

Fig. 6.3 An axon e_1 projecting into a dendrite e_2 through a synapse v

potential in the dendrites—i.e., between the boundary values of the unknowns u_1, u_2 in the diffusion equations. Indeed, a chemical synapse is not an electrical connector!

6.7 Nonlinear Parabolic Equations

Let us briefly discuss how the theory summarized in Sect. 6.2.2 can be adapted to the setting of equations on networks.

We restrict ourselves for the sake of simplicity to the case of

$$\text{a uniformly locally finite, weighted oriented graph } \boxed{\mathsf{G} = (\mathsf{V}, \mathsf{E}, \gamma)}$$

and consider the discrete p-Laplacian introduced in Sect. 2.4 (the case of a merely locally finite graph has been treated in [87]). For $1 < p < \infty$ one introduces the energy functional

$$\mathcal{E} : f \mapsto \frac{1}{p} \|\mathcal{I}^T f\|^p_{\ell^p_\mu},$$

which is well-defined on the separable, reflexive Banach space

$$w^{1,p,2}_\mu(\mathsf{V}) := \left\{ f \in \ell^2 : \mathcal{I}^T f \in \ell^p_\mu(\mathsf{E}) \right\}.$$

(Observe that for $p = 2$ we recover the quadratic form introduced in (6.23), up to the factor $\frac{1}{p}$.)

Lemma 6.3. *Let $p \in (1, \infty)$. Then the functional $\mathcal{E}$ is convex, coercive, and continuously Fréchet differentiable as a functional on $w^{1,p,2}_\mu(\mathsf{V})$ and lower semicontinuous as a functional on $\ell^2(\mathsf{V})$. It is coercive whenever restricted to the closure of $c_{00}(\mathsf{V})$ in $w^{1,p,2}_\mu(\mathsf{V})$.*

Proof. Since for all $p \in (1, \infty)$

$$\mathcal{E} \equiv \frac{1}{p} \| \cdot \|^p_{\ell^p_\mu(\mathsf{E})} \circ \mathcal{I}^T \qquad \text{on } w^{1,p,2}_\mu(\mathsf{V}), \tag{6.47}$$

the composition of a convex functional and a linear operator, the functional $\mathcal{E}$ is convex. The assertion on coerciveness is clear, by definition.

In view of (6.47), and because $\mathcal{I}^T$ is bounded from $w_\mu^{1,p,2}(\mathsf{V})$ to $\ell_\mu^p(\mathsf{E})$, in order to check continuous Fréchet differentiability of $\mathcal{E}$ it suffices to observe that the functional $\|\cdot\|_{\ell_\mu^p}^p$ on $\ell_\mu^p(\mathsf{E})$ is continuously Fréchet differentiable for $p \in (1,\infty)$ with

$$\begin{aligned}\mathcal{E}'(f)h &= \left\langle |\mathcal{I}^T f|^{p-2}\mathcal{I}^T f, \mathcal{I}^T h\right\rangle_{\ell_\mu^{p'},\ell_\mu^p} \\ &= \sum_{\mathsf{e}\in\mathsf{E}} \mu(\mathsf{e})|(\mathcal{I}^T f)(\mathsf{e})|^{p-2}(\mathcal{I}^T f)(\mathsf{e})(\mathcal{I}^T h)(\mathsf{e}), \qquad f,h \in w_\mu^{1,p,2}(\mathsf{V}),\end{aligned}$$

by the chain rule. Thus, $\mathcal{E}$ is in particular lower semicontinuous as a functional on $w_\mu^{1,p,2}(\mathsf{V})$, and hence lower semicontinuity as a functional on $\ell^2(\mathsf{V})$. □

Using Lemma 6.42 one thus shows that the operator associated with $\mathcal{E}$ is actually (minus) the discrete p-Laplacian $\mathcal{L}_p$ in (2.54)—in a weak sense, and also in a classical sense whenever applied to test functions, i.e., to sequences in the space $c_{00}(\mathsf{V})$. Then, Theorem 6.45 yields well-posedness of the abstract Cauchy problem associated with $-\mathcal{L}_p$.

Similar considerations hold if instead of the discrete p-Laplacian, the more usual (differential) p-Laplacian is considered. This is the operator formally defined as (minus) the subdifferential (or rather, as before, as the Fréchet derivative) of the energy functional

$$\mathcal{E} : u \mapsto \frac{1}{p}\|\nabla u\|_{L^p}^p.$$

This operator was introduced back in the 1960s by G. Aronsson, E. DiBenedetto, O.A. Ladyženskaja, N.N. Ural'ceva and others, who were of course only interested in functions on domains of $\mathbb{R}^d$. In order to consider its network version, it suffices to take as domain of $\mathcal{E}$

$$W_Y^{1,p}\left((0,1);\ell_\mu^2(\mathsf{E})\right) :$$

This is defined in a way analogous to (6.32) for Y as in (6.29) (or even for a general closed subspace of $\ell_\mu^2(\mathsf{E})\times\ell_\mu^2(\mathsf{E})$). Again, this functional is convex, Fréchet differentiable and $\mathcal{E}(0) = 0$, and we can obtain well-posedness of the associated parabolic equation right away.

6.8 Notes and References

Section 6.1. Analytic semigroups were introduced by Hille in [51]. Much attention has been devoted to them ever since: We refer to [3, 76] and [5, § 3.7] for modern, comprehensive treatments. Indeed, analyticity is a very relevant property. On the one hand, it is sufficiently frequent: For instance many—if not all—elliptic operators of even order generate analytic semigroups. On the other hand, analytic semigroups have several distinctive features—like the perturbation results in Lemma 6.5.

In view of Theorem 6.2 and Proposition 6.8, (6.7) and (6.8) yield an estimate on the semigroup's analyticity angle and growth bound. However, observe that the optimal angle of $\frac{\pi}{2}$ cannot be deduced from (6.7)–(6.8) even when it is certainly known to hold, e.g. in the case of a bounded generator. It is also for this reason that Theorem 6.18 is particularly valuable, in view of Remark 6.10.

Section 6.2. The assumptions under which most of the results in Sect. 6.2 are formulated are by no means sharp. The reader is referred to [3, 4, 33, 97] for proofs, further details, and recent accounts about forms, although the first significant advances in the theory of sesquilinear forms go back to [58]. Here we do not discuss Kato's theory of *closed forms* and follow instead the approach proposed by J.-L. Lions, cf. [32]: that celebrated monograph includes an introduction to this theory along with many applications to problems from physics and mechanics.

A version of Theorem 6.18 was known to Kisynski, who in [62] gives to Lions credit for it. Condition 6.5 has become rather popular after Crouzeix showed in [31] that it plays a fundamental role in the development of a new type of functional calculus. Whenever the damping operator C in (4.19) is at least as unbounded as A, one does not expect generation of a C_0-group, but rather of an analytic C_0-semigroup. Form methods have proved effective in order to discuss also such damped wave equations, generalizing results that were previously known for self-adjoint operators: it has been proved in [85] that if C comes from a form and A is merely bounded from $[D(C)]$ to H, then (a certain extension of) the reduction matrix generates an analytic C_0-semigroup.

The constants in (6.8) have been tracked down in [91], while the perturbation Lemmata 6.22–6.23 are taken from [85, 90]. They are the form analogs of Lemma 6.5 and further perturbation results for operators, cf. [34,35]. Theorem 6.11 has been proved in [7], where also the notion of boundary group has been introduced.

The converse of the assertion in Remark 6.10 is not true: Littman showed in [73] that—unlike the heat equation on the same spaces—the classical wave equation is not well-posed on $L^p(\mathbb{R}^d)$ unless $p = 2$ or $d = 1$.

A comprehensive collection of applications and further results related to Ouhabaz' invariance criterion can be found in [97, Chapter 2]. Two special but fundamental instances—characterization of the invariance of the positive cone of L^2 and of the unit ball of L^∞—have been obtained for symmetric forms by Beurling and Deny in [20] and are therefore known as *Beurling–Deny criteria.* These results have also a nonlinear pendant proved in [10], cf. Lemma 6.1.

Theorem 6.31 can be traced back to [57], but has been re-discovered several times under slightly different assumptions. An account of its interesting history along with a vector-valued version are found in [89]. Observe that if $\mu(U) < \infty$, then $L^\infty(U \times U) \hookrightarrow L^2(U \times U)$ and hence we can conclude that T_K is a Hilbert–Schmidt operator. Recalling that the composition of Hilbert–Schmidt operators is of trace class, we deduce from the semigroup law another proof of the fact that under the assumptions of Corollary 6.66 and $(e^{tA})_{t\geq 0}$ is automatically of trace class, provided G has finite volume.

Even whenever Proposition 6.30.(5) applies and we know that the analytic semigroup associated with a form extrapolates to an L^∞-space, the semigroup on L^∞ need not be analytic itself—that is, its analyticity angle is 0. Nevertheless, all further extrapolated semigroups on L^p, $p \in (2, \infty)$, are indeed analytic, cf. [3, § 7.2].

The notion of j-elliptic form has been introduced by Arendt and ter Elst in [8], from where most of the results in this section are taken. Very similar ideas had already appeared in [18] as a refinement of the theory of traces of Dirichlet forms in [45].

The theory of Arendt and ter Elst also allows for some useful spectral relations—in this case between Δ and $\mathbb{A}$—as it has been shown in [90]. This theory is particularly efficient in order to discuss the Dirichlet-to-Neumann operator, but we omit the easy details and refer to [90, § 4] for related results.

Classical references for the theory of nonlinear semigroups are [23, 81, 102] or a legendary uncompleted manuscript by P. Bénilan, M. Crandall, and A. Pazy, which is still circulating in mathematical departments as a samizdat. The recent lecture notes [29] offer an easy-to-read introduction to the theory of subdifferentials, whereas [9] is more focused on applications. In Theorems 6.45 and 6.46 we have summarized several different results obtained with different methods, and in particular some celebrated results by H. Brezis and T. Kato, cf. [102, Thm. IV.4.1, Thm. IV.4.3, and Thm. IV.8.2] or [23, Théo. 3.1, Théo. 3.2, Théo. 3.3, Théo. 3.6]. Lemma 6.1 combines [10, Théo 1.1 and Cor. 2.2], [23, Prop. 4.5], and [30, Cor. 3.7].

Section 6.3. Delay differential equations represent a broad and classical research topic, cf. [37]. The approach presented here follows that developed in [11]. Well-posedness of parabolic delay differential equations is proved using a technique based on the theory of *one-sided coupled operator matrices* due to K.-J. Engel, cf. [11, Thm. 3.35]. Many sophisticated results on asymptotics of delay evolution equations can be found in [11, Part III], including Proposition 6.50.

Section 6.4. We have already mentioned that algebraic methods were applied to graph-theoretical problems already by Kirchhoff in [61]. It seems that the firsts who suggested a functional analytical theory of networks were Beurling and Deny in [21]. (Likely because their actual goal was clearly more general and ambitious—viz, the development of the theory of forms associated with sub-Markovian semigroups on general metric measure spaces—their "elementary case" was possibly overlooked or considered as a mere toy model for several years.) Similar ideas, also in connection with potential theory, appeared in [40] and were then systematically developed by Yamasaki and his coauthors in a series of papers beginning with [110]. They were devoted to potential theory on infinite networks and in this context Yamasaki and his coauthors naturally introduced a non-standard kind of random walk—in our language, the discrete diffusion equation—along with some different but related versions of the discrete Sobolev spaces. A few years later, Mohar started to investigate the spectral properties of the adjacency matrix in [82]. Dodziuk studied in [38] the discrete Laplacian from the point of view of dynamical systems, and also considered the same operator in dependence of different node weights—specifically, of the deg-weight. Observe that $\frac{1}{\deg}\mathcal{L}$ is similar to $\mathcal{L}_{\text{norm}}$ by the unitary transformation $\mathcal{D}^{\frac{1}{2}}$. Soon after [39] many authors, including D. Cartwright, M. Picardello, P. Soardi, and W. Woess began to combine all these different approaches along with group theory and stochastics, thus obtaining a comprehensive and elegant theory of difference operators on discrete graphs, both on the elliptic and parabolic side, cf. [104, 109]. Finally, in the last few years new clusters of authors began interbreeding the approach of the 1990s with new mathematical physical flavors, in particular in three series of papers by P. Jorgensen and E. Pearse, by U. Smilansky, and by M. Keller and D. Lenz, and some further coauthors: They are conveniently surveyed in [55, 59, 103], respectively.

Non-autonomous problems like that considered in Remark 6.25 arise, e.g., whenever we discuss a linearized version of (5.6) and at the same time allow for synaptic pruning and/or plasticity, so that the entries of $\mathcal{A}^{\text{in}}$ may change over time.

In Remark 6.51 we have imposed condition (6.22) for the sake of simplicity. The more challenging case of measures ν that are allowed to be unbounded from above and/or from below is considered, e.g., in [60, 87]. It can be treated using the weighted, discrete Sobolev spaces $w^{1,2}_{\mu,\nu}(\mathsf{V})$.

In this generality, Theorems 6.52 and 6.54 about well-posedness and sub-Markov property of $(e^{t\mathcal{L}^D})_{t\geq 0}$, $(e^{t\mathcal{L}^N})_{t\geq 0}$ have been obtained in [50, 60] using the theory of Dirichlet forms of Beurling and Deny. Indeed, the special case of finite graphs had been fully discussed already in [20]. An equivalent statement of Theorem 6.54 is precisely that the Laplace–Beltrami matrix is associated with a Dirichlet form. In our proof of these results we follow instead the partially different methods discussed in [87], which incidentally also extend to nonlinear diffusion problems associated with the discrete p-Laplacians briefly introduced in Sect. 2.4. The existence of realizations of the Laplace–Beltrami matrix that generate semigroups that are not sub-Markovian has been proved in [53].

The possibility of approximating the solution of an evolution equation on a network by cutting the graph outside ever-growing "discrete balls", solving the problem on such finite graphs, and then going to the limit has been envisaged by several authors. Proposition 6.55 is taken from [87], but [59, Prop. 10] is very similar. Strong and spectral radius convergence has been observed already in [82] in the case of the adjacency matrix. Observe that the same Galerkin methods may be also applied to find solutions of discrete advective problems as in Sect. 6.4.2, as they are associated with perturbations of quadratic forms. In contrast, application of Galerkin methods to advective problems is much more delicate in the usual continuous setting.

Section 6.5. Many properties of the semigroup $(e^{t\Delta})_{t\geq 0}$ associated with the sesquilinear form a defined in (6.31) have also been obtained, by other methods and with different motivations, in [15, 64, 67], but special instances were already known since [12], see also [14] and references therein. A more precise estimate that takes into account the topology of the graph is presented in [36]. Theorem 6.67 and Example 6.73, which together state that the second derivative with continuity and Kirchhoff node conditions comes from a Dirichlet form (and hence that it generates a sub-Markovian semigroup), has been obtained in [67, 84] in the case of finite graphs, and in [27] in the infinite case. We have observed in Example 6.73 failure of the (sub-)Markovian property for the semigroup associated with $a_{\mathcal{W}}$ where $\mathcal{W} = -\mathcal{L}$—i.e., for the semigroup generated by the Krein–von Neumann extension of the second derivative: This is also a well-known fact in abstract potential theory, cf. [44, Thm. 2.3.2].

More general system of diffusion processes can be considered, describing non-local interactions that take place not only in the nodes, but also in the edges. More precisely, (Di) can be generalized to

$$\frac{\partial u_{\mathsf{e}}}{\partial t}(t,x) = \sum_{\mathsf{f}\in\mathsf{E}} \frac{\partial}{\partial x}\left(c^2_{\mathsf{ef}}(x)\frac{\partial u_{\mathsf{f}}}{\partial x}\right)(t,x) - \sum_{\mathsf{f}\in\mathsf{E}} p_{\mathsf{ef}} u_{\mathsf{f}}(t,x), \quad x \in (0,1),\ \mathsf{e}\in\mathsf{E},\ t\geq 0, \tag{6.48}$$

where $c^2(x), p(x)$ are (possibly non-diagonal) matrices for all $x \in [0,1]$. (We have already seen this equation as a mathematical model of ephaptic coupling of neurons, cf. Chap. 5.) It has to be complemented by (Cc) and the generalized Kirchhoff conditions

$$\sum_{\mathsf{e},\mathsf{f}\in\mathsf{E}} \sum_{\mathsf{w}\in\mathsf{V}} c^2_{\mathsf{ef}}(\mathsf{w}) \left(\iota^+_{\mathsf{ve}}\iota^+_{\mathsf{wf}} - \iota^-_{\mathsf{ve}}\iota^-_{\mathsf{wf}}\right) u'_{\mathsf{f}}(t,\mathsf{w}) = \sum_{\mathsf{w}\in\mathsf{V}} \omega_{\mathsf{vw}} u(\mathsf{w}), \quad t\geq 0,\ \mathsf{v}\in\mathsf{V}. \tag{gKc}$$

It has been shown in [28] that the generation result in Theorem 6.67 extends to such a general setting, under the additional assumptions that c is sufficiently smooth—$C^1\left([0,1];\ell^2_\mu(\mathsf{E})\times\ell^2_\mu(\mathsf{E})\right)$ will do —, that $p \in L^\infty\big((0,1)\times(0,1);\ell^2_\mu(\mathsf{E})\big)$, and that the ellipticity condition

$$\Gamma|\xi|^2_{\ell^2} \geq \operatorname{Re}\sum_{\mathsf{e},\mathsf{f}\in\mathsf{E}} c^2_{\mathsf{ef}}(x)\xi_{\mathsf{f}}\overline{\xi_{\mathsf{e}}} \geq \gamma|\xi|^2_{\ell^2} \quad \text{for all } \xi\in\ell^2(\mathsf{E}) \text{ and a.e. } x\in(0,1),$$

is satisfied for some $\Gamma,\gamma > 0$. More precisely, despite being highly non-local this problem is governed by an analytic C_0-semigroup on $L^2\big((0,1);\ell^2_\mu(\mathsf{E})\big)$. Remarkably, in spite of its parabolic nature (shown e.g. in [80]) (6.48) lacks most of the regularity properties that are typical of diffusion problems if c is not diagonal for all $x \in [0,1]$: e.g., the solution is not of class C^∞ and the governing C_0-semigroup is not (sub-)Markovian—i.e., no parabolic maximum principle holds, and the system is not dissipative with respect to the L^1 and L^∞-norms.

Existence of traveling waves as solutions of (Di) – (Cc) – (Kc) is known to be possible under some combinatorial assumptions on the graph underlying the neuronal network, cf. [14, §§ 16–17]: This suggests that a network may support spikes moving at constant velocity in the framework of (HH). In the case of a single interval of infinite length, traveling wave solutions of different neural models have studied by several authors, cf. [100, Chapter 6] and references therein. As already mentioned in Remark 6.99. It is however not clear how (HH) should be extended to a model of a whole network.

The possibility of interpreting some class of discrete diffusion equations as stochastic processes that approximate classical diffusion processes associated with differential or hybrid operators on metric graphs has been discussed in [108]. Conversely, a fascinating instance of emergence of diffusion equations on networks from a discrete setting was suggested by Freidlin and Wentzell in [43]: If one considers a family of dynamical systems, say indexed by $\epsilon > 0$, it may happen that as $\epsilon \to 0$ the system tends to have faster diffusion in certain directions than in other ones. (Formally speaking, weak convergence of distributions in a space of continuous functions is meant). Hence, the phase plane is subdivided in regions that as $\epsilon \to 0$ may collapse to edges of a metric graph (actually, of a tree), so that transition from one region to the others corresponds to crossing of ramification points. Similar ideas concerning the possibility of letting a system of diffusion equations on decoupled intervals, but with non-local Robin-type boundary conditions, converge to a discrete diffusion equation on a line graph have been analyzed in [22].

Finally, let us observe that, just like its underlying discrete graph G, also a metric graph $\mathfrak{G}$ might suffer from a growth so ill-behaved that $\mathfrak{G}$ develops a kind of boundary at infinity. More precisely, even in the unweighted case it might be that the space of compactly supported $C^1(\mathfrak{G})$-functions is not dense in $W^{1,2}(\mathfrak{G})$ even if G is infinite and has no node of degree 1. We have avoided to treat these aspects in detail, but the reader should keep in mind that bizarre non-uniqueness phenomena may arise that resemble the situation described in Theorem 6.52.

Section 6.5.1. Proposition 6.79 is the main result in [75]. He proved it—in a significantly more general context—making use of his theory of *local dissipativity*, cf. [74], which roughly speaking allows for an extension of Hilbert space methods to evolution equations taking place in spaces of continuous functions. Observe that the "wrong sign" in the node conditions in Proposition 6.79 appears in the domain of the operator rather than in the operator itself as in Corollary 6.96, and thus cannot be treated by a simple perturbation argument (Perturbation results for *domains* of generators do in fact exist, cf. [48,49], but are seemingly not applicable here.)

Up to minor notational changes, the more benign node conditions (Cc′) – (KRc′) have been studied already in [13, 47], and (more implicitly) already by Hölder, cf. [52, § 3]. This setting has become rather popular after Kuchment discussed it in [68, § 3] in the case of Hermitian $\mathcal{W}$ as well, in a slightly more general form, in [46], cf. Sect. 7.2.1. The case of non-Hermitian $\mathcal{W}$ has been discussed, among others, in [17,27,56,64,84]. An extension to the case of dynamic node conditions has been performed in [86, 101]. Theorems 6.72 has been proved independently in [27,56]. It seems that the Laplacian with anti-Kirchhoff node conditions presented in Remark 6.81 has been first proposed in [76]. Example 6.86 is taken from [86].

Local finiteness of the graph has been one of our standing assumptions throughout Sects. 6.4 and 6.5. If we drop it, the results of these sections become much more delicate and several counter-intuitive phenomena may happen, both in the continuous and in the discrete case. To the best of our knowledge, the latter case has been investigated in only in [60] and a few other papers. An example concerning the continuous case is given by the following, taken from [25].

Proposition 6.100. *Let* $\mathfrak{G}$ *be the metric graph constructed over a countable unweighted graph* G. *Let* Δ *be the operator on* $L^2(\mathfrak{G})$ *associated with the form* $a_{\mathcal{W}}$ *for* $\mathcal{W} = 0$ *with domain* $W^{1,2}_{Y}((0,1);\ell^2(\mathsf{E}))$ *as in* (6.32). *Then the associated semigroup* $(e^{t\Delta})_{t\geq 0}$ *is irreducible if and only if for one/all* $\mathsf{e} \in \mathsf{E}$ *and any other edge* f *there exists a path from* e *to* f *that does not contain any node of infinite degree.*

Section 6.6.1. We are not aware of literature specifically devoted to the parabolic properties of the Dirichlet-to-Neumann operators on graphs. In the case of domains of $\mathbb{R}^d$, the fact that the relevant C_0-semigroup is sub-Markovian—i.e., a counterpart of Theorem 6.90—has been proved in [98, Thm. 9.1] and independently by form methods in [41].

Section 6.6.2. The operator $\mathbb{A}$ appears in several models of neuroscience [24, 77] and elasticity [70, § II.7], with essentially the same explanation/derivation: For example, in the lumped soma model introduced in Chap. 5 one wants electric potential neither to vanish in the soma, nor to be reflected in the dendritic network, but rather to accumulate there (until an action potential is triggered).

Material in this section is mainly taken from [86, 92, 93], where the interplay with the investigations in [77] has also been discussed. Related mathematical results appear in [56, 101], whereas [26] is more devoted to modeling issues. There seems to be no consensus in the literature on how the biochemistry of even very small pools of neurons connected by chemical synapses should be modeled by means of differential equations: To the best of our knowledge, [54] is one of the few articles devoted to this topic. Also motivated by biological investigations, this time by those in [24], parabolic equations for hybrid operators have seemingly been mathematically investigated for the first time by Nicaise and later in a more detailed way by von Below, cf. in particular [14, 15, 94] and references therein. Such systems also fit into the abstract theory of so-called "interaction problems" introduced by Ali Mehmeti in [2].

Our investigation is based on the possibility of applying form method, which depends on the particular structure of $\mathbb{A}$. More general dynamic boundary conditions (and hence different entries in the operator matrix $\mathbb{A}$) might be chosen: This more general setting can still be studied by different methods, e.g. along the lines of [14, 66, 88, 106].

Feller and Wentzell intensively studied stochastic processes associated with the realization of $\mathbb{A}$ in $C(\overline{U})$, beginning with [42, 107]: Such boundary conditions (or their generalization as in (2.52)) are therefore sometimes also referred to as of *Feller* or *generalized Wentzell* type. They essentially describe the behavior of a Brownian motion in which any particle that hits the boundary is neither simply reflected or absorbed, but rather can spend some time there, cf. [105] for a modern and general overview. In this sense, it is indeed natural to study them in a space of continuous functions, like in Corollary 6.98. A characterizations of generators of Brownian motions on metric graphs in terms of Laplacians with Generalized Wentzell conditions has been proved in [65, Theorems 2.5 and 2.8].

Concerning differential operators with general node conditions as in (2.53), the main difficulty is that they need not come from a form—indeed, they are possibly not even quasi-dissipative. It is indeed more natural to discuss realizations of these operators in $C_0(\mathfrak{G})$: This has been done in [75] for $\alpha \equiv 0$, and in [14] for $\alpha \not\equiv 0$.

References

1. H. Abels, M. Wilkem, Convergence to equilibrium for the Cahn–Hilliard equation with a logarithmic free energy. Nonlinear Anal. Theory Methods Appl. **67**, 3176–3193 (2007)
2. F. Ali Mehmeti, Regular solutions of transmission and interaction problems for wave equations. Math. Meth. Appl. Sci. **11**, 665–685 (1989)
3. W. Arendt, Semigroups and evolution equations: functional calculus, regularity and kernel estimates, in *Handbook of Differential Equations: Evolutionary Equations*, vol. 1, ed. by C.M. Dafermos, E. Feireisl (North Holland, Amsterdam, 2004)
4. W. Arendt, Heat Kernels–Manuscript of the 9th Internet Seminar, 2006. (Freely available at http://www.uni-ulm.de/fileadmin/website_uni_ulm/mawi.inst.020/arendt/downloads/internetseminar.pdf)
5. W. Arendt, C.J.K. Batty, M. Hieber, F. Neubrander, *Vector-Valued Laplace Transforms and Cauchy Problems, Monographs in Mathematics*, vol. 96 (Birkhäuser, Basel, 2001)
6. W. Arendt, D. Dier, H. Laasri, E.M. Ouhabaz, Maximal regularity for evolution equations governed by non-autonomous forms. arXiv:1303.1166 (2013)
7. W. Arendt, O. El Mennaoui, M. Hieber, Boundary values of holomorphic semigroups. Proc. Amer. Math. Soc. **125**, 635–647 (1997)
8. W. Arendt, T. ter Elst, Sectorial forms and degenerate differential operators. J. Operator Th. **67**, 33–72 (2012)
9. V. Barbu, in *Nonlinear Differential Equations of Monotone Types in Banach Spaces*. Monographs in Mathematics (Springer, Berlin, 2010)
10. L. Barthélemy, Invariance d'un convexe fermé par un semi-groupe associé à une forme non-linéaire. Abstr. Appl. Analysis **1**, 237–262 (1996)
11. A. Bátkai, S. Piazzera, *Semigroups for Delay Equations* (AK Peters, Wellesley, 2005)
12. J. von Below, A characteristic equation associated with an eigenvalue problem on c^2-networks. Lin. Algebra Appl. **71**, 309–325 (1985)
13. J. von Below, Classical solvability of linear parabolic equations on networks. J. Differ. Equ. **72**, 316–337 (1988)
14. J. von Below, *Parabolic Network Equations* (Tübinger Universitätsverlag, Tübingen, 1994)
15. J. von Below, S. Nicaise, Dynamical interface transition in ramified media with diffusion. Comm. Partial Differ. Equations **21**, 255–279 (1996)
16. J. von Below, J.A. Lubary, The eigenvalues of the Laplacian on locally finite networks. Result. Math. **47**, 199–225 (2005)
17. J. von Below, D. Mugnolo, The spectrum of the Hilbert space valued second derivative with general self-adjoint boundary conditions. Lin. Alg. Appl. **439**, 1792–1814 (2013)
18. A. Ben Amor, J.F. Brasche, Sharp estimates for large coupling convergence with applications to Dirichlet operators. J. Funct. Anal. **254**, 454–475 (2008)

19. P. Bénilan, H. Brézis, Solutions faibles d'équations d'évolution dans les espaces de Hilbert. Ann. Inst. Fourier **22**, 311–329 (1972)
20. A. Beurling, J. Deny, Dirichlet spaces. Proc. Natl. Acad. Sci. USA **45**, 208–215 (1959)
21. A. Beurling, J. Deny, Espaces de Dirichlet. I: Le cas élémentaire. Acta Math. **99**, 203–224 (1959)
22. A. Bobrowski, From diffusions on graphs to Markov chains via asymptotic state lumping. Ann. Henri Poincaré A **13**, 1501–1510 (2012)
23. H. Brézis, *Operateurs Maximaux Monotones et Semi-Groupes de Contractions dans les Espaces de Hilbert* (North-Holland, Amsterdam, 1973)
24. H. Camerer, *Die Elektrotonische Spannungsbreitung im Soma, Dendritenbaum und Axon von Nervenzellen*, Ph.D. thesis, Universität Tübingen, 1980
25. S. Cardanobile, The L^2-strong maximum principle on arbitrary countable networks. Lin. Algebra Appl. **435**, 1315–1325 (2011)
26. S. Cardanobile, D. Mugnolo, Analysis of a FitzHugh–Nagumo–Rall model of a neuronal network. Math. Meth. Appl. Sci. **30**, 2281–2308 (2007)
27. S. Cardanobile, D. Mugnolo, Parabolic systems with coupled boundary conditions. J. Differ. Equ. **247**, 1229–1248 (2009)
28. S. Cardanobile, D. Mugnolo, R. Nittka, Well-posedness and symmetries of strongly coupled network equations. J. Phys. A **41**, 055102 (2008)
29. R. Chill, E. Fašangová, *Gradient Systems* (MatFyzPress, Prague, 2010)
30. F. Cipriani, G. Grillo, Nonlinear Markov semigroups, nonlinear Dirichlet forms and applications to minimal surfaces. J. Reine Ang. Math. **562**, 201–235 (2003)
31. M. Crouzeix, Operators with numerical range in a parabola. Arch. Math. **82**, 517–527 (2004)
32. R. Dautray, J.-L. Lions, *Mathematical Analysis and Numerical Methods for Science and Technology*, vol. 5 (Springer, Berlin, 1992)
33. E.B. Davies, *Heat Kernels and Spectral Theory*. Cambridge Tracts in Mathematics, vol. 92 (Cambridge University Press, Cambridge, 1989)
34. W. Desch, W. Schappacher, On relatively bounded perturbations of linear C_0-semigroups. Ann. Sc. Norm. Super. Pisa, Cl. Sci. **11**, 327–341 (1984)
35. W. Desch, W. Schappacher, Some perturbation results for analytic semigroups. Math. Ann. **281**, 157–162 (1988)
36. L. Di Persio, G. Ziglio, Gaussian estimates on networks with applications to optimal control. Networks Het. Media **6**, 279–296 (2011)
37. O. Diekmann, S.A. van Gils, S.M. Verduyn Lunel, H.-O. Walther, *Delay Equations: Functional-, Complex-, and Nonlinear Analysis*. Applied Mathematical Sciences, vol. 110 (Springer, Berlin, 1995)
38. J. Dodziuk, Difference equations, isoperimetric inequality and transience of certain random walks. Trans. Amer. Math. Soc. **284**, 787–794 (1984)
39. J. Dodziuk, W.S. Kendall, Combinatorial Laplacians and isoperimetric inequality, in *From Local Times to Global Geometry, Control and Physics*. Proceedings of Coventry 1984/85. Pitman Research Notes in Mathematics Series, vol. 150 (Pitman, Harlow, 1986), pp. 68–74
40. R.J. Duffin, The extremal length of a network. J. Math. Anal. Appl. **5**, 200–215 (1962)
41. H. Emamirad, I. Laadnani, An approximating family for the Dirichlet-to-Neumann semigroup. Adv. Diff. Equ. **11**, 241–257 (2006)
42. W. Feller, The parabolic differential equations and the associated semi-groups of transformations. Ann. Math. **55**, 468–519 (1952)
43. M.I. Freidlin, A.D. Wentzell, Diffusion processes on graphs and the averaging principle. Ann. Probab. **21**, 2215–2245 (1993)
44. M. Fukushima, *Dirichlet Forms and Markov Processes*. Mathematical Library, vol. 23 (North-Holland, Amsterdam, 1980)
45. M. Fukushima, Y. Oshima, M. Takeda, *Dirichlet Forms and Symmetric Markov Processes*. Studies in Mathematics, vol. 19 (de Gruyter, Berlin, 2010)
46. S.A. Fulling, P. Kuchment, J.H. Wilson, Index theorems for quantum graphs. J. Phys. A **40**, 14165–14180 (2007)

47. V.I. Gorbachuk, M.L. Gorbachuk, *Boundary Value Problems for Operator Differential Equations*. Mathematics and its Applications (Soviet Series), vol. 48 (Kluwer, Dordrecht, 1991)
48. G. Greiner, Perturbing the boundary conditions of a generator. Houston J. Math **13**, 213–229 (1987)
49. G. Greiner, K. Kuhn, Linear and semilinear boundary conditions: the analytic case, in *Semigroup Theory and Evolution Equations*, ed. by P. Clément, B. de Pagter, E. Mitidieri. Lecture Notes in Pure and Applied Mathematics, vol. 135 (Dekker, Delft, 1991), pp. 193–211
50. S. Haeseler, M. Keller, D. Lenz, R. Wojciechowski, Laplacians on infinite graphs: Dirichlet and Neumann boundary conditions. J. Spectral Theory **2**, 397–432 (2012)
51. E. Hille, Notes on linear transformations. ii: analyticity of semi-groups. Ann. Math. **40**, 1–47 (1939)
52. E. Hölder, Entwicklungssätze aus der Theorie der zweiten Variation: Allgemeine Randbedingungen. Acta Math. **70**, 193–242 (1939)
53. X. Huang, M. Keller, J. Masamune, R.K. Wojciechowski, A note on self-adjoint extensions of the Laplacian on weighted graphs. J. Funct. Anal. **265**, 1556–1578 (2013)
54. A. Hussein, D. Mugnolo, Quantum graphs with mixed dynamics: the transport/diffusion case. J. Phys. A **46**, 235202 (2013)
55. P.E.T. Jorgensen, E.P.J. Pearse, Resistance boundaries of infinite networks, in *Random Walks, Boundaries and Spectra*, ed. by D. Lenz, F. Sobieczky, W. Woess. Progress in Probability, vol. 64 (Springer, Berlin, 2011), pp. 111–142
56. U. Kant, T. Klauß, J. Voigt, M. Weber, Dirichlet forms for singular one-dimensional operators and on graphs. J. Evol. Equ. **9**, 637–659 (2009)
57. L.V. Kantorovich, B.Z. Vulikh, Sur la représentation des opérations linéaires. Compos. Math. **5**, 119–165 (1937)
58. T. Kato, *Perturbation Theory for Linear Operators*. Grundlehren der mathematischen Wissenschaften, vol. 132 (Springer, Berlin, 1980)
59. M. Keller, D. Lenz, Unbounded Laplacians on graphs: basic spectral properties and the heat equation. Math. Model. Nat. Phenom. **5**, 198–224 (2010)
60. M. Keller, D. Lenz, Dirichlet forms and stochastic completeness of graphs and subgraphs. J. Reine Angew. Math. **666**, 189–223 (2012)
61. G. Kirchhoff, Ueber die Auflösung der Gleichungen, auf welche man bei der Untersuchung der linearen Vertheilung galvanischer Ströme geführt wird. Ann. Physik **148**, 497–508 (1847)
62. J. Kisynski, Semi-groups of operators and some of their applications to partial differential equationsm, in *Control Theory and Topics in Functional Analysis*, vol. III (IAEA, Vienna, 1976), pp. 305–405
63. V. Kostrykin, R. Schrader, Laplacian on metric graphs: eigenvalues, resolvents and semigroups, in *Quantum Graphs and Their Applications*, ed. by G. Berkolaiko, S.A. Fulling, P. Kuchment. Proceedings of the Conference on Quantum Graphs and Their Applications held in Snowbird, 2005. Contemporary Mathematics, vol. 415 (American Mathematical Society, Providence, 2006)
64. V. Kostrykin, J. Potthoff, R. Schrader, Contraction semigroups on metric graphs, in *Analysis on Graphs and its Applications*, ed. by P. Exner, J. Keating, P. Kuchment, T. Sunada, A. Teplyaev. Proceedings of Symposia in Pure Mathematics, vol. 77 (American Mathematical Society, Providence, 2008), pp. 423–458
65. V. Kostrykin, J. Potthoff, R. Schrader, Brownian motions on metric graphs. J. Math. Phys. **53**, 095206 (2012)
66. M. Kramar, D. Mugnolo, R. Nagel, Semigroups for initial-boundary value problems, in *Evolution Equations 2000: Applications to Physics, Industry, Life Sciences and Economics*, ed. by M. Iannelli, G. Lumer. Proceedings Levico Terme, 2000. Progress in Nonlinear Differential Equations, vol. 55 (Birkäuser, Basel, 2003), pp. 277–297
67. M. Kramar Fijavž, D. Mugnolo, E. Sikolya, Variational and semigroup methods for waves and diffusion in networks. Appl. Math. Optim. **55**, 219–240 (2007)

68. P. Kuchment, Quantum graphs I: some basic structures. Wave. Random Media **14**, 107–128 (2004)
69. P. Kuchment, Quantum graphs: an introduction and a brief survey, in *Analysis on Graphs and its Applications*, ed. by P. Exner, J. Keating, P. Kuchment, T. Sunada, A. Teplyaev. Proceedings of Symposia in Pure Mathematics, vol. 77 (American Mathematical Society, Providence, 2008), pp. 291–314
70. J.E. Lagnese, G. Leugering, E.J.P.G. Schmidt, *Modeling, Analysis, and Control of Dynamic Elastic Multi-Link Structures*. Systems and Control: Foundations and Applications (Birkhäuser, Basel, 1994)
71. J.L. Lions, *Equations Différentielles Opérationnelles et Problèmes aux Limites* (Springer, Berlin, 1961)
72. J.L. Lions, *Quelques Méthodes de Résolution des Problèmes aux Limites non Linéaires* (Dunod, Paris, 1969)
73. W. Littman, The wave operator and L_p norms. J. Math. Mech **12**, 55–68 (1963)
74. G. Lumer, Problème de Cauchy pour opérateurs locaux et "changement de temps". Ann. Inst. Fourier **25**, 409–446 (1975)
75. G. Lumer, Connecting of local operators and evolution equations on networks, in *Potential Theory*, ed. by F. Hirsch. Proceedings of Potential Theory Copenhagen 1979 (Springer, Berlin, 1980), pp. 230–243
76. A. Lunardi, *Analytic Semigroups and Optimal Regularity in Parabolic Problems*. Progress in Nonlinear Differential Equations and their Applications, vol. 16 (Birkhäuser, Basel, 1995)
77. G. Major, J.D. Evans, J.J. Jack, Solutions for transients in arbitrarily branching cables: I. Voltage recording with a somatic shunt. Biophys. J. **65**, 423–449 (1993)
78. A. Manavi, H. Vogt, J. Voigt, Domination of semigroups associated with sectorial forms. J. Oper. Theory **54**, 9–25 (2005)
79. A. McIntosh, On representing closed accretive sesquilinear forms as $(A^{\frac{1}{2}}u, A^{*\frac{1}{2}}v)$, in *Collège de France Seminar*, ed. by H. Brézis, J.L. Lions. Research Notes in Mathematics, vol. 70 (Pitman, Boston, MA, 1982), pp. 252–267
80. D. Mercier, S. Nicaise, Existence results for general system of differential equations on one-dimensional networks and prewavelets approximation. Disc. Cont. Dyn. Syst. **4**, 273–300 (1998)
81. I. Miyadera, *Nonlinear Semigroups*. Translations of Mathematical Monographs, vol. 109 (American Mathematical Society, Providence, 1992)
82. B. Mohar, The spectrum of an infinite graph. Linear Alg. Appl. **48**, 245–256 (1982)
83. U. Mosco, Approximation of the solutions of some variational inequalities. Ann. Sc. Norm. Super. Pisa Cl. Sci. III Serie **21**, 373–394 (1967)
84. D. Mugnolo, Gaussian estimates for a heat equation on a network. Networks Het. Media **2**, 55–79 (2007)
85. D. Mugnolo, A variational approach to strongly damped wave equations, in *Functional Analysis and Evolution Equations: The Günter Lumer Volume*, ed. by H. Amann, W. Arendt, M. Hieber, F. Neubrander, S. Nicaise, J. von Below (Birkhäuser, Basel, 2008), pp. 503–514
86. D. Mugnolo, Vector-valued heat equations and networks with coupled dynamic boundary conditions. Adv. Diff. Equ. **15**, 1125–1160 (2010)
87. D. Mugnolo, Parabolic theory of the discrete p-Laplace operator. Nonlinear Anal., Theory Methods Appl. **87**, 33–60 (2013)
88. D. Mugnolo, Asymptotics of semigroups generated by operator matrices. Arabian J. Math. (2014 in press)
89. D. Mugnolo, R. Nittka, Properties of representations of operators acting between spaces of vector-valued functions. Positivity **15**, 135–154 (2011)
90. D. Mugnolo, R. Nittka, Convergence of operator-semigroups associated with generalised elliptic forms. J. Evol. Equ. **12**, 593–619 (2012)
91. D. Mugnolo, R. Nittka, O. Post, Convergence of sectorial operators on varying Hilbert spaces. Oper. Matrices **7**, 955–995 (2013)

92. D. Mugnolo, S. Romanelli, Dirichlet forms for general Wentzell boundary conditions, analytic semigroups, and cosine operator functions. Electronic J. Differ. Equ. **118**, 1–20 (2006)
93. D. Mugnolo, S. Romanelli, Dynamic and generalized Wentzell node conditions for network equations. Math. Meth. Appl. Sci. **30**, 681–706 (2007)
94. S. Nicaise, Some results on spectral theory over networks, applied to nerve impulse transmission, in *Polynômes Orthogonaux et Applications*, ed. by C. Brezinski, A. Draux, A. P. Magnus, P. Maroni, A. Ronveaux. Proceedings of the Laguerre Symposium held at Bar-le-Duc, 1984. Lecture Notes in Mathematics, vol. 1171 (Springer, Berlin, 1985), pp. 532–541
95. R. Nittka, *Elliptic and Parabolic Problems with Robin Boundary Conditions on Lipschitz Domains*, Ph.D. thesis, Universität Ulm, 2010
96. E.M. Ouhabaz, Invariance of closed convex sets and domination criteria for semigroups. Potential Analysis **5**, 611–625 (1996)
97. E.M. Ouhabaz, *Analysis of Heat Equations on Domains*. London Mathematical Society Monograph Series, vol. 30 (Princeton University Press, Princeton, 2005)
98. K. Sato, T. Ueno, Multi-dimensional diffusion and the Markov process on the boundary. Kyoto J. Math. **4**, 529–605 (1965)
99. K. Schmüdgen, *Unbounded Self-adjoint Operators on Hilbert Space*.Graduate Texts in Mathematics, vol. 265 (Springer, Berlin, 2012)
100. A. Scott, *Neuroscience: A Mathematical Primer* (Springer, New York, 2002)
101. C. Seifert, J. Voigt, Dirichlet forms for singular diffusion on graphs. Oper. Matrices **5**, 723–734 (2011)
102. R.E. Showalter, *Monotone Operator in Banach Space and Partial Differential Equations*, Mathematical Surveys and Monographs, vol. 49 (American Mathematical Society, Providence, 1997)
103. U. Smilansky, Discrete graphs: a paradigm model for quantum chaos. Séminaire Poincaré **14**, 89–114 (2010)
104. P.M. Soardi, Rough isometries and Dirichlet finite harmonic functions on graphs. Proc. Amer. Math. Soc. **119**, 1239–1248 (1993)
105. K. Taira, *Semigroups, Boundary Value Problems and Markov Processes* (Springer, Berlin, 2004)
106. J.L. Vázquez, E. Vitillaro, Heat equation with dynamical boundary conditions of reactive type. Comm. Partial Differ. Equations **33**, 561–612 (2008)
107. A.D. Venttsel', On boundary conditions for multidimensional diffusion processes. Theor. Probab. Appl. **4**, 164–177 (1960)
108. M. Weber, On occupation time functionals for diffusion processes and birth-and-death processes on graphs. Ann. Appl. Probability **11**, 544–567 (2001)
109. W. Woess, *Random Walks on Infinite Graphs and Groups*. Cambridge Tracts in Mathematics, vol. 138 (Cambridge University Press, Cambridge, 2000)
110. M. Yamasaki, Extremum problems on an infinite network. Hiroshima Math. J. **5**, 223–250 (1975)

Chapter 7
Evolution Equations Associated with Self-Adjoint Operators

The theory of forms presented in Chap. 6 was originally developed in order to extend the study of parabolic problems beyond the setting of the Spectral Theorem, in much the same way the Lax–Milgram Lemma extended the applicability of the Riesz–Fréchet Theorem. This program was successful: Nowadays many relevant results on linear parabolic problems have been extended to the non-self-adjoint case by form methods and further results depend on much deeper techniques, including sophisticated functional calculi that conveniently replace the Spectral Theorem and whose exposition goes beyond the scope of our book, cf. [50].

However, some properties still seem to be typical for self-adjoint operators. In this chapter we collect a few results that are either specific for the self-adjoint case or whose general validity does not seem to be known yet. In the remainder of the chapter we apply them to different classes of evolution equations on networks.

7.1 The Spectral Theorem and Dirichlet Forms

Throughout this section we assume that

V, H are separable, complex Hilbert spaces with V densely and continuously embedded in H.

We recall the *Spectral Theorem* for (possibly unbounded) self-adjoint operators.

Theorem 7.1. *Let A be a self-adjoint operator on H. Then there exist a σ-finite measure space X, a measurable function $q : X \to \mathbb{R}$, and a unitary operator $U : H \to L^2(X)$ such that $A = U^{-1}M_q U$, where M_q is the multiplication operator defined in Example 4.17, i.e.,*

$$D(A) = \{u \in H : Uu \in D(M_q)\}, \qquad Au = U^{-1}(q \cdot Uu).$$

D. Mugnolo, *Semigroup Methods for Evolution Equations on Networks*, Understanding Complex Systems, DOI 10.1007/978-3-319-04621-1_7,

Furthermore, the self-adjoint operator A is dissipative (and hence by definition negative semidefinite) if and only if $q(x) \le 0$ for a.e. $x \in X$.

If A is self-adjoint and dissipative, so that $-q$ is a positive function, the *square root* of $-A$ is

$$D\left(\sqrt{-A}\right) := \left\{u \in H : \sqrt{-q} \cdot (Uu) \in L^2(X)\right\},$$
$$\sqrt{-A} := U^{-1}\left(\sqrt{-q} \cdot (Uu)\right).$$

Lemma 7.2. *Let $a : V \times V \to \mathbb{C}$ be a continuous, H-elliptic sesquilinear form. If a is symmetric, then the associated operator A is self-adjoint.*

Conversely, let A be a self-adjoint and ω-quasi-dissipative operator on H. Then

$$a(u,v) := \left(\sqrt{-A-\omega\,\mathrm{Id}}\, u \middle| \sqrt{-A-\omega\,\mathrm{Id}}\, v\right)_H, \qquad u,v \in V := D(\sqrt{-A-\omega\,\mathrm{Id}}),$$

defines a continuous, H-elliptic, symmetric sesquilinear form $a : V \times V \to \mathbb{C}$.

If V is compactly embedded in H, then X can be chosen to be a subset of $\mathbb{N}$ (actually a finite subset, if $\dim H < \infty$) with the standard atomic measure and q is simply the sequence of eigenvalues of A. Hence we obtain the following.

Corollary 7.3. *Let $a : V \times V \to \mathbb{C}$ be a symmetric, continuous, H-elliptic sesquilinear form with associated operator A. If the embedding of V in H is compact, then the following assertions hold.*

(1) A has compact resolvent and hence its spectrum consists solely of eigenvalues of finite multiplicity.
(2) Ordering the eigenvalues in decreasing order and counting multiplicities, the k-th largest eigenvalue $\lambda_k(A)$ is given by Courant's minimax formula

$$-\lambda_k = \min_{\substack{E \subset V \\ \dim E = k}} \max_{\substack{u \in E \\ u \neq 0}} \frac{a(u,u)}{\|u\|_H^2}, \qquad k \in \mathbb{N}, \tag{7.1}$$

i.e., E runs over the k-dimensional subspaces of V. Moreover, $\lim_{n\to\infty} \lambda_n = -\infty$ if $\dim H = \infty$.
(3) There exists an orthonormal basis of H consisting of eigenvectors e_n of A associated with λ_n such that $e_n \in V$ and for which

$$D(A) = \left\{u \in H : \sum_{n=1}^{\dim H} \lambda_n^2 (u|e_n)_H^2 < \infty\right\}, \qquad Au = \sum_{n=1}^{\dim H} \lambda_n (u|e_n)_H e_n.$$

The fraction in (7.1) is usually referred to as *Rayleigh quotient*.

While all quasi-dissipative self-adjoint operators come from a form, there exist self-adjoint operators that are not quasi-dissipative—an example is the momentum operator with periodic boundary conditions, cf. Definition 2.44. We rather study them exploiting directly the spectral theorem.

Evolution equations involving multiplication operators can usually be explicitly solved, e.g.,

$$e^{tM_q}u(x) = e^{tq(x)}u(x), \qquad x \in X.$$

Furthermore, these solutions are well-behaved under isomorphic transformations U: e.g.,

$$e^{tU^{-1}M_qU} = U^{-1}e^{tM_q}U, \qquad t \geq 0.$$

Clearly, unitarity of the semigroup is also preserved if (and only if) U is additionally unitary.

Remark 7.4. The spectral theorem suggests how to solve an abstract Cauchy problem associated with a self-adjoint operator A: It suffices to define an exponential function by

$$e^{tA}u := U^{-1}(e^{tq} \cdot Uu), \qquad t \geq 0. \tag{7.2}$$

In particular, these formulae show that $(e^{tA})_{t\geq 0}$ is contractive (resp., uniformly exponentially stable) if and only if $q(x) \leq 0$ (resp., if and only if $q(x) \leq q_0$ for some $q_0 < 0$) for a.e. $x \in X$.

If furthermore $A - \omega\,\mathrm{Id}$ is a negative definite operator, then (7.2) defines a bounded analytic C_0-semigroup of angle $\frac{\pi}{2}$ such that

$$\|e^{tA}\|_{\mathcal{L}(X)} \leq e^{\omega \operatorname{Re} t} \qquad \text{for all } t \in \mathbb{C} \text{ s.t. } \operatorname{Re} t > 0.$$

Corollary 7.5. *Under the assumptions of Corollary 7.3 if we denote by P the orthogonal projector onto the eigenspace associated with $s(A)$, then*

$$\|e^{-s(A)t}e^{tA} - P\|_{\mathcal{L}(H)} \leq Me^{-\epsilon t} \qquad \textit{for all } t \geq 0,$$

where ϵ is the largest eigenvalue of A different from $s(A)$.

Proof. It follows directly from Corollary 7.3 that for all $u \in H$

$$e^{tA}u = \sum_{n=1}^{\dim H} e^{t\lambda_n}(u|e_n)_H e_n, \qquad t \geq 0. \tag{7.3}$$

Now, $s(A)$ is an eigenvalue and $A - s(A)\,\mathrm{Id}$ is a non-invertible self-adjoint operator. By (7.3), the semigroup generated by $A - s(A)\,\mathrm{Id}$ leaves invariant its eigenspace associated with the operator 0, hence it follows by (7.3) that

$$\left(e^{-s(A)t}e^{tA} - P\right)u = \sum_{n=m_1+1}^{\dim H} e^{t(\lambda_n - s(A))}(u|e_n)_H e_n, \qquad t \geq 0, \qquad (7.4)$$

where m_1 is the multiplicity of the largest eigenvalue $s(A)$. □

Example 7.6. Let G be a finite oriented graph. Let us assume for simplicity that G is not weighted, i.e., $\mu \equiv 1$.

(1) It is clear by definition that the discrete Laplacian $\mathcal{L}$ is symmetric and hence also self-adjoint, as it is bounded, and furthermore $-\mathcal{L}$ is dissipative. Since G is finite, by Lemma 2.13 $s(\mathcal{L}) = 0$, hence by Corollary 7.5 $(e^{-t\mathcal{L}})_{t\geq 0}$ converges exponentially to a linear combination of the characteristic function of each connected component.
(2) Similarly, it is clear by definition that the signless Laplacian $\mathcal{Q}$ is self-adjoint. By Lemma 2.13 $s(\mathcal{Q}) = 0$, hence by Corollary 7.5 $(e^{-t\mathcal{Q}})_{t\geq 0}$ converges exponentially to a linear combination of the β eigenvectors associated with the eigenvalue 0. Each of these eigenvectors is a function that assigns a value ± 1 to each node of the k-th bipartite connected component V_k, $1 \leq k \leq \beta$, depending on whether the node belongs to V_k^+ or V_k^-. Hence, a convenient method to check whether a given connected graph is bipartite or not is to check whether $(e^{-t\mathcal{Q}}\mathbf{1})_{t\geq 0}$ converges to 0 or rather to a ± 1-function.

More generally, self-adjointness allows for the definition of a "natural" functional calculus for A, by acting on the function q: In this way properties of $f(A)$ can be read off $f \circ q$. Besides the cases of the square root $\sqrt{-\cdot}$ and the exponential $e^{t\cdot}$, arbitrary continuous functions may be applied—e.g. $\cos(t\sqrt{-\cdot})$, yielding that each self-adjoint operator generates a C_0-cosine operator function. As a further instance we explicitly mention *Stone's Theorem.*

Theorem 7.7. *Let A be a densely defined operator on a Hilbert space. Then the following assertions are equivalent.*

(a) iA is the generator of a C_0-group of unitary operators.
(b) A is self-adjoint.

Example 7.8. The (time-dependent) *Schrödinger equation* is

$$i\hbar\frac{\partial\psi}{\partial t}(t,x) = \mathfrak{H}\psi(t,x), \qquad t \in \mathbb{R}, \qquad (7.5)$$

where the Hamiltonian $\mathfrak{H}$ is a self-adjoint operator acting on some Hilbert space. In the easiest, one-dimensional instance one studies the "free Hamiltonian" $\mathfrak{H} = \Delta$ on $L^2(\mathbb{R})$. Then it can be proved as usual that $\mathfrak{H}$ comes from a symmetric form and is thus self-adjoint: One concludes from Stone's Theorem that the initial value problem associated with (7.8) is governed by $(e^{it\Delta})_{t\in\mathbb{R}}$, hence it is in particular well-posed. Indeed, it is easily seen that (4.6) can be modified to yield for all $u \in L^2(\mathbb{R})$ the formula

$$e^{it\Delta}u(x) = (G_1(it,\cdot) * u)(x) = \frac{1}{(4\pi it)^{\frac{1}{2}}} \int_{\mathbb{R}} e^{-\frac{\|x-y\|^2}{4it}} u(y)dy, \qquad t \neq 0,\ x \in \mathbb{R}, \tag{7.6}$$

i.e., one simply replaces real time t by purely imaginary time it in the Gaussian kernel G_1 (concerning $i^{\frac{1}{2}}$, we take the square root in the first quadrant of the complex plane). We extend this formula letting

$$e^{i0\Delta} := \mathrm{Id} :$$

This defined a unitary C_0-group $(e^{it\Delta})_{t\in\mathbb{R}}$.

Let us in the following assume that

$$\boxed{(\Omega, d, \tilde{\mu}) \text{ is a metric measure space and } H := L^2(\Omega, \tilde{\mu}).}$$

Definition 7.9. A sesquilinear mapping $a : V \times V \to \mathbb{C}$ is called a *Dirichlet form* if it is continuous, H-elliptic and symmetric and if additionally the associated semigroup is sub-Markovian.

It is called *local* if for all open, disjoint $U, V \subset \Omega$

$$a(u, v) = 0, \qquad \text{whenever } u, v \in L^2(\Omega) \text{ with } \operatorname{supp} u \subset U, \operatorname{supp} v \subset V. \tag{7.7}$$

Example 7.10. Because for all $u \in W^{1,2}$-functions the support of u' is not larger than the support of u, the form $a_{\mathcal{W}}$ in (6.33) is certainly local whenever the matrix $\mathcal{W}$ is diagonal. By Theorems 6.71 and 6.72, $a_{\mathcal{W}}$ is also a Dirichlet form.

The form a in (6.23) is also Dirichlet, by Theorem 6.54, but it is non-local, as one sees even in the simple case of

$$\mathsf{V} = \{\mathsf{v}_1, \mathsf{v}_2\}, \qquad \mathsf{E} = \{(\mathsf{v}_1, \mathsf{v}_2)\}, \qquad f(\mathsf{v}_1) = 0,\ f(\mathsf{v}_2) = 1,\ g(\mathsf{v}_1) = 1,\ g(\mathsf{v}_2) = 0.$$

However, whenever G is unweighted a clearly satisfies the following weaker locality property (for $\delta = 2$): There exists $\delta > 0$ such that for all open, disjoint $U, V \subset \Omega$ with $d(U, V) \geq \delta$

$$a(u, v) = 0, \qquad \text{whenever } u, v \in L^2(\Omega) \text{ with } \operatorname{supp} u \subset U, \operatorname{supp} v \subset V.$$

Definition 7.11. Let $(T(t))_{t\geq 0}$ be a C_0-semigroup on $L^2(\Omega)$.

- If there exist $b, c > 0$ such that for all open $U, V \subset \Omega$

$$|(T(t)u|v)_{L^2}| \leq c e^{-\frac{\operatorname{dist}(U,V)^2}{bt}} \|u\|_{L^2} \|v\|_{L^2},$$
$$\text{for all } u, v \in L^2(\Omega) \text{ s.t. } \operatorname{supp} u \subset U, \operatorname{supp} v \subset V,\ t > 0.$$

  then $(T(t))_{t\geq 0}$ is said to satisfy *Davies–Gaffney estimates.*

- Let $(T(t))_{t\geq 0}$ have an integral kernel $(K_t)_{t\geq 0}$ as in (6.36). If there exist constants $\tilde{b}, \tilde{c} > 0$ and $d \in \mathbb{N}$ such that

$$0 \leq K_t(x,y) \leq \tilde{c}t^{-\frac{d}{2}} e^{-\frac{|x-y|^2}{\tilde{b}t}} \qquad \text{for all } x,y \in \Omega,\ t > 0, \tag{7.8}$$

then $(T(t))_{t\geq 0}$ is said to satisfy *d-dimensional Gaussian estimates*.

The latter name is due to fact that the term on the right hand side is, up to rescaling factors, just the Gaussian kernel introduced in Example 4.15.

Proposition 7.12. *The semigroup associated with a local Dirichlet form satisfies Davies–Gaffney estimates.*

Pointwise estimates of Gaussian type may appear stronger than integrated estimates like those of Davies–Gaffney. However, the following holds.

Proposition 7.13. *If $(T(t))_{t\geq 0}$ satisfies Davies–Gaffney estimates and it is ultracontractive of dimension d, then it satisfies d-dimensional Gaussian estimates.*

Proposition 7.14. *Let an operator A on $L^2(\Omega)$ be self-adjoint and dissipative. Then* (ACP2) *enjoys finite speed of propagation if and only if $(e^{tA})_{t\geq 0}$ satisfies Davies–Gaffney estimates.*

7.2 Self-Adjoint Operators on Networks and Evolution Equations

We assume throughout this section that

$\mathfrak{G}$ is the metric graph associated with a locally finite, weighted oriented graph $\mathsf{G} = (\mathsf{V}, \mathsf{E}, \mu)$.

The difference operators we have considered in Sect. 6.4 satisfy the range condition in the Hille–Yosida Theorem, hence they are self-adjoint if and only if they are symmetric—their symmetry can in turn be checked directly, already looking at the formal definitions in Sect. 2.1. For this reason, in this chapter we mostly focus on differential operators.

7.2.1 Diffusion Equation

We come back to the study of diffusion equations on graphs and investigate the spectrum of Δ, the second derivative standard node conditions on $L^2(\mathfrak{G})$: Knowing its eigenvalues and its eigenvectors we may then by (7.3) solve the heat equation in terms of a series. If $\mathcal{W}$ is self-adjoint, then the sesquilinear form in (6.31) is symmetric and we conclude that the associated Laplacian Δ is self-adjoint. We simultaneously discuss the Laplacian $\tilde{\Delta}$ with anti-Kirchhoff node conditions introduced in Remark 6.81, in order to emphasize the symmetry between the spectra of these both operators (Figs. 7.1 and 7.2). Throughout this section we restrict to the case of equilateral metric graphs.

Theorem 7.15. *Let $\mathfrak{G}$ be equilateral. Let us denote by κ the number of its connected components and by β the number of its connected components that are additionally bipartite. Then for $\lambda > 0$ the following assertions hold.*

- *If u is an eigenvector of Δ with node conditions* (Cc) – (Kc) *with associated eigenvalue $-\lambda$, then the corresponding vector $u_{|\mathsf{V}} \in \mathbb{C}^{\mathsf{V}}$ of node values is a (right) eigenvector of the transition matrix $\mathcal{T}$: more precisely,*

$$\mathcal{T} u_{|\mathsf{V}} = \cos\sqrt{\lambda}\, u_{|\mathsf{V}}. \tag{7.9}$$

  Conversely, if $\cos\sqrt{\lambda}$ *is an eigenvalue of $\mathcal{T}$ of multiplicity* $\text{mult}(\cos\sqrt{\lambda})$ *and ψ is an associated eigenvector, then λ is an eigenvalue of $-\Delta$ with standard boundary conditions and $u_{|\mathsf{V}} := \psi$ is the vector of node values of some eigenvector associated with the eigenvalue λ. The multiplicities are*

$$\text{mult}(\lambda) = \begin{cases} \kappa & \text{if } \lambda = 0, \\ \text{mult}(\cos\sqrt{\lambda}) & \text{if } \sin\sqrt{\lambda} \neq 0, \\ |\mathsf{E}| - |\mathsf{V}| + 2\kappa & \text{if } \cos\sqrt{\lambda} = 1,\ \lambda > 0, \\ |\mathsf{E}| - |\mathsf{V}| + 2\beta & \text{if } \cos\sqrt{\lambda} = -1,\ \lambda > 0. \end{cases}$$

- *Let $\tilde{u} := \partial_{c^2} u \in \mathbb{C}^{\mathsf{V}}$, where u is an eigenvector of $\tilde{\Delta}$ with associated eigenvalue $-\lambda$. Then*

$$\mathcal{T}\tilde{u} = -\cos\sqrt{\lambda}\, \tilde{u}. \tag{7.10}$$

  Conversely, if $\lambda > 0$ and $-\cos\sqrt{\lambda}$ is an eigenvalue of $\mathcal{T}$ admitting the eigenvector $\psi \in \mathbb{C}^n$, then λ is an eigenvalue of $-\tilde{\Delta}$ with anti-Kirchhoff conditions and $\psi = \partial_{c^2} u$ for some eigenfunction u belonging to λ. The multiplicities of the eigenvalues are

$$\text{mult}(\lambda) = \begin{cases} |\mathsf{E}| - |\mathsf{V}| + \beta & \text{if } \lambda = 0, \\ \text{mult}(-\cos\sqrt{\lambda}) & \text{if } \sin\sqrt{\lambda} \neq 0, \\ |\mathsf{E}| - |\mathsf{V}| + 2\kappa & \text{if } \cos\sqrt{\lambda} = -1,\ \lambda > 0, \\ |\mathsf{E}| - |\mathsf{V}| + 2\beta & \text{if } \cos\sqrt{\lambda} = 1,\ \lambda > 0. \end{cases}$$

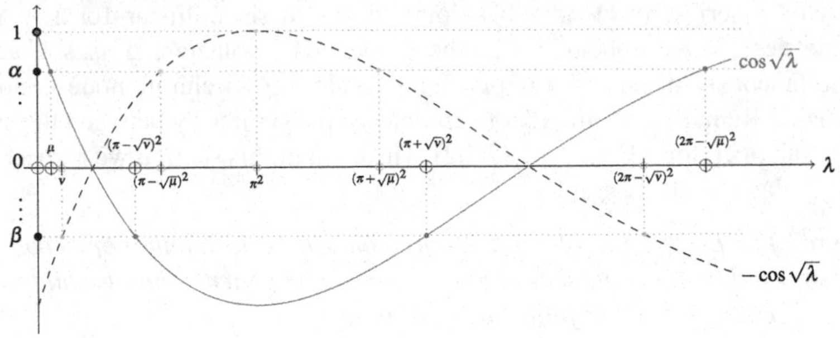

Fig. 7.1 On the abscissa, the eigenvalues of $-\Delta$ are plotted (*solid line*) in correspondence with the associated eigenvalues of $\mathcal{T}$ on the ordinate axis. This plot reflects the case of a non-bipartite graph, for which the spectrum of $\mathcal{T}$ is not symmetric with respect to 0. The eigenvalues of $-\widetilde{\Delta}$, (minus) the Laplacian with anti-Kirchhoff node conditions, are also plotted (*dashed line*)

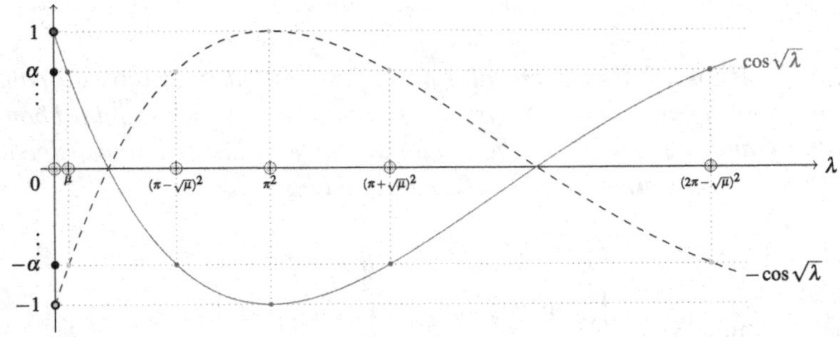

Fig. 7.2 This plot corresponds to the bipartite case, when the spectrum of $\mathcal{T}$ is symmetric about 0, cf. Remark 2.17

Example 7.16. The most relevant limitation to the application of Theorem 7.15 is of course that only equilateral graphs are allowed. Under this restrictive assumption, there are quite a few consequences that are easy to deduce but nevertheless worth to remark. For example, the formulae for the multiplicities show that simplicity of all eigenvalues implies that the graph is connected and $|\mathsf{V}| - |\mathsf{E}| = 1$—i.e., by (A.3), that the graph G is a tree. This is true both in the case of standard node conditions (Cc) – (Kc) and in the case of anti-Kirchhoff conditions. If we also know that all eigenvalues of the normalized Laplacian of the graph are simple, then we can conclude that all eigenvalues of the second derivative with both standard and anti-Kirchhoff node conditions have multiplicity one—not surprisingly, this is e.g. the case for path graphs.

The above description of the spectrum of Δ has further interesting consequences: Let for instance G be finite and connected. The heat equation on $\mathfrak{G}$ with node conditions (Cc) – (Kc), i.e., the abstract Cauchy problem associated with Δ, is governed by a C_0-semigroup. By dissipativity of Δ and because $\mathbf{1} \in D(\Delta)$ with $\Delta\mathbf{1} = 0$, $s(\Delta) = 0$ and by Corollary 7.5 the rate of convergence of $(e^{t\Delta})_{t\geq 0}$

towards the projector onto the first eigenspace (i.e., the space of constant functions) is directly influenced by the second largest eigenvalue $\lambda_2(\Delta)$ of Δ, hence by the second largest eigenvalue $\lambda_2(\mathcal{T})$ of $\mathcal{T}$. Further properties hold; e.g., it follows from Remark 2.17 that the $(e^{t\Delta})_{t\geq 0}$ converges exponentially to an equilibrium with the same rate (namely, $e^{-\lambda_2 t} = e^{-\frac{\pi^2}{4}t}$) on all metric graphs over complete bipartite graphs.

Remark 7.17. By Theorem 7.15 and (2.23), spectral properties of the normalized Laplacian $\mathcal{L}_{\text{norm}}$ can be easily transferred to properties of Δ. One of the most interesting consequences of this general principle becomes clear when one observes that in the last few years many interlacing results have been obtained for the eigenvalues of $\mathcal{L}$ under graph operations. For example, it is proved in [15, Theorem 1.2] that if

- Γ_1 is a graph on n vertices,
- $\tilde{\Gamma}_2$ is a subgraph of Γ_1,
- t is the number of isolated vertices of $\tilde{\Gamma}_2$,
- Γ_2 is the complement of $\tilde{\Gamma}_2$ in Γ_1, i.e., the graph obtained by deleting from Γ_1 the edges of $\tilde{\Gamma}_2$,
- $\mu_j^{(1)}$ and $\mu_j^{(2)}$ denote the eigenvalues of the normalized Laplacian of Γ_1, Γ_2, respectively,

then for $k = 1, \ldots, n$

$$\mu_{k-t+1}^{(1)} \leq \mu_k^{(2)} \leq \begin{cases} \mu_{k+t-1}^{(1)}, & \text{if } \Gamma_2 \text{ is bipartite,} \\ \mu_{k+t}^{(1)}, & \text{otherwise,} \end{cases}$$

where

$$\mu_{-t+1}^{(1)} = \ldots = \mu_0^{(1)} := 0 \quad \text{and} \quad \mu_{n+1}^{(1)} = \ldots = \mu_{n+t}^{(1)} := 2.$$

(An analogous relation holds if a graph is added, instead of subtracted [15, Corollary 1.4]). Yet more refined results related to more subtle structures (like coverings and spanning subgraphs) have been obtained in [16, 22, 40]. In view of Theorems 7.15 it is possible to translate all these interlacing results for the spectrum of $\mathcal{L}_{\text{norm}}$ into interlacing results for spectral subsets of Δ.

In the case G has finite volume, the embedding of the form domain V in $H = L^2(\mathfrak{G})$ is even a Hilbert–Schmidt operator, and in particular Δ has purely point spectrum and (7.2) reads

$$e^{t\Delta}u = \sum_{n\in\mathbb{N}} e^{t\lambda_n}(u|e_n)_H e_n, \qquad t \geq 0.$$

We know from Lemma 6.20 that $(e^{t\Delta})_{t\geq 0}$ is then of trace class. One can now recover relevant information, as the following *trace formula* due to J.-P. Roth shows.

Theorem 7.18. *Let* G *be finite, with* |V| *nodes and* |E| *edges. Then for all* $t > 0$

$$\operatorname{Tr} e^{t\Delta} := \sum_{n\in} e^{\lambda_n t} = \frac{\mathrm{vol}_\mu(\mathsf{G})}{2\sqrt{\pi t}} + \frac{|\mathsf{V}|-|\mathsf{E}|}{2} + \frac{1}{2\sqrt{\pi t}} \sum_{\mathsf{C}\in\mathfrak{C}} \sigma(\mathsf{C})\,\mathrm{len}_\mu(\mathrm{Gen}(\mathsf{C}))e^{-\frac{\mathrm{vol}_\mu(\mathsf{C})^2}{4t}},$$

where the series are absolutely convergent.

Here $\mathfrak{C}$ denotes the set of all circuits of $\overline{\mathsf{G}}$, where $\overline{\mathsf{G}}$ was introduced before Definition 2.36, and $\sigma(\mathsf{C})$ is the number defined in (2.35) based on the scattering matrix $\mathcal{S}$ of $\overline{\mathsf{G}}$. Recall that for any circuit C we denote by $\mathrm{len}_\mu(\mathrm{Gen}(\mathsf{C}))$ its length and by Gen(C) its gene—i.e., the circuit that yields C after some finite number of reiterations—, cf. Remark A.18 and Definition A.3, respectively.

Example 7.19. Since repetitions are allowed, $(\mathsf{e},\bar{\mathsf{e}},\mathsf{e},\bar{\mathsf{e}},\mathsf{e},\bar{\mathsf{e}})$ defines in a natural way a closed path in $\overline{\mathsf{G}}$, its gene being the circuit $\mathsf{e},\bar{\mathsf{e}}$. On the other hand, the circuit defined by $(\mathsf{e},\bar{\mathsf{e}},\mathsf{e},\bar{\mathsf{e}},\mathsf{f},\bar{\mathsf{f}},\mathsf{e},\bar{\mathsf{e}})$ agrees with its own gene.

Hence for a circuit associated with $(\mathsf{e}_1,\ldots,\mathsf{e}_n)$ we have in particular

$$\sigma(\mathsf{C}) = \sigma_{\mathsf{e}_n\mathsf{e}_1}\sigma_{\mathsf{e}_1\mathsf{e}_2}\cdots\sigma_{\mathsf{e}_{n-1}\mathsf{e}_n}.$$

Clearly, summing over $\mathfrak{C}$ and not only over the set of all genes is essential in order to have an infinite sum on the right hand side (as one expects, since also the sum on the left hand side consists of infinitely many terms).

Roth also found a fundamental solution of the heat equation. It shows in turn that diffusion on metric graphs consists of a weighted overlapping of diffusions along infinitely many paths.

Theorem 7.20. *For any two* $x,y\in\mathfrak{G}$ *let* $\mathfrak{P}_{x,y}$ *be the set of all paths from* x,y. *Then*

$$e^{t\Delta}u(x) = \int_{\mathfrak{G}} K_t(x,y)u\,dy \qquad \text{for all } t>0 \text{ and } x\in\mathfrak{G},$$

where

$$K_t(x,y) := \sum_{\mathsf{P}\in\mathfrak{P}_{x,y}} \sigma(\mathsf{P})G_1(t,\mathrm{dist}_\mu(x,y)), \qquad t>0,\ x,y\in\mathfrak{G}. \tag{7.11}$$

Here G_1 is the one-dimensional Gaussian kernel in (4.5), dist_μ is the metric distance introduced in Definition 3.14 and, again, $\sigma(\mathsf{P})$ is defined as in (2.35).

In the special case of two points x,y on an unweighted path graph the only elements of $\mathfrak{P}_{x,y}$ that yield a non-zero contribution are those for which no reflection ever takes place, since by Definition 2.36 $\sigma_{\mathsf{e}\bar{\mathsf{e}}} = 0$ whenever $\mathsf{e}_{\mathrm{term}}$ is an inessential node, i.e., whenever it has exactly two incident edges. We conclude that only the shortest path between x,y yields a contribution, and we hence recover the usual setting of Example 4.15 as a special case.

Example 7.21. It is remarkable that non-trivial qualitative information about the spectrum of elliptic operators on networks becomes available by a direct application of the minimax formula (7.1) in Lemma 7.3, without necessarily applying Theorem 7.15 to determine eigenvalues explicitly. First of all, it is clear that the form domain $W_Y^{1,2}\left((0,1);\ell^2_\mu(\mathsf{E})\right)$ is contained for Y as in (6.29)—indeed, for *every* closed subspace of $\ell^2_\mu(\mathsf{E})\times\ell^2_\mu(\mathsf{E})$!—between $W_0^{1,2}\left((0,1);\ell^2_\mu(\mathsf{E})\right)$ and $W^{1,2}\left((0,1);\ell^2_\mu(\mathsf{E})\right)$. Accordingly, the operator A associated with $a_{\mathcal{W}}$ with form domain $W_Y^{1,2}\left((0,1);\ell^2_\mu(\mathsf{E})\right)$—i.e., the Laplacian with standard node conditions—is always included, in the sense of self-adjoint operators, between the Laplacians with Dirichlet and Neumann boundary conditions on $L^2\left((0,1);\ell^2_\mu(\mathsf{E})\right)$. The same interlacing holds for the k-th eigenvalue λ_k, for each $k\in\mathbb{N}$.

It is perhaps more interesting to compare Laplacians with standard node conditions on different networks. Let us mention three easy examples.

- Whenever we consider the Laplace–Beltrami matrix $\mathcal{L}$ on a given finite weighted oriented graph G, adding an edge obviously does not change the state space $\ell^2_{\deg_\gamma}(\mathsf{V})$, but it does certainly enlarge the numerator of the Rayleigh quotient. This modification of the graph hence yields a smaller $\lambda_2(-\mathcal{L})$. The same assertion holds for $\mathcal{Q}, \mathcal{L}_{\rm norm}, \mathcal{Q}_{\rm norm}$.
- Let us now consider a metric graph $\mathfrak{G}$ and $\mathsf{V}_0\subset\mathsf{V}$. For the hybrid Laplacian $\mathbb{\Delta}$ on $\mathfrak{G}$ with standard node conditions on V_0^C and dynamic ones on V_0, the denominator of the quotient in (7.1) certainly decreases if one replaces the standard node condition on some $\mathsf{v}\in\mathsf{V}_0^C$ by a dynamic one in the same node; hence, the new hybrid Laplacian thus obtained has a smaller $\lambda_2(\mathbb{\Delta})$.
- The dependence of the eigenvalues of the Laplacian Δ with standard node conditions on deleting of edges is rather subtle. However, it is not difficult to see that the quotient in (7.1) is well-behaved under certain different graph operations. If e.g. two different nodes $\mathsf{v},\mathsf{w}\in\mathsf{V}$ are identified yielding a new weighted oriented graph G', then it is clear that $W^{1,2}(\mathfrak{G}')$ is a closed subspace of $W^{1,2}(\mathfrak{G})$. Again, Courant's minimax principle says that $\lambda_2(\Delta)$ becomes smaller than it was in $\mathfrak{G}$.

In view of Corollary 7.5, all these observations can be translated into assertions on the rate of convergence of diffusion equations on different networks. In this way, the connectivity of the network is enhanced, and the corresponding diffusive system convergence more quickly to equilibrium (recall that all λ_2 is a strictly negative number, since all our operators are dissipative).

Are there any further self-adjoint realizations of the Laplacian, or more general elliptic operators, on a metric graph? Certainly. Indeed, by Lemma 7.2 any operator associated with a densely defined, elliptic continuous form that is symmetric is also self-adjoint; and by Lemma 6.63 the standard form $a_{\mathcal{W}}$ is symmetric if and only if $\mathcal{W}$ is a self-adjoint operator on $\ell^2_{\deg_\mu}(\mathsf{V})$. This is not the end of the story, however.

Indeed, we have already mentioned in Sect. 6.5 that the node conditions (Cc′) – (KRc′) may well be defined not only for Y defined as in (6.29), but for a general closed subspace of $\ell^2_\mu(\mathsf{E}) \times \ell^2_\mu(\mathsf{E})$: It is immediate to check that also in this case $a_{\mathcal{W}}$, formally defined in the same way, is symmetric. The following sharper result is due to P. Kuchment.

Theorem 7.22. *Let Y be a closed subspace of $\ell^2_\mu(\mathsf{E}) \times \ell^2_\mu(\mathsf{E})$ and a self-adjoint bounded linear operator $\mathcal{W}$ on Y. Then the operator associated with $a_{\mathcal{W}}$ is self-adjoint.*

If additionally G is finite, then conversely for any self-adjoint realization of an elliptic operator $\nabla(c\nabla)$ there exist a closed subspace Y of $\ell^2_\mu(\mathsf{E})\times\ell^2_\mu(\mathsf{E})$ and a self-adjoint operator (i.e., a Hermitian matrix) $\mathcal{W} \in \mathcal{L}(Y)$ such that the node conditions of $\nabla(c\nabla)$ can be written as (Cc′) – (KRc′).

The very convenient variational structure of this class of problems can be easily exploited to investigate spectral properties of the vector-valued elliptic problem

$$\begin{cases} (cu')'(x) = \lambda u(x), \qquad x \in (0,1) \\ \begin{pmatrix} u(1) \\ u(0) \end{pmatrix} \in Y, \\ \begin{pmatrix} (cu')(1) \\ -(cu')(0) \end{pmatrix} + \mathcal{W}\begin{pmatrix} u(1) \\ u(0) \end{pmatrix} \in Y^\perp. \end{cases} \tag{EP$_{Y,\mathcal{W}}$}$$

Proposition 7.23. *Let Y be a closed subspace of $\ell^2_\mu(\mathsf{E}) \times \ell^2_\mu(\mathsf{E})$ and a self-adjoint bounded linear operator $\mathcal{W}$ on Y. The spectrum of the operator $\nabla(c\nabla)$ associated with $a_{\mathcal{W}}$ is real; it is contained in $(-\infty, 0]$ if R is positive semidefinite. If H is finite-dimensional, then $\nabla(c\nabla)$ has pure point spectrum. Then the following assertions hold.*

1) 0 is an eigenvalue of (EP$_{Y,\mathcal{W}}$) if and only if

$$H_Y := \{(A,B) \in \ell^2_\mu(\mathsf{E}) \times \ell^2_\mu(\mathsf{E}) : A = B\} \cap Y \neq \{0\} \quad \text{and} \quad \mathcal{W} = 0,$$

and in this case the multiplicity of 0 agrees with $\dim H_Y \leq \dim \ell^2_{\deg_\mu}(\mathsf{V})$.

2) If $\mathcal{W}$ is positive semidefinite, then $\lambda > 0$ is an eigenvalue of (EP$_{Y,\mathcal{W}}$) if and only if the space $H_{Y,\mathcal{W}}$ of all solutions $(A,B) \in \ell^2_{\deg_\mu}(\mathsf{V}) \times \ell^2_{\deg_\mu}(\mathsf{V})$ of the system

$$\begin{cases} P_{Y^\perp}\begin{pmatrix} A \\ A\cos\sqrt{\lambda} + B\sin\sqrt{\lambda} \end{pmatrix} = 0 \\ P_Y\left(\sqrt{\lambda}\begin{pmatrix} B \\ A\sin\sqrt{\lambda} - B\cos\sqrt{\lambda} \end{pmatrix} - R\begin{pmatrix} A \\ A\cos\sqrt{\lambda} + B\sin\sqrt{\lambda} \end{pmatrix}\right) = 0 \end{cases} \tag{7.12}$$

has nonzero dimension; and in this case the multiplicity of λ *agrees with* $\dim H_{Y,R} \leq 2 \dim \ell^2_{\deg_\mu}(\mathsf{V})$.

This result complements the explicit formula for the resolvent in Proposition 6.83. Observe in particular that if $\mathcal{W}$ is self-adjoint, then by Theorem 7.22 the operator A is self-adjoint and therefore iA generates a C_0-group. Now, Propositions 4.51, 7.23 and 6.83 can be combined to study spectral properties of the bi-Laplacian on a metric graph that was introduced in Sect. 2.2.4.

7.2.2 Three Schrödinger-Type Equations

By Stone's theorem, the initial-value problem associated with each time-dependent Schrödinger-type equation on $\mathfrak{G}$

$$i\hbar \frac{\partial \psi}{\partial t}(t,x) = \mathfrak{H}\psi(t,x), \qquad t \in \mathbb{R},\ x \in \mathfrak{G}, \tag{7.13}$$

is well-posed if the relevant observable—the Hamiltonian $\mathfrak{H}$—is self-adjoint. Conversely, in the mathematical formulation of quantum mechanics each physically meaningful self-adjoint operator on a Hilbert space can be viewed as an observable of a given system.

One possible "physically meaningful" observable is the momentum. Classical momentum can be quantized to yield the operator $i\nabla$ in $\mathbb{R}^3$, or rather $i\frac{d}{dx}$ on $\mathfrak{G}$—this is the operator $\tilde{A}$ with node conditions defined by a matrix $\mathcal{U}$ introduced in Definition 2.44.

Proposition 7.24. *Let* $\mathcal{U} \in \mathcal{L}(\ell^2(\mathsf{E}))$. *Then the operator* $\tilde{A}$ *on* $\mathfrak{G}$ *is self-adjoint on* $L^2(\mathfrak{G})$ *if and only if* $\mathcal{U}$ *is unitary. In this case the unitary group* $(e^{it\tilde{A}})_{t\geq 0}$ *is given by*

$$e^{it\tilde{A}}u(x) := \mathcal{U}^k u(t+x-k) \qquad \text{if } t+x \in [k,k+1),\ k \in \mathbb{N},\ x \in (0,1),\ t \geq 0.$$

Thus, (i times) the first derivative with standard node conditions $\overleftarrow{A}$ is usually not self-adjoint.

Another relevant equation of Schrödinger type arises if the momentum operator is replaced by another observable—the Dirac operator D from Sect. 2.2.3, whose notation we adopt in the following. The following characterizes its self-adjointness.

Proposition 7.25. *A Dirac operator with boundary conditions defined by* $\mathcal{Z}_1, \mathcal{Z}_2 \in \mathcal{L}\big(\ell^2_\mu(\mathsf{E}) \times \ell^2_\mu(\mathsf{E})\big)$ *is self-adjoint if* $(\mathcal{Z}_1\ \mathcal{Z}_2)$ *is surjective and if moreover* $\mathcal{Z}_1\mathcal{Z}_2^*$ *is self-adjoint. The converse implication holds if* G *is finite.*

Finally, let us now consider (7.13) for the free Hamiltonian $\mathfrak{H} = \Delta$. We know from Example 7.8 that $(G_1(i\ \cdot, \cdot))_{t>0,x\in\mathbb{R}}$ yields the integral kernel of $(e^{it\Delta})_{t\in\mathbb{R}}$. The proof of the fact that $(K_t)_{t>0}$ in (7.11) is a fundamental solution of the heat equation on $\mathfrak{G}$ in [19, 69] consists essentially of three parts: It has to be checked that

- for each $t > 0$ K_t has the required regularity properties,
- for each $t > 0$ K_t satisfies the standard node conditions, and
- $K_\cdot(\cdot)$ solves (Di).

In the case of the Schrödinger equation one can mimic the proofs of the above assertions and thus easily deduce the following, where we adopt the same notations.

Proposition 7.26. *The unitary group that solves the Schrödinger equation on* $\mathfrak{G}$ *with continuity and Kirchhoff node conditions* (Cc) – (Kc) *consists of integral operators whose kernel is given by*

$$H_t(x, y) := K_{it}(x, y) = \sum_{\mathsf{P}\in\mathfrak{P}_{x,y}} \sigma(\mathsf{P}) G_1(it, \mathrm{dist}_\mu(x, y)), \qquad t > 0,\ x, y \in \mathfrak{G}. \tag{7.14}$$

7.2.3 Wave Equations

Let us now consider the second order abstract Cauchy problem (ACP2) associated with $M_c\Delta - M_p$, where $c \geq 0$ and Δ is the Laplacian with standard node conditions, i.e.,

$$\frac{\partial^2 u_{\mathsf{e}}}{\partial t^2}(t, x) = \frac{\partial}{\partial x}\left(c_{\mathsf{e}} \frac{\partial u_{\mathsf{e}}}{\partial x}\right)(t, x) - p_{\mathsf{e}}(x) u_{\mathsf{e}}(t, x), \qquad t \geq 0,\ x \in (0, \mu(\mathsf{e})),\ \mathsf{e} \in \mathsf{E}, \tag{Wa}$$

complemented by (Cc) – (KRc). This is a system of wave equations (or, more precisely, of *Klein-Gordon equations* whenever $p \neq 0$). We already know that the associated form a is of Lions type, hence (ACP2) is well-posed. Now, a is a local, symmetric Dirichlet form whenever $p \geq 0$ and $\mathcal{W}$ is diagonal with positive entries: By Propositions 7.12 and 7.14 the following holds.

Proposition 7.27. *Let* $\mathcal{W}$ *be diagonal with non-negative entries and* $p_{\mathsf{e}} \geq 0$ *for all* $\mathsf{e} \in \mathsf{E}$. *Then* (ACP2) *associated with the elliptic operator* $\nabla(c\nabla) - M_p$ *with node conditions* (Cc) – (KRc) *enjoys finite speed of propagation.*

The abstract wave equation on $\mathfrak{G}$ over a finite graph G with node conditions (Cc) – (Kc), i.e., the second order abstract Cauchy problem associated with Δ, is governed by a C_0-cosine operator function $(C(t, \Delta))_{t\geq 0}$. Because Δ comes from an accretive form, it is dissipative and hence by the Spectral Theorem $(C(t, \Delta))_{t\geq 0}$ is bounded. Also by the Spectral Theorem, Proposition 4.53 applies and we conclude that periodicity of $(C(t, \Delta))_{t\geq 0}$ is equivalent to (4.21). In view of Theorem 7.15, this imposes a strong condition on the connectivity of G, which is however satisfied if G is a path or a cycle.

7.2.4 Beam Equations

As a direct consequence of the self-adjointness of the Laplacian with standard node conditions, one deduces the following.

Proposition 7.28. *Consider the bi-Laplacian with standard node conditions*

$$D(\Delta^2) := \left\{ u : (0,1) \in W^{4,2}\Big((0,1); \ell^2_\mu(\mathsf{E})\Big) \text{ s.t.} \begin{array}{l} \exists u_{|\mathsf{V}}, u''_{|\mathsf{V}} \in \mathbb{C}^{\mathsf{V}} \quad \text{with} \\ (\mathcal{I}^-)^\top u_{|\mathsf{V}} = u(0),\ (\mathcal{I}^+)^\top u_{|\mathsf{V}} = u(1), \\ (\mathcal{I}^-)^\top u''_{|\mathsf{V}} = u''(0),\ (\mathcal{I}^+)^\top u''_{|\mathsf{V}} = u''(1), \\ \mathcal{I}^+ \mathcal{M} u'(1) - \mathcal{I}^- \mathcal{M} u'(0) + \mathcal{W} u_{|\mathsf{V}} = 0, \\ \text{and } \mathcal{I}^+ \mathcal{W} u'''(1) - \mathcal{I}^- \mathcal{W} u'''(0) + \mathcal{W} u''_{|\mathsf{V}} = 0 \end{array} \right\},$$

$$\Delta^2 u := \frac{d^4 u}{dx^4}.$$

Then $-\Delta^2$ generates a C_0-cosine operator function on $L^2(\mathfrak{G})$, thus the associated beam equation is well-posed.

Indeed, as we know the semi-explicit expression for $(e^{it\Delta})_{t\in\mathbb{R}}$ given by (7.14), we also know by Proposition 4.51 a semi-explicit expression for $(C(t,-\Delta^2))_{t\geq 0}$.

7.3 Quantum Graphs

Consider a Hamiltonian system associated with a function $\mathfrak{H}$ of time t, momentum $p : \mathbb{R} \to \mathbb{R}^n$ and position $q : \mathbb{R} \to \mathbb{R}^d$—it is assumed that momentum and position completely describe the state of the system. For example, in the classical description of a particle in three dimensions, one has

$$\mathfrak{H}(p,q,t) = \frac{(p|p)_{\mathbb{R}^3}}{2m} + V(q,t), \qquad t \in \mathbb{R}.$$

It is well known from classical theoretic mechanics that the particle's dynamics is described by the $2d$-dimensional system

$$\begin{cases} \frac{dp}{dt}(t) = -\frac{\partial \mathfrak{H}}{\partial q}(p(t), q(t), t), & t \in \mathbb{R}, \\ \frac{dq}{dt}(t) = \frac{\partial \mathfrak{H}}{\partial p} p(t), q(t), t), & t \in \mathbb{R}, \end{cases}$$

where the Hamiltonian function represents the energy of the system. As any other dynamical system, the above Hamiltonian system may or may not have a chaotic behavior. Because of conservation of energy, the Hamiltonian function agrees at any time with the energy E, i.e.,

$$E(p, q, t) = \frac{(p|p)_{\mathbb{R}^3}}{2m} + V(q, t), \qquad t \in \mathbb{R}.$$

When trying to extend the classical framework to quantum mechanics, there is a standard way of *quantizing* the above Hamiltonian system, i.e., to obtain a Hamiltonian operator that formally resembles the Hamiltonian function and which can be used to set up a Schrödinger equation (7.5) in a suitable Hilbert space (typically, $L^2(\mathbb{R}^3)$). This quantization is commonly performed by formally replacing the momentum q by its quantum analog, the *momentum operator* $-i\hbar\nabla$ and the potential energy by the corresponding multiplication operator M_V, so that the Hamiltonian operator is eventually given by $\mathfrak{H} := \frac{1}{2m}(-i\hbar\nabla) \cdot (-i\hbar\nabla) + M_V$, or rather

$$\mathfrak{H}\psi := -\frac{\hbar^2}{2m}\Delta\psi + V\psi.$$

In this way, it is possible to formally translate the dynamics governing classical systems into a quantum mechanical formalism, and vice versa. For reasons we are not able to discuss in this brief account, it is relevant to investigate the properties of those quantum mechanical systems whose classical analog has chaotic behavior. This branch of mathematical physics is commonly referred to as "quantum chaos".

One of the most intriguing open questions in this field is related to the so-called *Bohigas–Giannoni–Schmit conjecture*. In their paper [10] they make a case for the following assertion[1]:

> Spectra of time-reversal-invariant systems whose classical analogs are K systems show the same fluctuation properties as predicted by GOE.

What is truly remarkable about this conjecture is apparently that it links two rather different subjects—quantum chaos and random matrices. Ever since, more and more numerical evidence has been collected in the literature for an even stronger

[1] Here, GOE denotes the Gaussian orthogonal ensemble, the set of all symmetric matrices which becomes a measure space whenever it is endowed with a certain Gaussian measure. One similarly defines the ensembles of hermitian matrices GUE and of self-dual matrices GSE. We refer e.g. [57] for a good introduction to random matrix theory.

claim, which is what nowadays usually goes under the name of BGS-conjecture and which we can summarize as follows:

> Spectra of quantum systems whose classical analogs are chaotic show the same fluctuation properties as predicted by at least one of the random matrix ensembles GOE, GUE or GSE.

Failure of the BGS-conjecture is commonly regarded as utterly unlikely in the quantum chaos community, but investigations in this subject had traditionally to deal with models that can only be dealt with numerically: typically, spectra of Laplacians with Dirichlet boundary conditions on two dimensional domains with peculiar geometries—so-called *quantum billiards*.

Then, in 1997, T. Kottos and U. Smilansky made a surprising discovery. They observed in [45] that even a very simple differential operator—to be precise, a Schrödinger operator with magnetic potential A—defined on a *metric graph* displays analogous fluctuation properties as predicted by GOE (if $A = 0$) or GUE (if $A \neq 0$), under the sole assumption that the edge lengths are rationally independent. This is true already for very simple graphs—the example of a metric graph over the complete graph with four nodes is done in [45]. This operator has a well-behaved spectrum that can be determined rather explicitly. Similar properties have been observed in [12] for the Dirac operator and GSE.

The discovery of Kottos and Smilansky has promptly aroused much interest in the quantum chaos community: the expression "quantum graph" was born.[2] Meanwhile it has been extended to refer to the whole field of the analysis of partial differential operators on metric graphs (and, more recently, of difference operators on discrete graphs), but the amount of work devoted to spectral theoretical aspects is still preponderant. Two usual references for spectral theory and ordinary differential equations on graphs are [35,49].

7.4 Notes and References

Section 7.1. The formulation of the Spectral Theorem in Theorem 7.1 was suggested by P. Halmos in [39] with the aim of emphasize the similarities with the eponymous result for Hermitian matrices.

Corollary 7.5 is a special case of [33, Corollary V.3.3], which holds under slightly milder assumptions. Many properties of semigroups that we have seen in dependence of positivity have natural self-adjoint counterparts. Indeed, historically the theory of positive C_0-semigroups has been developed largely with the aim of dropping the assumption of self-adjointness from many results on long-time behavior—e.g., for shift semigroups.

Theorem 7.7 was proved by M.H. Stone in [73] and explains why it is mostly self-adjoint observables that are studied in quantum mechanics: The normed square

---

[2] In fact *re*born: the same name was earlier sporadically used in relation to Feynman diagrams.

of a wavefunction $U(t)\psi_0$ can only be interpreted as a probability (for a quantum system with initial configuration ψ_0) if the C_0-group $(U(t))_{t\in\mathbb{R}}$ is unitary.

If A is self-adjoint and $-A$ is dissipative, so that A comes from an accretive, symmetric form, then $a(\psi,\psi)$ represents the system's energy at a certain configuration ψ: this explains why techniques for analyzing an evolution equation based on properties of a form a evaluated along its diagonal are sometimes referred to as *energy methods*.

Although Davies–Gaffney estimates had already been suggested in earlier articles, they have been become popular with [29]. Proposition 7.12 was shown in [74]. Propositions 7.13 and 7.14 appear in [27, 72]. Because the Gaussian semigroup is the prototypical example of a semigroup governing a well-behaved parabolic problem, Gaussian estimates are quite naturally to expect. Theorem 7.13 and some related properties have been studied in in [27, 72]. It seems to be unknown whether Proposition 7.14 holds if the symmetry assumption is dropped and a is only required to be of Lions type. A concise but efficient overview of the interplay between heat kernel estimates and properties of metric measure spaces is given in [26]. It is known that semigroup with Gaussian estimates also enjoy several further properties: For example, if a an analytic C_0-semigroup on L^2 admits Gaussian estimates (and hence it necessarily extrapolates to L^p, $p\in(1,\infty)$), then all extrapolates semigroups on L^p have the same analyticity angle—in particular also the extrapolated semigroup on L^1 which (unlike in the case of general analytic semigroups that are contractive with respect to both the L^2 and L^∞-norm) is always strongly continuous and analytic. Furthermore, all their generators have coincident spectrum—but this is also true in the case that the generator in L^2 has compact resolvent, which is usually much easier to check. We refer to [1, Sect.. 7.4] for further related properties.

A classical way of solving the one-dimensional wave equation

$$\frac{\partial^2 u}{\partial t^2}(t,x)=\frac{\partial^2 u}{\partial x^2}(t,x),\qquad t\geq 0,\ x\in\mathbb{R},\tag{7.15}$$

consists in factoring it as

$$\left(\frac{\partial}{\partial t}-\frac{\partial}{\partial x}\right)\left(\frac{\partial}{\partial t}+\frac{\partial}{\partial x}\right)u(t,x)=0,\qquad t\geq 0,\ x\in\mathbb{R},$$

which is justified if u is smooth enough and hence the assumptions of Schwarz' Theorem are satisfied. This suggests to define a new unknown

$$v(t,x):=\left(\frac{\partial}{\partial t}+\frac{\partial}{\partial x}\right)u(t,x),\qquad t\geq 0,\ x\in\mathbb{R},\tag{7.16}$$

using which (7.15) becomes an advection equation

$$\left(\frac{\partial}{\partial t}-\frac{\partial}{\partial x}\right)v(t,x)=0,\qquad t\geq 0,\ x\in\mathbb{R}.\tag{7.17}$$

Thus, we can first solve (7.17) and then use the solution as an inhomogeneous data in (7.16). In this way we can deduce a few properties of the wave equation from the corresponding solutions of the advection equation, which in turn can be explicitly solved by Example 4.16.

In particular, this approach yields (4.20) and hence the fact that the one-dimensional wave equation exhibits finite speed of propagation. Indeed, this is more generally true of the wave equation associated with the d-dimensional Laplacian, by Proposition 7.14. Accordingly, if the initial data u_0, u_1 are compactly supported, so is the solution u at any time. In this case, a spectacular result due to J.A. Goldstein and R.J. Duffin in [32, 36] states that if d is odd (and only then) both the kinetic energy $\|u_t\|^2_{L^2}$ and the potential energy $\|\nabla u\|^2_{L^2}$ become constant, and in fact equal, within finite time.

One can replicate the same idea also for wave equations on networks. In this case, it is tempting to replace the second derivative on $\mathbb{R}$ by the second derivative with standard node conditions on a metric graph and to exploit the factorization (2.47). However, $\overleftarrow{A}_K, \overleftarrow{A}_C$ are in general not generators of C_0-semigroups (it is easy to see that they need not have the necessary number of boundary conditions). A tricky workaround has been developed in [42].

Section 7.2.1. The core of Theorem 7.15 has been proved by J. von Below in [5], cf. also [60], and variously extended to infinite graphs and to Laplacians with further, non-standard boundary conditions in [8]—from where also Proposition 7.23 is taken—and other papers over the last three decades, and in particular in [14, 18, 64]. It is particularly significant that, in view of Theorem 7.15, there is a monotonic dependence of changes on $\lambda_2(\mathcal{T})$ on $\lambda_2(\Delta)$. Much is known about $\lambda_2(\mathcal{T})$, in particular whenever G is a regular graph, cf. e.g. [24,58]; also the behavior of $\lambda_2(\mathcal{T})$ and hence of $\lambda_2(\Delta)$ under graph operations has been thoroughly investigated, cf. [8] and references therein.

If G is weighted in a non-trivial way, an analog of Theorem 7.15 is less immediate to obtain; in particular, the knowledge of the spectrum $\mathcal{T}$ is poorer. Some information on $\lambda_2(\Delta)$, the second largest eigenvalue of Δ, has been obtained both by analytic and numerical methods in [51,56], cf. also the earlier investigations in [61]. In these papers, particular attention is devoted to the shifts of $\lambda_2(\Delta)$ induced by deleting or inserting edges, and to comparisons with the (well-known) values of $\lambda_2(\Delta)$ in some elementary graphs, most notably paths, cycles and complete graphs. Even the very elementary case of (non-simple) graphs G_m obtained connecting two nodes with m edges of length $1 = \mu_1, \mu_2, \ldots, \mu_m$, respectively, has a non-obvious behavior. The authors of [51] prove that the second largest eigenvalue of the Laplacian with standard node conditions on G_2 can be either larger or smaller than on G_1, depending on μ_2; and that the second largest eigenvalue of G_{n-1} is smaller than the second largest eigenvalue of G_n for any $n \geq 3$, *no matter how large* $\mu_3, \ldots, \mu_n$ *are*. In view of Example 7.16, this leads to surprising statement that the rate of convergence to equilibrium of a diffusion process on a network can become *slower* by *enhancing* connectivity. A phenomenon that seems to be closely related to this has been first predicted for traffic networks in [13] and is

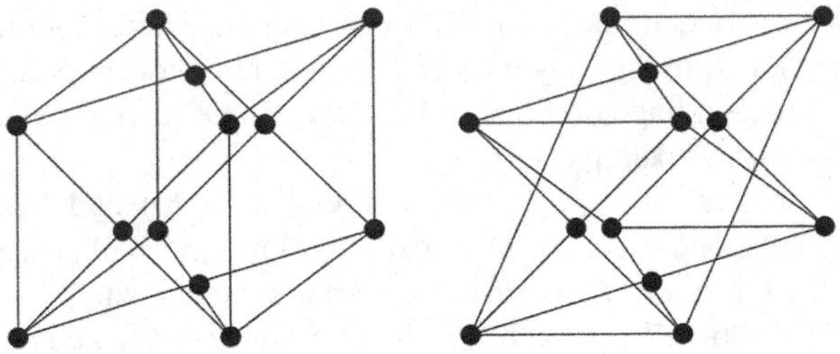

Fig. 7.3 An example of two isospectral metric graphs (courtesy of J. von Below)

nowadays referred to as *Braess' paradox*. It has been actually observed in quantum mechanical experiments, cf. [62]. This seems highly counter-intuitive in particular in view of Rayleigh's principle: *The effective resistance between any two nodes of an electric circuit cannot be decreased by cutting edges.* This suggests that the effective resistance is not directly related to the lowest non-trivial eigenvalue of the discrete Laplacian.

It has been observed in [8,34,65] that the multiplicities of the null spaces of $\Delta, \tilde{\Delta}$ carry relevant topological information: In the case of an unweighted, finite G

$$\dim \operatorname{Ker} \tilde{\Delta} - \dim \operatorname{Ker} \Delta = \operatorname{Tr} e^{t\tilde{\Delta}} - \operatorname{Tr} e^{t\Delta} = |\mathsf{E}| - |\mathsf{V}| - \kappa^-, \tag{7.18}$$

where κ^- is the number of connected components that are *not* bipartite.

Can one hear the shape of a metric graph? Following M. Kac [41], this question is quite natural. One answer relies directly upon Theorem 7.15: It suffices to observe that the Laplacians on two different metric graphs $\mathfrak{G}_1, \mathfrak{G}_2$ are clearly isospectral if so are the transition matrices $\mathcal{T}$ of the underlying discrete graphs $\mathsf{G}_1, \mathsf{G}_2$, provided $\mathsf{G}_1, \mathsf{G}_2$ are both connected and either both bipartite or both non-bipartite. Since $\mathcal{T}$ are isospectral $\mathsf{G}_1, \mathsf{G}_2$ if so are the adjacency matrices of $\mathsf{G}_1, \mathsf{G}_2$, and because indeed isospectral adjacency matrices on regular connected, non-bipartite graphs are known, von Below could answer the above question in the negative in [6] (Fig. 7.3).

But is it possible to modify the setting in order to recover the shape of a graph? Indeed, it was proved in [38] that this is possible if the weights μ of the edges are not pairwise rationally dependent. It has been shown in [8] that, remarkably, there seems to be no relevant difference between information encoded in the spectrum of Δ and $\tilde{\Delta}$: this depends on some interesting symmetry relation between the eigenvalues of these operators, even if their multiplicities are generally different. A beautiful general method for constructing isospectral graphs, relating spectral issues and (geometric) symmetries of a metric graph, has been presented in [2].

It was sketched in Example 7.16 how it is possible to obtain estimates for the speed of convergence toward equilibrium of diffusion processes on metric graphs that depend on graph-theoretical results. This idea has been elaborated e.g. in [5, 8, 47]. If for example $\mathcal{W} = 0$ and $\mu \equiv 1$ (i.e., all edges have the same length and conditions (Kc) – (Cc) are imposed, then among all metric graphs with given node set the fastest convergence is attained when the underlying graph is complete, cf. [47, Sect. 5]. Further estimates can be obtained in terms of certain parameters that are relevant in graph theory, like the diameter and the edge connectivity parameter of a graph.

A precise description of the eigenspaces that evidences their symmetric features is obtained in [17]. In the case of infinite graphs there are several potential theoretical reasons that motivate the study of the Laplacian in $L^\infty(\mathfrak{G})$ instead of $L^2(\mathfrak{G})$: This rich theory has been studied in a long series of papers that begins with [7]. Spectral theory of second order differential operators on metric graphs is thoroughly discussed in [9, Chaps. 3–5].

Some simple instances of (7.11) have been proved in [61,69] for finite graphs. Its later non-trivial generalization to the infinite case has been introduced (in a slightly different formulation) by C. Cattaneo in [19,20].

It was showed by A. Selberg in [71] that the trace of $e^{t\Delta_M}$, where Δ_M denotes the Laplace–Beltrami operator on a Riemannian manifold M, carries relevant information about the geometry of M and in particular about the lengths of its geodesics, cf. also [25]. The formula in Theorem 7.18 arises in close analogy with Selberg's theory upon regarding cycles in a graph as the analogs of closed geodesics of a manifold; it is due to J.-P. Roth [68, 69], cf. also [46, 61]. A comparable trace formula has been obtained in[52], and later for metric graphs with more general boundary conditions in [11,44]. More precisely, the trace formulae in the latter two articles are expressed in terms of an identity of two measures: Roth's formula in Theorem 7.18 can then be deduced integrating them against the Gaussian kernel. The idea of describing a diffusion process as a weighted sum of random walks, like in Theorem 7.20, is old and was already at the basis of the *Feynman–Kac formula*. There, the solution of a diffusion-type equation is expressed in terms of a conditional expectation with respect to a probability measure associated some Brownian motion: In the case of a network, the set or Brownian motions from x to y is simply $\mathfrak{P}_{x,y}$.

One-dimensional Gaussian estimates for the semigroup generated by the Laplacian with Kirchhoff–Robin node conditions have been proved in [59] for finite metric graphs by the so-called Davies' trick, which amounts to prove uniform L^∞-(quasi)contractivity estimates for a class of perturbed semigroups. The proof relies substantially on the identification suggested in Lemma 3.22 and only works in the finite case: The reason for this is that in general the isomorphism Φ in Lemma 3.22 fails to yield a rough isometry in the infinite case. A simpler proof proposed in [63] applies Propositions 7.12 and 7.13 to Example 7.10, thus it additionally allows to treat the case of diffusion on infinite metric graphs. Furthermore, *two*-dimensional Gaussian estimates are proved in [63] for the metric graph over $\mathbb{Z}^2$: This means that diffusion on this metric graph behaves like on an individual interval for short times (intuitively, until some particles initially confined in an edge hit the edge's boundaries), and then it starts resembling diffusion in a domain of $\mathbb{R}^2$. Here we have followed Pang's approach, generalized in [66] to much more general infinite planar graphs that arise as skeletons of tessellations of the plane.

Also $(e^{-t\mathcal{L}})_{t\geq 0}$, the semigroup that governs the discrete heat equation, can be trivially represented as a family of kernel operators. Kernel estimates for this semigroup are delicate and depend essentially on metric properties of the underlying graph. Results in this direction have been obtained among others in [30, 37]. In particular, T. Delmotte has proved in [31] estimates on the kernel of the random walk on a graph by a rather sophisticated method based on proving a Harnack inequality for the time-continuous problem and then extending it to the time-discrete case by a Moser-type iteration argument.

In this context, Theorem 7.22 has first appeared in [48], but cf. the historical notes at the end of Chap. 6.

Section 7.2.2. Proposition 7.25 is taken from [12]. J. Bolte and his coauthors have systematically investigated properties of the Dirac equation on a graph.

Unlike quantum mechanics, quantum fields theory on graphs seems to be still in its early stage. Some investigations have been performed by B. Bellazzini, M. Mintchev and coauthors beginning with [3, 4].

Section 7.2.3. The study of wave equations seems to have been one of the earliest active subfields of the theory of PDEs on networks. There are apparent motivations for this interest, mostly coming from mechanical engineering: We refer to [28, 54] for results—mostly focused on controllability—and for possible higher dimensional extensions. An early investigation of the wave equation on $\mathfrak{G}$ has been performed in [69, Sect. 6]. If $\mathcal{W} = 0$, and hence the node conditions are of purely Kirchhoff type, an explicit formula for the cosine operator function generated by Δ on $\mathfrak{G}$ was obtained by C. Cattaneo and L. Fontana in [21, Thm. 3] see also [67]. (However, in a recent article [55], D. Lenz and K. Pankrashkin shed light on a little known article (cf. [43], or [55, Prop. 3] for a summary in English), only available in Russian, where a D'Alembert-type formula for the solution has seemingly been obtained at least in the case of a *finite* weighted oriented graph: A general D'Alembert-type formula for graphs that need not even be locally finite has been obtained in [55].) A solution formula for the case of general $\mathcal{W}$ could probably be obtained combining the ideas of [21,23]. However, the formula in [21] looks extremely complicated even on networks with simple connectivities like a star or a complete graph. Actually, the formula hardly gives any hint on the behavior of solutions. It is therefore desirable to prove by abstract methods at least some property one naturally expects.

Finite speed of propagation for the wave equation on a metric graph as in Proposition 7.27 has been observed in [70] by different methods. In a different but related setting, it has been proved in [53] that only wave equations with node conditions that give rise to local forms can enjoy finite speed of propagation.

Also the second order abstract Cauchy problem associated with the discrete Laplacian $\mathcal{L}$ is well-posed, by Theorem 6.52. One would expect finite speed of propagation there, too, but the form associated with $\mathcal{L}$ is not local and therefore Proposition 7.14 does not apply directly.

Section 7.2.4. The analysis of metric graph of beams is common in the literature, though less than waves, and the literature on this topic includes several contributions by, among others, K. Ammari, S. Avdonin, V. Komornik, G.R. Leugering, S. Nicaise, E.J.P.G. Schmidt, E. Zuazua, cf. [28, 54]

References

1. W. Arendt, Semigroups and evolution equations: Functional calculus, regularity and kernel estimates, in *Handbook of Differential Equations: Evolutionary Equations – Vol. 1*, ed. by C.M. Dafermos, E. Feireisl (North Holland, Amsterdam, 2004)
2. R. Band, T. Shapira, U. Smilansky, Nodal domains on isospectral quantum graphs: the resolution of isospectrality? J. Phys. A **39**, 13999–14014 (2006)
3. B. Bellazzini, M. Mintchev, Quantum fields on star graphs. J. Phys. A **39**, 11101–11117 (2006)
4. B. Bellazzini, M. Mintchev, P. Sorba, Bosonization and scale invariance on quantum wires. J. Phys. A **40**, 2485–2507 (2007)
5. J. von Below, A characteristic equation associated with an eigenvalue problem on c^2-networks. Lin. Algebra Appl. **71**, 309–325 (1985)
6. J. von Below, Can one hear the shape of a network? in *Partial Differential Equations on Multistructures (Proc. Luminy 1999)*, ed. by F. Ali Mehmeti, J. von Below, S. Nicaise. Lecture Notes in Pure and Applied Mathematics, vol. 219 (Marcel Dekker, New York, 2001), pp. 19–36
7. J. von Below, J.A. Lubary, The eigenvalues of the Laplacian on locally finite networks. Result. Math. **47**, 199–225 (2005)
8. J. von Below, D. Mugnolo, The spectrum of the Hilbert space valued second derivative with general self-adjoint boundary conditions. Lin. Alg. Appl. **439**, 1792–1814 (2013)
9. G. Berkolaiko, P. Kuchment, *Introduction to Quantum Graphs*. Mathematical Surveys and Monographs, vol. 186 (American Mathematical Society, Providence, 2013)
10. O. Bohigas, M.J. Giannoni, C. Schmit, Characterization of chaotic quantum spectra and universality of level fluctuation laws. Phys. Rev. Lett. **52**, 1–4 (1984)
11. J. Bolte, S. Endres, The trace formula for quantum graphs with general self-adjoint boundary conditions. Ann. Henri Poincaré A **10**, 189–223 (2009)
12. J. Bolte, J. Harrison, Spectral statistics for the Dirac operator on graphs. J. Phys. A **36**, 2747–2769 (2003)
13. D. Braess, Über ein Paradoxon aus der Verkehrsplanung. Unternehmensforschung **12**, 258–268 (1968)
14. J. Brüning, V. Geyler, K. Pankrashkin, Spectra of self-adjoint extensions and applications to solvable Schrödinger operators. Rev. Math. Phys. **20**, 1–70 (2008)
15. S. Butler, Interlacing for weighted graphs using the normalized Laplacian. Electronic J. Lin. Alg. **16**, 87 (2007)
16. S.K. Butler, Eigenvalues and Structures of Graphs. PhD thesis, University of California, San Diego, 2008
17. R. Carlson, Eigenvalue cluster traces for quantum graphs with equal edge lengths. Rocky Mount. J. Math. **42**, 467–490 (2012)
18. C. Cattaneo, The spectrum of the continuous Laplacian on a graph. Monats. Math. **124**, 124–215 (1997)
19. C. Cattaneo, The spread of the potential on a homogeneous tree. Ann. Mat. Pura Appl., IV Ser. **175**, 29–57 (1998)
20. C. Cattaneo, The spread of the potential on a weighted graph. Rend. Semin. Mat. Torino **57**, 221–229 (1999)
21. C. Cattaneo, L. Fontana, D'Alembert formula on finite one-dimensional networks. J. Math. Anal. Appl. **284**, 403–424 (2003)
22. G. Chen, G. Davis, F. Hall, Z. Li, K. Patel, M. Stewart, An interlacing result on normalized Laplacians. SIAM J. Discr. Math. **18**, 353–361 (2004)
23. R. Chill, V. Keyantuo, M. Warma, Generation of cosine families on $L^p(0, 1)$ by elliptic operators with Robin boundary conditions, in *Functional Analysis and Evolution Equations*, ed.by H. Amann et al. (Birkhäuser, Basel, 2008), pp. 113–130
24. F.R.K. Chung, *Spectral Graph Theory*. Regional Conference Series in Mathematics, vol. 92 (American Mathematical Society, Providence, 1997)

25. Y. Colin de Verdière, Spectre du laplacien et longueurs des géodésiques périodiques. ii. Compos. Math. **27**, 159–184 (1973)
26. T. Coulhon, Heat kernel estimates, sobolev-type inequalities and riesz transform on non-compact riemannian manifolds, in *Analysis and geometry of metric measure spaces (Proc. Montréal 2011)*. CRM Proceedings and Lecture Notes, vol. 56 (American Mathematical Society, Providence, 2013), pp. 55–66
27. T. Coulhon, A. Sikora, Gaussian heat kernel upper bounds via the Phragmén–Lindelöf theorem. Proc. London Math. Soc. **96**, 507–544 (2008)
28. R. Dáger, E. Zuazua, *Wave propagation, observation and control in 1-d flexible multi-structures*. Mathématiques et Applications, vol. 50 (Springer, Berlin, 2005)
29. E.B. Davies, Heat kernel bounds, conservation of probability and the Feller property. J. Anal. Math. **58**, 99–119 (1992)
30. E.B. Davies, Large deviations for heat kernels on graphs. J. London Math. Soc. **2**, 65–72 (1993)
31. T. Delmotte, Parabolic Harnack inequality and estimates of Markov chains on graphs. Rev. Mat. Iberoam. **15**, 181–232 (1999)
32. R.J. Duffin, Equipartition of energy in wave motion. J. Math. Anal. Appl. **32**, 386–391 (1970)
33. K.-J. Engel, R. Nagel, *One-Parameter Semigroups for Linear Evolution Equations*. Graduate Texts in Mathematics, vol. 194 (Springer, New York, 2000)
34. S.A. Fulling, P. Kuchment, J.H. Wilson, Index theorems for quantum graphs. J. Phys. A **40**, 14165–14180 (2007)
35. S. Gnutzmann, U. Smilansky, Quantum graphs: Applications to quantum chaos and universal spectral statistics. Adv. Phys **55**, 527–625 (2006)
36. J.A. Goldstein, An asymptotic property of solutions of wave equations. Proc. Am. Math. Soc. **23**, 359–363 (1969)
37. A. Grigor'yan, A. Telcs, Sub-gaussian estimates of heat kernels on infinite graphs. Duke Math. J. **109**, 451–510 (2001)
38. B. Gutkin, U. Smilansky, Can one hear the shape of a graph? J. Phys. A **34**, 6061–6068 (2001)
39. P.R. Halmos, What does the spectral theorem say? Am. Math. Mon **70**, 241–247 (1963)
40. D. Horak, J. Jost, Interlacing inequalities for eigenvalues of discrete Laplace operators. Ann. Global Anal. Geom. **43**, 177–207 (2013)
41. M. Kac, Can one hear the shape of a drum? Am. Math. Mon **73**, 1–23 (1966)
42. B. Klöss, The flow approach for waves in networks. Oper. Matrices **6**, 107–128 (2012)
43. A.V. Kopytin, On representation of solutions to the wave equation on graphs with commensurable edges. Tr. Mat. Fak. Voronezh. Gos. Univ. **6**, 67–77 (2001)
44. V. Kostrykin, J. Potthoff, R. Schrader, Heat kernels on metric graphs and a trace formula, in *Adventures in Mathematical Physics*. Contemporary Mathematics, vol. 447 (American Mathematical Society, Providence, 2007), pp. 175–198
45. T. Kottos, U. Smilansky, Quantum chaos on graphs. Phys. Rev. Lett. **79**, 4794–4797 (1997)
46. T. Kottos, U. Smilansky, Periodic orbit theory and spectral statistics for quantum graphs. Ann. Physics **274**, 76–124 (1999)
47. M. Kramar Fijavž, D. Mugnolo, E. Sikolya, Variational and semigroup methods for waves and diffusion in networks. Appl. Math. Optim. **55**, 219–240 (2007)
48. P. Kuchment, Quantum graphs I: Some basic structures. Waves Random Media **14**, 107–128 (2004)
49. P. Kuchment, Quantum graphs: an introduction and a brief survey, in *Analysis on Graphs and its Applications*, ed. by P. Exner, J. Keating, P. Kuchment, T. Sunada, A. Teplyaev. Proceedings of Symposia in Pure Mathematics, vol. 77 (American Mathematical Society, Providence, 2008), pp. 291–314
50. P.C. Kunstmann, L. Weis, Maximal L_p-regularity for parabolic equations, Fourier multiplier theorems and H^∞-functional calculus, in *Functional Analytic Methods for Evolution Equations*. Lecture Notes in Mathematics, vol. 1855 (Springer, Berlin, 2004), pp. 65–311
51. P. Kurasov, G. Malenová, S. Naboko, Spectral gap for quantum graphs and their connectivity. J. Phys. A **46**, 275309 (2013)

52. P. Kurasov, M. Nowaczyk, Inverse spectral problem for quantum graphs. J. Phys. A **38**, 4901 (2005)
53. P. Kurasov, A. Posilicano, Finite speed of propagation and local boundary conditions for wave equations with point interactions. Proc. Am. Math. Soc. **133**, 3071–3078 (2005)
54. J.E. Lagnese, G. Leugering, E.J.P.G. Schmidt, *Modeling, Analysis, and Control of Dynamic Elastic Multi-Link Structures*. Systems and Control: Foundations and Applications (Birkhäuser, Basel, 1994)
55. D. Lenz, K. Pankrashkin, New relations between discrete and continuous transition operators on (metric) graphs. arXiv:1305.7491
56. G. Malenová, Spectra of quantum graphs. Master's thesis, Czech Technical University in Prague, 2013
57. M.L. Mehta, *Random Matrices*. Pure and Applied Mathematics, vol. 142 (Elsevier, Amsterdam, 2004)
58. B. Mohar, The Laplacian spectrum of graphs. Graph Theory Comb Appl **2**, 871–898 (1991)
59. D. Mugnolo, Gaussian estimates for a heat equation on a network. Netw. Het. Media **2**, 55–79 (2007)
60. S. Nicaise, Approche spectrale des problemes de diffusion sur les réseaux, in *Séminaire de Théorie du Potentiel Paris*, vol. 8 (Springer, Berlin, 1987), pp. 120–140
61. S. Nicaise, Spectre des réseaux topologiques finis. Bull. Sci. Math. II. Sér. **111**, 401–413 (1987)
62. M.G. Pala, S. Baltazar, P. Liu, H. Sellier, B. Hackens, F. Martins, V. Bayot, X. Wallart, L. Desplanque, S. Huant, Transport inefficiency in branched-out mesoscopic networks: An analog of the Braess paradox. Phys. Rev. Lett. **108**, 076802 (2012)
63. M.M.H. Pang, The heat kernel of the Laplacian defined on a uniform grid. Semigroup Forum **78**, 238–252 (2008)
64. K. Pankrashkin, Spectra of Schrödinger operators on equilateral quantum graphs. Lett. Math. Phys. **77**, 139–154 (2006)
65. O. Post, First order approach and index theorems for discrete and metric graphs. Ann. Henri Poincaré A **10**, 823–866 (2009)
66. R. Pröpper, Heat kernel bounds for the Laplacian on metric graphs of polygonal tilings. Semigroup Forum **86**, 262–271 (2013)
67. V. Pryadiev, Description of the solution of an initial-boundary value problem for the wave equation on a one-dimensional spatial network in terms of the green function of the corresponding boundary value problem for an ordinary differential equation. J. Math. Sci. **147**, 6470–6482 (2007)
68. J.-P. Roth, Spectre du laplacien sur un graphe. C. R. Acad. Sci. Paris Sér. I Math. **296**, 793–795 (1983)
69. J.-P. Roth, Le spectre du laplacien sur un graphe, in *Colloque de Théorie du Potentiel - Jacques Deny (Proc. Orsay 1983)*, ed. by G. Mokobodzki, D. Pinchon. Lecture Notes in Mathematics, vol. 1096 (Springer, Berlin, 1984), pp. 521–539
70. R. Schrader, Finite propagation speed and free quantum fields on networks. J. Phys. A **42**, 495401 (2009)
71. A. Selberg, Harmonic analysis and discontinuous groups in weakly symmetric Riemannian spaces with applications to Dirichlet series. J. Indian Math. Soc. **20**, 47–87 (1956)
72. A. Sikora, Riesz transform, Gaussian bounds and the method of wave equation. Math. Z. **247**, 643–662 (2004)
73. M.H. Stone, On one-parameter unitary groups in Hilbert space. Ann. Math. **33**, 643–648 (1932)
74. K.-T. Stur, Analysis on local Dirichlet spaces. II. Upper Gaussian estimates for the fundamental solutions of parabolic equations. Osaka J. Math. **32**, 275–312 (1995)

Chapter 8
Symmetry Properties

The aim of this final chapter is to discuss how possible symmetries in an (oriented) weighted graph G influence the behavior of evolution equations—either on G or on the metric graph $\mathfrak{G}$ over it. We will use the word "symmetry" in a rather broad sense, to mean different notions of structural regularity of graphs of the Laplacian on $\mathbb{R}^d$.

The common thread in our discussion is the search for reductions of an equation. By this we mean the possibility of decomposing it into components that are possibly easier to solve, typically because their relevant functional setting are lower dimensional than the original one. This usually amounts to showing that all operators of a semigroup (or a cosine operator function) commute with a certain operator that can be associated with some relevant symmetry of the system—like orthogonal matrices are associated with rotational symmetry.

8.1 Commutation Properties

In the classical case of a diffusion equation on a spherically symmetric domain—e.g., a ball or $\mathbb{R}^d$—it is well-known that solutions of the equation take the same value upon rotating their argument, provided the initial value is rotationally symmetric as well. It turns out that this observation is a special instance of the general theory of Lie groups of symmetries.

The theory of continuous symmetries for partial differential equations is a broad and interesting topic, but it is most effective whenever the equation is defined on the whole Euclidean space. Indeed, Lie groups of space-dependent transformations typically do not respect boundary conditions—or, in our case, node conditions. In this section we are going to sketch the most elementary traits of this theory.

Definition 8.1. Let U be an open domain of $\mathbb{R}^d$ and $K \in \mathbb{N}$. Let $k \in \{0, \dots, K\}$. Two functions $f, g \in C^K(\overline{U})$ are said to be *k-jet-equivalent at* $x_0 \in U$ if their

D. Mugnolo, *Semigroup Methods for Evolution Equations on Networks*, Understanding Complex Systems, DOI 10.1007/978-3-319-04621-1_8,

difference $f - g$ vanishes at x_0 along with all its partial derivatives of any order up to k.

The *k-th jet space* at x_0, denoted by $J^k_{x_0}$, is the quotient space of $C^K(\overline{U})$ with respect to the k-jet-equivalence relation at x_0. The generic element of $J^k_{x_0}$ is therefore of the form $\mathbf{x_0} \equiv (x_0, j_k f(x_0))$ for some $f \in C^K(\overline{U})$ and is called *k-jet* of f at x_0. The resulting function $j_k f : \mathbb{R}^d \to \mathbb{R}^{d_k}$ is called the *k-jet* of f. Finally,

$$J^k := J^k(U) := \{J^k_{x_0} : x_0 \in U\} \equiv U \times \mathbb{R}^{d_k}$$

is called the *k-th jet bundle*.

(Here the coefficients d_k, $k \in \mathbb{N}$, are defined by

$$d_0 := 1, \qquad d_k := 1 + \sum_{h=1}^{k} \binom{d+h-1}{h}, \qquad k \geq 1.)$$

Definition 8.2. For a bounded open domain $U \subset \mathbb{R}^d$ we call each mapping from J^0 to J^0 a *point transformation.*

A family $\mathcal{T} := (T_\epsilon(\mathbf{x}))_{\epsilon \in I_\mathbf{x}, \mathbf{x} \in J^0}$ ($I_\mathbf{x} \subset \mathbb{R}$ for $\mathbf{x} \in J^0$) of point transformations is a *point transformation group* if both following conditions hold.

- $\mathcal{T}$ satisfies the group law

$$T_{\epsilon_1+\epsilon_2}(\mathbf{x}) = T_{\epsilon_2}(T_{\epsilon_1}(\mathbf{x})) \quad \text{and} \quad T_0(\mathbf{x}) = \mathbf{x}$$

  for all $\mathbf{x} \in J^0$ and all $\epsilon_1, \epsilon_2 \in \mathbb{R}$ for which the identity makes sense;
- the dependence of $\mathcal{T}$ on $\mathbf{x}$ and ϵ is jointly continuously differentiable.

Let $\mathcal{T}$ be a point transformation group. The *k-jet of* $\mathcal{T}$ is the family $j_k\mathcal{T}$ of mappings from J^k to J^k defined by

$$(j_k T_\epsilon)(x, j_k f) := (\tilde{x}, j_k \tilde{f}), \quad \text{whenever } (\tilde{x}, \tilde{f}) = T_\epsilon(x, f), \qquad \text{for all } \epsilon \in \tilde{I}_\mathbf{x},$$

for some family $(\tilde{I}_\mathbf{x})_{\mathbf{x} \in J^0}$ of open intervals such that $\tilde{I}_\mathbf{x} \subset I_\mathbf{x}$ and $\inf \tilde{I}_\mathbf{x} = 0$ for each $\mathbf{x} \in J^0$.

Just like in the case of C_0-semigroups we can define the *generator* of a point transformation group $\mathcal{T}$ formally as its strong right derivative at 0.

Definition 8.3. Given a function $H : U \times C(U) \times \ldots \times C(U) \to \mathbb{R}$, a *point symmetry group* of the differential equation

$$H\left(x, u, \frac{\partial u}{\partial x_1}, \ldots, \frac{\partial u}{\partial x_d}, \frac{\partial^2 u}{\partial x_1^2}, \frac{\partial^2 u}{\partial x_1 \partial x_2}, \ldots \frac{\partial^2 u}{\partial x_d^2}, \ldots, \frac{\partial^K u}{\partial x_d^K}\right) = 0 \tag{8.1}$$

is a point transformation group such that

$$H\big((j_K T_\epsilon)(x, j_K u)\big) = 0 \quad \text{for all } \epsilon \in \tilde{I}_{\mathbf{x}} \qquad \text{whenever } H(x, j_K u) = 0. \quad (8.2)$$

Example 8.4. It can be shown (cf. [20, Example 2.41]) that the only point transformation groups of the one-dimensional heat equation

$$\frac{\partial u}{\partial t} = \frac{\partial^2 u}{\partial x^2}, \qquad t \geq 0,\ x \in \mathbb{R},$$

are those given by

$$\begin{aligned}
T_\epsilon^{(1)} &: (t, x, u) \mapsto (t + \epsilon, x, u), && \epsilon \in \mathbb{R},\\
T_\epsilon^{(2)} &: (t, x, u) \mapsto (e^{2\epsilon} t, e^\epsilon x, u), && \epsilon \in \mathbb{R},\\
T_\epsilon^{(3)} &: (t, x, u) \mapsto \left(\tfrac{t}{1-4\epsilon t}, \tfrac{x}{1-4\epsilon t}, \sqrt{1-4\epsilon t}\; e^{\frac{-\epsilon x^2}{1-4\epsilon t}} u\right), && \epsilon < \tfrac{1}{4t},\\
T_\epsilon^{(4)} &: (t, x, u) \mapsto (t, x + \epsilon, u), && \epsilon \in \mathbb{R},\\
T_\epsilon^{(5)} &: (t, x, u) \mapsto (t, x + 2\epsilon t, e^{-\epsilon x - \epsilon t^2} u), && \epsilon \in \mathbb{R},\\
T_\epsilon^{(6)} &: (t, x, u) \mapsto (t, x, e^\epsilon u), && \epsilon \in \mathbb{R},
\end{aligned}$$

Now, one sees that only $(T_\epsilon^{(1)})_{\epsilon\in\mathbb{R}}$ and $(T_\epsilon^{(6)})_{\epsilon\in\mathbb{R}}$ are independent of the space variable and are hence effective regardless of the geometry of the domain; or more specifically of the boundary—or the node—conditions. (Specific domains or metric graphs may support further families, though; e.g., $(T_\epsilon^{(4)})_{\epsilon\in\mathbb{R}}$ is clearly compatible with a heat equation on the one-dimensional torus, hence also with the metric graph over a cycle graph.)

But there exists also further groups that map solutions of the heat equations into solutions, although they do not carry a differentiable structure like a Lie group: for instance, the cyclic group generated by the involution

$$(t, x, u) \mapsto (t, -x, u)\,.$$

Because continuous symmetries are so rare in graphs and other discrete structures, we make a virtue of necessity and focus instead on different symmetries, of which there is often abundance in networks: discrete ones, as the involution considered at the end of Example 8.4.

By a *discrete* symmetry we loosely mean a finite or countable group of mappings, or even a single mapping (for instance a point transformation, if jets can be defined at all) such that some relation analogous to (8.2) holds. Actually, it is not always easy to make sense of (8.2) in the context of equations on metric graphs and/or whenever boundary conditions are imposed—let alone whenever H in (8.1) is replaced by a matrix acting on a space of sequences and hence a jet formalism is not available.

For all these reasons, we go over to a much more elementary notion of symmetry. Roughly speaking, we first content ourselves with operators that intertwine with the flow that drives an evolution equation.

Definition 8.5. Let E, F be normed spaces and R, S, T be linear operators from E to F, on E, and on F, respectively, such that $D(RS) \cap D(TR)$ dense in E. We say that R *intertwines between* S *and* T if

$$RSx = TRx \qquad \text{for all } x \in D(TR) \cap D(SR).$$

If $E = F$ and $S = T$, then R and T are said to *commute*.

(Here $D(TR) = \{x \in D(R) : Rx \in D(T)\}$ and $D(RS) = \{x \in D(S) : Sx \in D(R)\}$.)

In many cases, R will have some natural physical interpretation. Then, finding commutation relations is often a key to a deeper understanding of the behavior of an equation. If in particular R is self-adjoint, then iR generates a unitary group that may turn out to be a point transformation group for a relevant evolution equation: S, T may e.g. belong to a larger family—a C_0-semigroup, for instance, or a C_0-cosine operator function.

In the cases relevant for us, at least one of the operators is bounded, so that Definition 8.5 simplifies a bit. Let us begin with a simple but crucial observation.

Lemma 8.6. *Let E, F be normed spaces, S, T be bounded linear operators on E, F, respectively, and R be a closed, densely defined operator from E to F. Then R intertwines between S and T if and only if the graph of R*

$$\operatorname{Graph} R := \left\{ \begin{pmatrix} x \\ Rx \end{pmatrix} \in D(R) \times F \right\}$$

is invariant under the operator matrix $\operatorname{diag}(S, T)$, *which is a bounded linear operator on $E \times F$.*

Since an operator between Banach spaces E, F is closed if and only if its graph is a closed subspace of $E \times F$, we are now in the position to check whether an operator commutes with the solution operator of a first order evolution equation by simply applying our favorite characterization of invariance of closed convex subsets—actually, closed subspaces—under C_0-semigroups. To this aim, we first need the following, which can be checked directly.

Lemma 8.7. *Let E, F be normed spaces and R be a closed, densely defined operator from E to F. Then both* $\mathrm{Id}_F + RR^*$ *and* $\mathrm{Id}_E + R^*R$ *have bounded inverse and the orthogonal projector onto* $\operatorname{Graph} R$ *is given by*

$$P_{\operatorname{Graph} R} := \begin{pmatrix} (\mathrm{Id}_E + R^*R)^{-1} & R^*(\mathrm{Id}_F + RR^*)^{-1} \\ R(\mathrm{Id}_E + R^*R)^{-1} & \mathrm{Id}_F - (\mathrm{Id}_F + RR^*)^{-1} \end{pmatrix}. \tag{8.3}$$

Consequently, the following holds.

Corollary 8.8. *Let $V, \tilde{V}, H, \tilde{H}$ be Hilbert spaces with V and $\tilde{V}$ densely and continuously embedded in H and $\tilde{H}$, respectively. Let $a : V \times V \to \mathbb{C}$ and $\tilde{a} : \tilde{V} \times \tilde{V} \to \mathbb{C}$ be continuous sesquilinear forms that are H-elliptic and $\tilde{H}$-elliptic, respectively. Let us denote by $A, \tilde{A}$ the respective associated operators, and let R be a bounded linear operator from H to $\tilde{H}$.*

Then R intertwines between the C_0-semigroups on $H, \tilde{H}$ generated by $A, \tilde{A}$, respectively, i.e.,

$$e^{tA}Rx = Re^{t\tilde{A}}x \qquad \text{for all } t \geq 0 \text{ and all } x \in H$$

if and only if

$$R(V) \subset \tilde{V}, \quad R^*(\tilde{V}) \subset V,$$

and furthermore

$$a(L_1x + R^*L_2y, R^*RL_1x - R^*L_2y) = \tilde{a}(RL_1x + RR^*L_2y, RL_1x - L_2y) \quad \text{for all } x \in V,\ y \in \tilde{V}. \tag{8.4}$$

Here

$$L_1 := (\mathrm{Id}_H + R^*R)^{-1}, \qquad L_2 := (\mathrm{Id}_{\tilde{H}} + RR^*)^{-1}$$

and hence

$$\mathrm{Id}_{\tilde{H}} - L_2 = RR^*L_2, \qquad \mathrm{Id}_H - L_1 = R^*RL_1.$$

If in particular R is unitary, then

$$L_1 = \frac{1}{2}\,\mathrm{Id}_H \qquad \text{and} \qquad L_2 = \frac{1}{2}\,\mathrm{Id}_{\tilde{H}} \tag{8.5}$$

and we obtain the following.

Lemma 8.9. *Under the assumptions of Corollary 8.8, let R be unitary. Then R intertwines between $(e^{tA})_{t\geq 0}, (e^{t\tilde{A}})_{t\geq 0}$ if and only if $R(V) = \tilde{V}$ and*

$$a(x + R^*y, x - R^*y) = \tilde{a}(Rx + y, Rx - y), \qquad \text{for all } x \in V,\ y \in \tilde{V}. \tag{8.6}$$

If additionally a and $\tilde{a}$ are symmetric, then (8.6) can be further simplified to the condition

$$a(x, x) = \tilde{a}(Rx, Rx), \qquad \text{for all } x \in V. \tag{8.7}$$

Corollary 8.8 is a consequence of the following special case of Theorem 6.28.

Corollary 8.10. *Let V, H be Hilbert spaces with V densely and continuously embedded in H. Let $a : V \times V \to \mathbb{C}$ be a continuous H-elliptic sesquilinear form. If C is a closed subspace of H and we denote by P_C the orthogonal projector onto it, then the following assertions are equivalent.*

(a) C *is invariant under* $(e^{tA})_{t\geq 0}$.
(b) $P_C V \subset V$ *and* $\operatorname{Re} a(P_C u, u - P_C u) \geq 0$ *for all* $u \in V$.
(c) $P_C V \subset V$ *and* $\operatorname{Re} a(u, v) = 0$ *for all* $u \in V \cap C$ *and all* $v \in V \cap C$.

If a is additionally symmetric, then the assertions (a)–(c) are also equivalent to the following ones.

(d) P_C *commutes with* $(e^{tA})_{t\geq 0}$.
(e) $P_C V \subset V$ *and* $a(P_C u, P_C u) \leq a(u, u)$ *for all* $u \in V$.

Proof of Corollary 8.8. The operator matrix $\operatorname{diag}(A, \tilde{A})$ generates the C_0-semigroup $\left(\operatorname{diag}(e^{tA}, e^{t\tilde{A}})\right)_{t\geq 0}$ on the Hilbert space $H \times \tilde{H}$. It comes from a form defined by

$$\left(\begin{pmatrix} x_1 \\ x_2 \end{pmatrix}, \begin{pmatrix} y_1 \\ y_2 \end{pmatrix}\right) \mapsto a(x_1, y_1) + \tilde{a}(x_2, y_2),$$

with form domain $V \times \tilde{V}$. Now, use (8.3) and apply Theorem 6.28. □

In this section we focus on the invariance of closed subspaces *under semigroups*, but let us mention that Corollary 4.32 and the formula for the resolvent operator of the reduction matrix in (4.17), together with Corollary 8.10, have the following straightforward consequence.

Corollary 8.11. *Let the assumptions of Corollary 8.10, and in particular let C be a closed subspace of the Hilbert space H.*

(1) Let additionally a be a form of Lions type, so that by Theorem 6.18 the associated operator A generates a C_0-cosine operator function $(C(t, A))_{t\geq 0}$. Then the assertions (a)–(c) in Corollary 8.10 are also equivalent to the following.

(a′) C *is invariant under* $(C(t, A))_{t\geq 0}$.

(2) Let additionally a be symmetric, so that by Theorem 7.7 the associated operator A generates a unitary C_0-group $(e^{itA})_{t\geq 0}$. Then the assertions (a)–(c) in Corollary 8.10 are also equivalent to the following.

(a″) C *is invariant under* $(e^{itA})_{t\in\mathbb{R}}$.

Example 8.12. When one tries to apply Corollary 8.10, the main difficulty is in most cases to determine the orthogonal projector onto C. Luckily, several orthogonal projector onto relevant closed subspaces are explicitly known. Consider for instance the *cut space* of a finite, oriented graph G, i.e., the range of the transpose of the incidence matrix $\mathscr{I}$ introduced in Definition 2.2: Then the orthogonal projector of $\ell^2(\mathsf{E})$ onto it can be found in [12, Thm. 14.8.1].

8.2 Graph Symmetries

Let us extend the notions of automorphisms—introduced in Definition A.6—to nodes and edges of a *weighted* graph.

We assume in the remainder of this chapter that[1]

$\mathfrak{G} = (\mathsf{V}, \mathfrak{E})$ is the metric graph over a locally finite, connected, weighted oriented graph $\mathsf{G} = (\mathsf{V}, \mathsf{E}, \rho)$.

Definition 8.13. A permutation O on V is called an *automorphism* of G if O preserves (weighted) node adjacency, i.e., $\alpha_{O\mathsf{v}\,O\mathsf{w}} = \alpha_{\mathsf{vw}}$ for all $\mathsf{v}, \mathsf{w} \in \mathsf{V}$, where $\mathscr{A} = (\alpha_{\mathsf{vw}})$ is the adjacency matrix of Definition 2.8.

Given an automorphism O, an *induced edge automorphism* $\tilde{O}$ is defined by[2]

$$\tilde{O}\mathsf{e} := (O\mathsf{v}, O\mathsf{w}) \qquad \text{whenever } \mathsf{e} = (\mathsf{v}, \mathsf{w}) \in \mathsf{E}. \tag{8.8}$$

We denote by $\mathrm{Aut}(\mathsf{G})$ and $\mathrm{Aut}'(\mathsf{G})$ the groups of all automorphisms and induced edge automorphisms of G, respectively.

By Remark 2.9, a permutation on V is—in the sense of the above definition—an automorphism with respect to some orientation of G if and only if it is an automorphism with respect to all orientations. Observe moreover that, by definition, ρ is constant along the orbits of any induced edge automorphism $\tilde{O}$.

The groups $\mathrm{Aut}(\mathsf{G}), \mathrm{Aut}'(\mathsf{G})$ will be in most cases isomorphic: The proof of Lemma A.8 carries over verbatim to the weighted case. However, the automorphism group of a weighted oriented graph does in general differ from the automorphism group of the associated unweighted graph.

Example 8.14. Let G be the oriented cycle on three edges, cf. Fig. 8.1.

---

[1] Unlike in the previous chapters we prefer to discuss simultaneously the operators on the graph and on the metric graph. For this reason we avoid to stress the difference between a resistance-like edge weight μ and a conductance-like edge weight γ and rather formulate all results for a generic weight ρ. This seems to be harmless, since no embedding in the Euclidean space will play a role in this chapter and therefore it is not relevant to make precise whether ρ should be thought of as directly or inversely proportional to the length of an edge.

[2] By the following notation we mean that the (unique) edge that has v, w as endpoints is mapped into the (unique) edge that has $O\mathsf{v}, O\mathsf{w}$ as endpoints. The edge orientations given originally are respected, and so are the weights, by definition of node automorphism: i.e., $\rho\left(\tilde{O}(\mathsf{v}, \mathsf{w})\right) = \rho(\mathsf{v}, \mathsf{w})$ whenever v, w are adjacent.

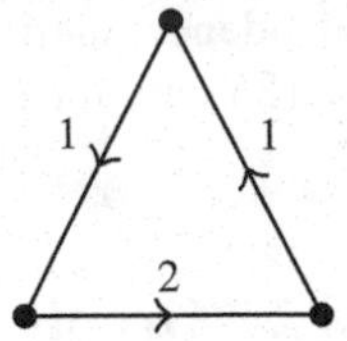

Fig. 8.1 A weighted graph with weighted automorphism group of order 2 and unweighted automorphism group of order 3

As long as no weight is assigned to the edges, all nodes (and all edges) can be mutually identified and Aut(G) is therefore isomorphic to the symmetric group S_3. If however we consider a weight function $(2, 1, 1)$, then the node that is *not* endpoint of the edge with weight 2 has to be fixed. Hence, Aut(G) becomes in this case isomorphic to C_2: The remaining nodes can only be either fixed or switched.

Each node permutation O induces two mappings on the vector spaces $\mathbb{C}^{\mathsf{V}}$ and $\mathbb{C}^{\mathsf{E}}$ defined by

$$f : \mathsf{v} \mapsto f(O\mathsf{v}) \qquad \text{and} \qquad u : \mathsf{e} \mapsto u(\tilde{O}\mathsf{e}), \tag{8.9}$$

respectively. We will not distinguish between the node/edge permutations and their associated operators as in (8.9), i.e., we adopt the notation.

$$(Of)(\mathsf{v}) := f(O\mathsf{v}) \quad \text{and} \quad (\tilde{O}u)(\mathsf{e}) := u(\tilde{O}\mathsf{e}), \qquad \mathsf{v} \in \mathsf{V},\ \mathsf{e} \in \mathsf{E}. \tag{8.10}$$

Concerning the metric graph $\mathfrak{G}$ associated with G, O induces also an operator that acts on functions defined on $\mathfrak{G}$ by

$$(\Omega u)(x) := \begin{cases} u_{\tilde{O}\mathsf{e}}\left(\frac{x}{\rho(\tilde{O}\mathsf{e})}\right), & \text{if } x \in (0, \rho(\mathsf{e})),\ \mathsf{e} \in \mathsf{E}, \\ u(O\mathsf{v}), & \text{if } x \in \mathsf{V}, \end{cases} \tag{8.11}$$

(recall that by definition a point of $\mathfrak{G}$ is either an element of V or an element of some metric edge). If we are instead interested in L^2-functions over the metric graphs, which are not defined in the Lebesgue null set V, we have simply

$$(\Omega u)_{\mathsf{e}} := u_{\tilde{O}\mathsf{e}}, \qquad u \in L^2\big((0,1); \ell^2_\rho(E)\big),\ \mathsf{e} \in \mathsf{E}.$$

The following observation will prove useful in view of Remark 8.9: We omit its straightforward proof.

Lemma 8.15. *If O is an automorphism on $\mathsf{G} = (\mathsf{V}, \mathsf{E}, \rho)$, then the mappings $O, \tilde{O}, \Omega$ are unitary operators on $\ell^2(\mathsf{V})$, $\ell^2_\rho(\mathsf{E})$, and $L^2\big((0,1; \ell^2_\rho(E)\big)$, respectively.*

Applying a node automorphism O to a function on V, in the sense of (8.10), means performing permutations of its nodal values that are more or less apparently compatible with its symmetries: rotating it, switching nodes, etc. In the rest of this section we devote our attention to the following apparently related question for the discrete diffusion equation (4.22):

Is it clear that the same results from

- first applying an automorphism to the initial data of (4.22) and then measuring the temperature f of the system after a time t, or else
- first letting the system evolve, then at time t applying an automorphism to the values of the temperature function f, and finally reading off the values of f in the nodes?

If G is finite and O is an automorphism of G, then for all $t \geq 0$ the semigroup $(e^{-t\mathscr{L}})_{t\geq 0}$ generated by the Laplace–Beltrami matrix $\mathscr{L}$ is given by (4.1) and

$$e^{-t\mathscr{L}}O = Oe^{-t\mathscr{L}} \qquad \text{for all } t \geq 0 :$$

Indeed, O clearly does not only commute with the adjacency matrix $\mathscr{A}$, but also with the degree matrix $\mathscr{D}$ and hence by Proposition 2.12 with $\mathscr{L}$. Thus, for all $t \geq 0$

$$\begin{aligned} e^{-t\mathscr{L}}O &= \sum_{k=0}^{\infty} \frac{(-1)^k t^k}{k!} \mathscr{L}^k O \\ &= \sum_{k=0}^{\infty} \frac{(-1)^k t^k}{k!} \mathscr{L}^{k-1} O \mathscr{L} \\ &= \ldots \\ &= O \sum_{k=0}^{\infty} \frac{(-1)^k t^k}{k!} \mathscr{L}^k = Oe^{-t\mathscr{L}} . \end{aligned}$$

Hence, the answer to the above question is positive. To treat the general case we present another proof that is based on Corollary 8.10 and contains an idea that will be recurrent in the rest of this chapter.

Proposition 8.16. *If $O \in Aut(\mathsf{G})$, then*

$$e^{-t\mathscr{L}^N}O = Oe^{-t\mathscr{L}^N} \quad \textit{and} \quad e^{-t\mathscr{L}^D}O = Oe^{-t\mathscr{L}^D} \qquad \textit{for all } t \geq 0,$$

where $\mathscr{L}^N, \mathscr{L}^D$ are the self-adjoint realizations of the Laplace–Beltrami matrix studied in Sect. 6.4.1.

Proof. The form a in (6.23) satisfies the condition (8.7) because

$$a(f, f) = \|\mathscr{I}^T f\|^2_{\ell^2_\rho(\mathsf{E})} = \|\mathscr{I}^T Of\|^2_{\ell^2_\rho(\mathsf{E})} = a(Of, Of) \quad \text{for all } f, g \in w^{1,2}_\rho(\mathsf{V}),$$

where the central identity is justified by the fact that $f \in w^{1,2}_\rho(\mathsf{V})$ if and only if $Of \in w^{1,2}_\rho(\mathsf{V})$, as node permutations can be interpreted as rearrangements of series: Due to absolute convergence, summability properties are not changed. Also in view of the fact that and using the fact that ρ is constant along induced orbits, this completely proves the claimed commutation relation for the semigroup $(e^{-t\mathcal{L}^N})_{t\geq 0}$ associated with a.

The same considerations apply to elements of $\mathring{w}^{1,2}_\rho(\mathsf{V})$ and thus yield the claim for $(e^{-t\mathcal{L}^D})_{t\geq 0}$, too. □

Remark 8.17. Let A be self-adjoint. One of the reasons why one is interested in discovering bounded linear operators R that commute with a semigroup $(e^{tA})_{t\geq 0}$—and then by Proposition 4.28.(1) with the resolvent operators $R(\mu, A)$, $\mu \in \rho(A)$, hence with A itself by standard properties of functional calculus for self-adjoint operators—is the following: If λ is an eigenvalue of A with associated eigenvalue $x \in D(A)$, then Rx satisfies

$$ARx = RAx = \lambda Rx.$$

Hence, if $Rx \neq \alpha x$ for any $\alpha \in \mathbb{C}$, one concludes that the geometric multiplicity of λ as an eigenvalue of A is at least 2.

A computation analogous to that in the proof of Proposition 8.16 holds for the signless incidence matrix $\mathcal{I}$ from Remark 2.4, thus each automorphism commutes with the Dirichlet and Neumann-like realizations of the signless Laplace–Beltrami matrix, too. Additionally, a corresponding commutation result can be proved for the semigroups generated by the advection and Kirchhoff matrices introduced in Sect. 6.4.2, this time checking condition (8.6) instead of (8.7), since the associated forms are not symmetric.

What about the operators that act on functions defined on the metric graph, like Ω introduced by means (8.11) and (8.8)? In the following we work in the setting of Sect. 6.5, also adopting the notation therein. In particular, $\nabla(c^2\nabla)$ is the elliptic operator with standard node conditions (Cc) – (KRc), cf. Definition 2.40.

We will assume in the remainder of this chapter that

$$\boxed{c \in L^\infty(0,1;\ell^\infty_\rho(\mathsf{E}))\text{ with } c_{\mathsf{e}}(x) \geq c_0 \text{ for some } c_0 > 0, \text{ all } \mathsf{e} \in \mathsf{E} \text{ and a.e. } x \in (0,1)}$$

and that

$$\boxed{\mathcal{W} \text{ is a bounded linear operator on } Y := \operatorname{Rg}\begin{pmatrix}(\mathcal{I}^+)^T\\(\mathcal{I}^-)^T\end{pmatrix} \simeq \ell^2_{\deg_\rho}(\mathsf{V}).}$$

Proposition 8.18. *Let $O \in Aut(\mathsf{G})$. Assume that O commutes with $\mathcal{W}$ and that c is constant along the orbits of $\tilde{O}$, i.e.,*

$$c_{\mathsf{e}}(x) = c_{\tilde{O}\mathsf{e}}(x) \qquad \text{for all } \mathsf{e} \in \mathsf{E} \text{ and a.e. } x \in (0,1).$$

Then

$$e^{t\nabla(c^2\nabla)}\Omega = \Omega e^{t\nabla(c^2\nabla)} \qquad \text{for all } t \geq 0.$$

Proof. The form domain $W_Y^{1,2}((0,1);\ell^2_\rho(\mathsf{E}))$ defined in (6.32) is left invariant under Ω: Indeed, if $u \in W_Y^{1,2}((0,1);\ell^2_\rho(\mathsf{E}))$, then clearly $\Omega u \in W^{1,2}(0,1;\ell^2_\rho(E))$ and moreover the continuity condition (Cc) is preserved, as one has

$$(\Omega u)_{\mathsf{e}}(\mathsf{v}) = u_{\tilde{O}\mathsf{e}}(O\mathsf{v}) = u_{\tilde{O}\mathsf{f}}(O\mathsf{v}) = (\Omega u)_{\mathsf{f}}(\mathsf{v}) \qquad \text{for all } \mathsf{e},\mathsf{f} \in \mathsf{E}_{\mathsf{v}} \text{ and all } \mathsf{v} \in \mathsf{V}.$$

Now, for all $u, v \in W_Y^{1,2}((0,1);\ell^2_\rho(\mathsf{E}))$ one has to check that (8.7) holds for the form $a_{\mathscr{W}}$ in (6.33). The corresponding equation reads

$$\sum_{\mathsf{e}\in\mathsf{E}} \int_0^1 c_{\mathsf{e}}(x)|u'_{\mathsf{e}}(x)|^2\rho(\mathsf{e})dx - \sum_{\mathsf{e}\in\mathsf{E}} \int_0^1 c_{\mathsf{e}}(x)|\Omega u'_{\mathsf{e}}(x)|^2\rho(\mathsf{e})dx$$
$$= (\mathscr{W}(\Omega u)_{|\mathsf{V}} \mid (\Omega u)_{|\mathsf{V}})_{\ell^2_{\deg_\rho}(\mathsf{V})} - (\mathscr{W} u_{|\mathsf{V}} \mid u_{|\mathsf{V}})_{\ell^2_{\deg_\rho}(\mathsf{V})} :$$

The right hand side—an inner product in Y—vanishes, simply because by assumption $\mathscr{W}$ and O commute and hence Corollary 8.8 applies. The left hand side vanish, too, by a change of variable $\mathsf{e} \mapsto \tilde{O}\mathsf{e}$ and using the fact that ρ is constant along induced orbits. □

Observe that under the assumptions of Proposition 8.18 the operator Ω leaves invariant the domain of $\nabla(c^2\nabla)$, and in particular functions that satisfy the standard node conditions (Cc) – (Kc) are mapped into functions that satisfy the same conditions. Indeed, one has

$$\Omega\nabla(c^2\nabla) = \nabla(c^2\nabla)\Omega.$$

Let us denote in the following by Aut($\mathfrak{G}$) the group of all unitary operators on $L^2((0,1);\ell^2_\rho(\mathsf{E}))$ that commute with $(e^{t\Delta})_{t\geq 0}$, where Δ is the second derivative with standard node conditions. We can recover a version of Frucht's Theorem A.9 for metric graphs. For the sake of simplicity we restrict to the unweighted case of $\rho \equiv 1$.

Corollary 8.19. *Let Γ be a (possibly infinite) group. Then there exists a metric graph $\mathfrak{G}$ such that Γ is isomorphic to a subgroup of Aut($\mathfrak{G}$).*

Proof. To begin with, apply Frucht's Theorem A.9 to find some graph G such that Aut (G) is isomorphic to Γ. Such a G can be chosen to be connected and with more than three nodes, hence by Lemma A.8 the group Aut(G) of all automorphisms and the group Aut′(G) of all induced edge automorphisms of G are isomorphic.

The operator Ω defined in (8.11) is unitary, since $\tilde{O}$ is a permutation, and the chain of identifications

$$O \mapsto \tilde{O} \mapsto \Omega$$

allow us to define a group

$$\mathfrak{b} := \{\Omega : O \in \mathrm{Aut}(\mathsf{G})\}$$

of unitary operators on $L^2\big((0,1);\ell^2_\rho(\mathsf{e})\big)$, which by construction is isomorphic to $\mathrm{Aut}(\mathsf{G})$ and hence to Γ.

It follows from Theorem 8.18 that each such Ω commutes with $\big(e^{t\Delta}\big)_{t\geq 0}$: This shows that $\mathfrak{b}$ is a subgroup of $\mathrm{Aut}(\mathfrak{G})$ and concludes the proof. □

8.3 Shortings of Nodes and Edges

The results in Sect. 8.2 show that the diffusion semigroup $(e^{-t\mathcal{L}})_{t\geq 0}$ respects the symmetries of the graph G, in the sense that if f is *symmetric with respect to an automorphism* O (meaning that $Of = f$, i.e., f is constant along the orbits of O), then also the solution of the abstract Cauchy problem associated with $-\mathcal{L}$ with initial data f is symmetric with respect to O for all $t \geq 0$. In the following we will instead especially focus on the operation of averaging a node function over certain subsets of V. In view the electrostatic interpretation of our network models (cf. Sect. 2.1.4.1), we will refer to it as *shorting*. This seems to be a more general—and, we argue, more flexible—notion of symmetry.

Definition 8.20. Given an oriented metric graph $\mathsf{G} = (\mathsf{V}, \mathsf{E}, \rho)$ and a node weight function $\nu : \mathsf{V} \to (0,\infty)$, a *node shorting operator* of G is any orthogonal projector of $\ell^2_\nu(\mathsf{V})$ onto the closed subspace

$$\{f \in \ell^2_\nu(\mathsf{V}) : f(\mathsf{v}) = f(\mathsf{w}) \text{ for all } \mathsf{v},\mathsf{w} \in \mathsf{V}_0\},$$

where V_0 is some subset of V; while for $\mathsf{E}_0 \subset \mathsf{E}$ an *edge shorting operator* of G is any orthogonal projector of $\ell^2_\rho(\mathsf{E})$ onto the closed subspace

$$\{u \in \ell^2_\rho(\mathsf{E}) : u(\mathsf{e}) = u(\mathsf{f}) \text{ for all } \mathsf{e},\mathsf{f} \in \mathsf{E}_0\}.$$

Likewise, given an edge subset E_0, we call the orthogonal projector of $L^2\big((0,1);\ell^2_\rho(\mathsf{E})\big)$ onto its closed subspace

$$\Big\{u \in L^2\big((0,1);\ell^2_\rho(\mathsf{E})\big) : u_\mathsf{e}(x) = u_\mathsf{f}(x) \text{ for a.e. } x \in (0,1) \text{ and all } \mathsf{e},\mathsf{f} \in \mathsf{E}_0\Big\}$$

a *shorting operator* of the metric graph $\mathfrak{G}$.

If we take an automorphism O and consider the partition of V induced by the orbits of O, Propositions 8.16–8.18 show that the shorting operator with respect to

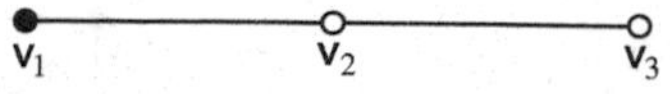

Fig. 8.2 Shorting two nodes of a path graph does generally not yield an invariant subsystem

each of its cells commutes with the respective C_0-semigroups. We will see in this section that partitions that come from automorphisms' orbits are not the only one that admit well-behaved shortings. Thus, such shorting procedures are rewarding: They may be used to reduce the complexity of the problem of solving a difference of differential equation in a non-obvious way.

Let us now identify classes of node subsets V_0 whose associated shorting operators commute with the C_0-semigroup generated by a matrix or a differential operator. To warm up, let us consider the easy case of $\mathsf{V}_0 = \mathsf{V}$. While a possible commutation of the operator that shorts *all nodes* with a semigroup that governs an abstract Cauchy problem does not yield much information, the associated shorting operator P is easy to find: If G has finite surface with respect to the node weight ν, cf. Definition A.17, then the all-node-shorting P is given by

$$Pf(\mathsf{v}) = \frac{1}{|\mathsf{V}|_\nu} \sum_{\mathsf{w}\in\mathsf{V}} f(\mathsf{w})\nu(\mathsf{w}), \qquad \mathsf{v} \in \mathsf{V},\ f \in \ell^2_\nu(\mathsf{V}). \tag{8.12}$$

This P projects onto the one-dimensional space spanned by $\mathbf{1}$, hence it certainly commutes with a semigroup if it is a spectral projector for the semigroup's generator A, i.e., if $\mathbf{1}$ is an eigenvector of A.

Example 8.21. (1) Let G be finite or, more generally, let G have finite volume with respect to the weight function ρ. Then $\mathbf{1} \in w^{1,2}_{\rho,\deg_\rho}(\mathsf{V})$ and the Laplace–Beltrami matrix $\mathscr{L}$ on $\ell^2_{\deg_\rho}(\mathsf{V})$ is not injective and $\mathbf{1} \in \operatorname{Ker}\mathscr{L}$ by Proposition 6.57. Hence, averaging a function $f : \mathsf{V} \to \mathbb{C}$ over *all* nodes (i.e., shorting all nodes) one obtains a new system which is trivially left invariant under the discrete heat equation, i.e., under $(e^{-t\mathscr{L}})_{t\geq 0}$, by Corollary 8.8 and because $a(\mathbf{1},\mathbf{1}) = 0$ for the associated sesquilinear form defined in (6.23); in particular this phenomenon is independent of the automorphism group of G. The same holds for $(e^{-t\mathscr{L}_{\text{norm}}})_{t\geq 0}$.

(2) By Proposition 6.58, also $(e^{-t\mathscr{Q}})_{t\geq 0}$—i.e., the semigroup generated by the signless Laplace–Beltrami matrix—commutes with the orthogonal projector defined in (8.12), provided G has some bipartite connected component.

(3) On the other hand, even in the unweighted case just shorting two arbitrary nodes is not sufficient to obtain an invariant subsystem under $(e^{-t\mathscr{L}})_{t\geq 0}$: Take e.g. a path of length 2 with $\rho \equiv 1$ and consider the orthogonal projector P of $\ell^2(\mathsf{V}) \equiv \mathbb{C}^3$ onto the space $\{f : \{\mathsf{v}_1, \mathsf{v}_2, \mathsf{v}_3\} \to \mathbb{C} : f(\mathsf{v}_2) = f(\mathsf{v}_3)\}$, (Fig. 8.2). Then, for $f(\mathsf{v}_n) := n$, $n = 1, 2, 3$, one has

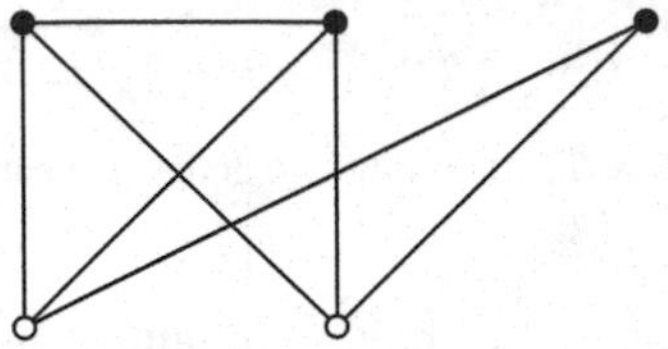

Fig. 8.3 A perturbation of a bipartite graph with respect to the partition $\mathsf{V} = \mathsf{V}_1 \dot\cup \mathsf{V}_2$, where $\mathsf{V}_1, \mathsf{V}_2$ comprise the lower and upper nodes, respectively

$$\left(\frac{3}{2}\right)^2 = a(Pf,Pf) > a(f,f) = 2.$$

Example 8.21.(3) suggests that shortings cannot generally be expected to be respected under time evolution of a diffusive problem unless they are compatible with the (weighted) connectivity of the graph.

One possible example of such a compatibility is given in the following, taken from [4], but similar results can actually be obtained under weaker assumptions—say, if G simply has a layer structure.

Proposition 8.22. *Let G be finite, unweighted (i.e., $\rho \equiv 1$), and orientedly bipartite (say, with $\mathsf{V} = \mathsf{V}_1 \dot\cup \mathsf{V}_2$). Let $\mathscr{W} = (\omega_{\mathsf{vw}})$ be a $\mathsf{V} \times \mathsf{V}$ matrix. Then the shorting operator defined in* (8.12) *commutes with $(e^{t\mathscr{W}})_{t\geq 0}$ if and only if there exist numbers $a_{11}, a_{12}, a_{21}, a_{22}$ such that*

$$\begin{aligned} a_{11} \deg(\mathsf{v}) &= \sum_{\mathsf{w}\in\mathsf{V}_1} \omega_{\mathsf{vw}}, \quad a_{12} \deg(\mathsf{v}) = \sum_{\mathsf{w}\in\mathsf{V}_2} \omega_{\mathsf{vw}} \text{ for all } \mathsf{v} \in \mathsf{V}_1, \quad \text{and} \\ a_{21} \deg(\mathsf{v}) &= \sum_{\mathsf{w}\in\mathsf{V}_1} \omega_{\mathsf{vw}}, \quad a_{22} \deg(\mathsf{v}) = \sum_{\mathsf{w}\in\mathsf{V}_2} \omega_{\mathsf{vw}} \text{ for all } \mathsf{v} \in \mathsf{V}_2. \end{aligned} \tag{8.13}$$

Example 8.23. (1) If $\mathscr{W} = \mathscr{A} = (\alpha_{\mathsf{vw}})$, the adjacency matrix of G, then Proposition 8.22 trivially applies, with $a_{11} = a_{22} = 0$ and $a_{12} = a_{21} = 1$.

(2) One may also apply Proposition 8.22 to prove once again that the shorting operator P defined in (8.12) commutes with the semigroup $(e^{-t\mathscr{L}})_{t\geq 0}$ generated by the discrete Laplacian or with the semigroup $(e^{t\mathscr{Q}})_{t\geq 0}$ generated by the signless Laplacian. Indeed condition (8.13) is satisfied with $\alpha_{11} = \alpha_{22} = 1$ and moreover $\alpha_{12} = \alpha_{21} = -1$ (for $\mathscr{L}$) or $\alpha_{12} = \alpha_{21} = 1$ (for $\mathscr{W} = \mathscr{Q}$).

We stress that the discrete Laplacian satisfies condition (8.13) also if new edges are added inside either of the partitions, cf. Fig. 8.3, as this graph perturbation does not affect the sums $\sum_{\mathsf{w}\in\mathsf{V}_1} \omega_{\mathsf{vw}}, \sum_{\mathsf{w}\in\mathsf{V}_2} \omega_{\mathsf{zw}}$ for $\mathsf{v} \in \mathsf{V}_1$ and $\mathsf{z} \in \mathsf{V}_2$.

(3) If $\mathscr{W} = \mathscr{L}$, then condition (8.13) may be satisfied even if the graph is not bipartite: An example is given by the graph in Fig. 8.3.

Example 8.23.(2) suggests that invariance of the all-node-shorting under time evolution of diffusive system is rather robust under certain graph perturbations. It is also noteworthy that in the case of the graph in Fig. 8.3 the all-node-shorting operator is not the only one that commutes with $\mathscr{L}$ and hence with $(e^{-t\mathscr{L}})_{t\geq 0}$: So does e.g. also the shorting operator defined by

$$Pf(\mathsf{v}) := \frac{1}{|\mathsf{V}_i|} \sum_{\mathsf{w} \in \mathsf{V}_i} f(\mathsf{w}) \qquad \text{for all } \mathsf{v} \in \mathsf{V}_i,\ i = 1, 2. \tag{8.14}$$

By definition, P averages the values of f over all nodes that belong to each cell of the partition (i.e., over the white and over the black nodes, respectively); it defines a doubly stochastic matrix, hence by Schur's test a bounded linear operator on $\ell^2(\mathsf{V})$.

We are soon going to see that this observation can be generalized and put into an abstract framework. It turns out that introducing a special class of partitions is critic to achieve this objective.

Definition 8.24. Let $\nu : \mathsf{V} \to (0, \infty)$. A partition of the node set V of G—i.e., a family $(\mathsf{V}_i)_{i \in I}$ of disjoint subsets of V whose union is V—is called *inward* or *outward almost equitable with respect to a node weight* ν *with cells* $(\mathsf{V}_i)_{i \in I}$ if for all $i, j \in I, i \neq j$,

$$\text{there are } c_{ij}^{\text{in}} \in \mathbb{R} \text{ s.t. } \sum_{\mathsf{w} \in \mathsf{V}_j} \iota_{\mathsf{we}}^{+} \rho\big((\mathsf{v}, \mathsf{w})\big) = c_{ij}^{\text{in}} \nu(\mathsf{v}) \qquad \text{for all } \mathsf{v} \in \mathsf{V}_i,\ \mathsf{e} := (\mathsf{v}, \mathsf{w}), \tag{8.15}$$

or

$$\text{there are } c_{ij}^{\text{out}} \in \mathbb{R} \text{ s.t. } \sum_{\mathsf{w} \in \mathsf{V}_j} \iota_{\mathsf{we}}^{-} \rho\big((\mathsf{v}, \mathsf{w})\big) = c_{ij}^{\text{out}} \nu(\mathsf{v}) \qquad \text{for all } \mathsf{v} \in \mathsf{V}_i,\ \mathsf{e} := (\mathsf{v}, \mathsf{w}), \tag{8.16}$$

respectively. It is called *almost equitable with respect to a node weight* ν (or simply *almost equitable* if $\nu \equiv 1$) if it is both inward and outward almost equitable with respect to the same node weight ν, i.e., if for all $i, j \in I, i \neq j$,

$$\text{there are } c_{ij} \in \mathbb{R} \text{ s.t. } \sum_{\mathsf{w} \in \mathsf{V}_j} \rho\big((\mathsf{v}, \mathsf{w})\big) = c_{ij} \nu(\mathsf{v}) \qquad \text{for all } \mathsf{v} \in \mathsf{V}_i. \tag{8.17}$$

It is called *equitable with respect to a node weight* ν (or simply *equitable* if $\nu \equiv 1$) if (8.17) holds for *all* $i, j \in I$, and not only whenever $i \neq j$.

Here and in the following we will assume for the sake of notational simplicity that a partition is indexed in $\mathbb{N}$, i.e.,

$$\boxed{(\mathsf{V}_i)_{i \in I} \equiv (\mathsf{V}_i)_{i = 1, 2, \ldots}} \tag{8.18}$$

Due to orientation-independence of the weights ρ, cf. Definition A.14, condition (8.17) and hence the very existence of an (almost) equitable partition does not depend on the orientation of G.

Remark 8.25. (1) In the unweighted case ($\rho \equiv 1, \nu \equiv 1$) existence of an inward equitable/outward equitable/equitable node partition amounts to saying that each node in V_i precedes/follows/is adjacent to exactly c_{ij} nodes in a cell V_j, respectively, for all $i, j \in I$. More generally, for $\nu \equiv 1$ (but general ρ) a partition is almost equitable if the weighted degree (i.e., the combined weight of all edges) towards the cell V_j is the same for all the nodes in V_i.

(2) If $\nu \equiv \deg_\rho$, then a partition is almost equitable if in each node $\mathsf{v} \in \mathsf{V}_i$ the combined weight of all edges towards the cell V_j—*in proportion* to the combined weight of *all* edges incident in v—only depends on V_i, and not on the individual node. In a graph with equitable partition the degrees of the nodes that belong to the same cell are necessarily the same (in this sense, equitable partition are proper generalizations of the partitions induced by orbits of groups of automorphisms). One thus sees that a partition is equitable with respect to $\nu \equiv 1$ if and only if it is equitable with respect to $\nu \equiv \deg_\rho$. The same equivalence fails to holds in the case of a partition that is merely *almost* equitable, cf. Fig. 8.6.

(3) There is an easy reason for introducing equitable partitions: Assume for the sake of simplicity that $\nu \equiv 1$ and define for a given partition $(\mathsf{V}_i)_{i\in I}$ of V an $I \times \mathsf{V}$ matrix $\mathscr{S} = (\sigma_{i\mathsf{v}})$ by

$$\sigma_{i\mathsf{v}} := \begin{cases} |\mathsf{V}_i|^{-\frac{1}{2}} & \text{if } \mathsf{v} \in \mathsf{V}_i, \\ 0 & \text{otherwise.} \end{cases}$$

Consider an $I \times I$ matrix

$$\mathscr{R} := \mathscr{S}\mathscr{A}\mathscr{S}^T.$$

Then it has been proved in [11, 22] that the partition is equitable if and only if $\mathscr{R}\mathscr{S} = \mathscr{S}\mathscr{A}$, and in this case for all $\lambda \in \mathbb{C}$ and all $x \in \mathbb{C}^I$

$$\mathscr{R}x = \lambda x \qquad \text{if and only if} \qquad \mathscr{A}\mathscr{S}^T x = \lambda \mathscr{S}^T x.$$

In particular, each eigenvalue of the so-called *quotient matrix* $\mathscr{R}$ is also an eigenvalue of $\mathscr{A}$.

(This result can be further refined if one considers the class of so-called *walk-regular graphs*, cf. [11, § 4] or [2, S VIII.3], for which the spectra of $\mathscr{R}$ and $\mathscr{A}$ agree—not counting multiplicities. An example is precisely the Petersen graph with respect to partition we have just considered.)

(4) Among others, the proof of the assertion reported in (3) uses the fact that $\mathscr{S}\mathscr{S}^T = \mathrm{Id}$. Instead, $\mathscr{S}^T\mathscr{S}$ is an orthogonal projector that takes the average of a function over all nodes that belong to a cell. More precisely,

$$Pf(\mathsf{v}) := \mathscr{S}^T\mathscr{S}f(\mathsf{v}) = \frac{1}{|\mathsf{V}_i|}\sum_{\mathsf{w}\in\mathsf{V}_i} f(\mathsf{w}), \qquad f \in \mathbb{C}^{\mathsf{V}},\ \mathsf{v} \in \mathsf{V}_i. \tag{8.19}$$

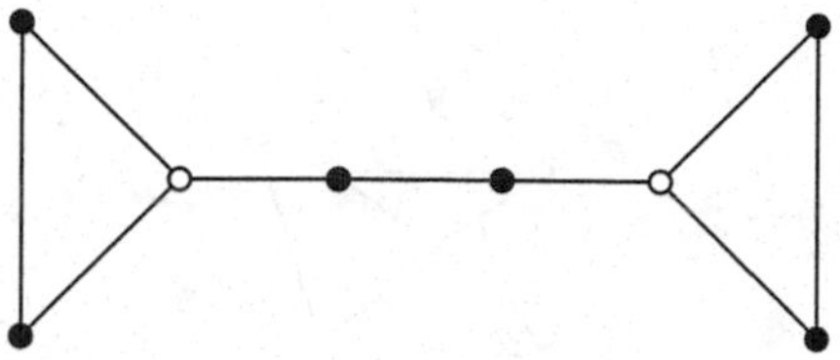

Fig. 8.4 A graph with an equitable partition that does not come from an automorphism. The two cells comprise the nodes painted in *white* and *black*, respectively. Condition (8.17) is satisfied with $c_{11} = c_{12} = 1, c_{21} = 2, c_{22} = 0$

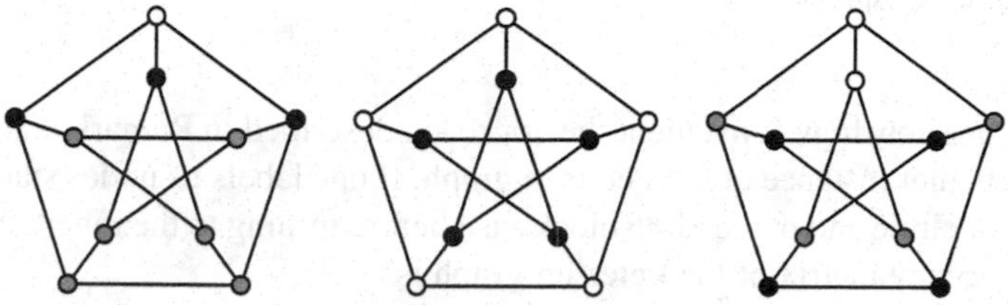

Fig. 8.5 Three equitable partitions of the unweighted Petersen graph: In each case, the cells comprise the nodes painted in the same color. In the former case, condition (8.17) is satisfied with $c_{12} = 3, c_{21} = 1, c_{23} = 2, c_{32} = 1, c_{33} = 2, c_{11} = c_{22} = 0$. (The same partition is equitable also with respect to the node weight $\nu = \deg_\rho$: Then, $c_{12} = 1$, $c_{21} = \frac{1}{3}$, $c_{23} = \frac{2}{3}$, $c_{32} = \frac{1}{3}$, $c_{33} = \frac{2}{3}$, $c_{11} = c_{22} = 0$.) With respect to the latter partition, condition (8.17) is satisfied with $c_{11} = c_{22} = 2$ and $c_{12} = c_{21} = 1$

Example 8.26. (1) A trivial equitable partition admitted by *every* graph is the one whose cells are all singletons.

For any graph, the one-cell partition of V—i.e., the partition for which all nodes belong to the same cell—is a further almost equitable partition. It is equitable if and only if the graph is regular (in the weighted sense of Definition 2.5).

(2) It is clear that if Γ is a subgroup of the automorphism group Aut(G), then the orbits of Γ define an equitable partition of G. However, not all equitable partitions of a graph G are the orbits of some subgroup of Aut(G). A simple, unweighted counterexample is shown in Fig. 8.4, taken from [10].

(3) Since each node has three neighbors, the Petersen graph in Example A.7 has two trivial equitable partitions $\mathcal{V}^{(1)}, \mathcal{V}^{(2)}$: those consisting of the whole node set and of singletons only, respectively.

Three further non-trivial equitable partitions $\mathcal{V}^{(3)}, \mathcal{V}^{(4)}, \mathcal{V}^{(5)}$ with respect to the edge weight $\rho \equiv 1$ and to the node weight $\nu \equiv 1$ are presented in Fig. 8.5. We stress that the second partition is induced by the graph automorphism group S_5, cf. Example A.7, but the other are not.

(By Theorem A.10, the Petersen graph is not Eulerian. Observe however that for all $k = 2, 3, 4$ each cell V_i in $\mathcal{V}^{(k)}$ induces an Eulerian subgraph, as c_{ii} is even for all $i \in I$.)

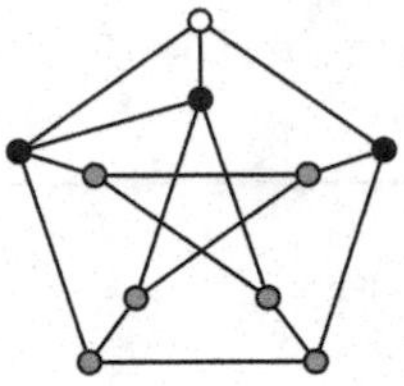

Fig. 8.6 This perturbation of the Petersen graph, for $\rho \equiv 1$, is a an example of a graph with an almost equitable partition that is not equitable (both with respect to $\nu \equiv 1$ and with respect to $\nu \equiv \deg_\rho$); and also of a graph with a partition that is almost equitable partition with respect to $\nu \equiv 1$ but not with respect to $\nu = \deg_\rho$.

Let us show how convenient the approach described in Remark 8.25.(3) is in the particular instance of the Petersen graph. If one labels its nodes starting with the outer circle and proceeding clockwise before turning to the inner circle, then the adjacency matrix of the Petersen graph is

$$\mathscr{A} = \begin{pmatrix} 0&1&0&0&1&1&0&0&0&0 \\ 1&0&1&0&0&0&1&0&0&0 \\ 0&1&0&1&0&0&0&1&0&0 \\ 0&0&1&0&1&0&0&0&1&0 \\ 1&0&0&1&0&0&0&0&0&1 \\ 1&0&0&0&0&0&0&1&1&0 \\ 0&1&0&0&0&0&0&0&1&1 \\ 0&0&1&0&0&1&0&0&0&1 \\ 0&0&0&1&0&1&1&0&0&0 \\ 0&0&0&0&1&0&1&1&0&0 \end{pmatrix},$$

whereas for the equitable partition denoted by $\mathscr{V}^{(3)}$

$$\mathscr{S} = \begin{pmatrix} 1 & 0 & 0 & 0 & 0 & 0 & 0 & 0 & 0 & 0 \\ 0 & \frac{1}{\sqrt{3}} & 0 & 0 & \frac{1}{\sqrt{3}} & \frac{1}{\sqrt{3}} & 0 & 0 & 0 & 0 \\ 0 & 0 & \frac{1}{\sqrt{6}} & \frac{1}{\sqrt{6}} & 0 & 0 & \frac{1}{\sqrt{6}} & \frac{1}{\sqrt{6}} & \frac{1}{\sqrt{6}} & \frac{1}{\sqrt{6}} \end{pmatrix}.$$

Then a direct computation yields

$$\mathscr{S}\mathscr{A}\mathscr{S}^T = \begin{pmatrix} 0 & \sqrt{3} & 0 \\ \sqrt{3} & 0 & \sqrt{2} \\ 0 & \sqrt{2} & 2 \end{pmatrix},$$

thus we can easily say that $-2, 1, 3$ are eigenvalues of $\mathscr{A}$. (Indeed, one can even show that there are no further eigenvalues, since $\mathscr{V}^{(3)}$ is compatible with the walk-regular structure of the Petersen graph.)

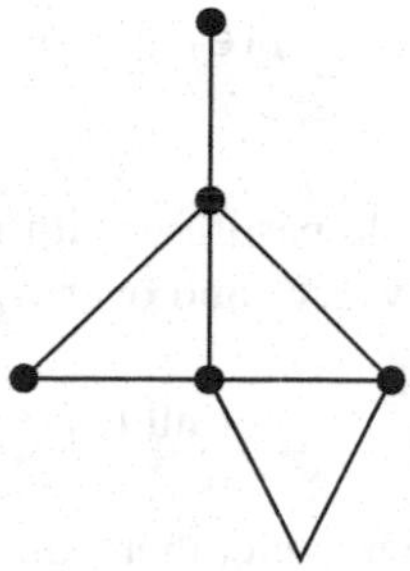

Fig. 8.7 An unweighted graph without any non-trivial almost equitable partition

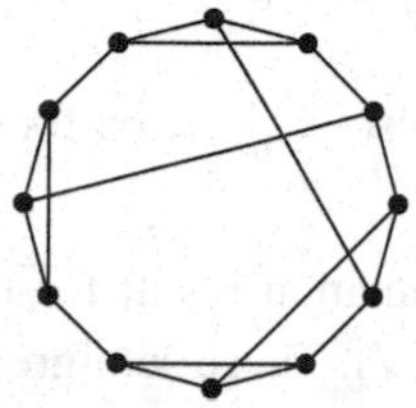

Fig. 8.8 The Frucht graph, an example of an unweighted graph that has only the trivial automorphism—i.e., each orbit is a singleton. However, being cubic it admits an equitable partition that does not consists solely of singletons

(4) More generally, if one takes a graph with an equitable partition and adds or deletes one edge between nodes that belong to the same cell as in Fig. 8.6, then the resulting graph has an almost equitable, non-equitable partition.
(4) There exists graphs without any non-trivial almost equitable partitions—and hence, *a fortiori* without any non-trivial automorphisms. An example (with respect to the edge weight $\rho \equiv 1$ and to the node weight $\nu \equiv 1$) is depicted in Fig. 8.7, taken from [5]
(6) In view of (2) a graph that is regular, but not node transitive, has an equitable partition whose sole cell contains all nodes, but which cannot be the orbit of any subgroup of Aut(G). An example is the so-called *Frucht graph* in Fig. 8.8, but actually by Frucht's Theorem A.9 there are uncountably many graphs with this property.

Remark 8.27. Any node partition induces an equivalence relation on V. If $\nu : \mathsf{V} \to (0, \infty)$, then we adopt the notation

$$|\mathsf{V}_i|_\nu := \sum_{\mathsf{v}\in\mathsf{V}_i} \nu(\mathsf{v}), \qquad i \in I.$$

Additionally, each node partition $(\mathsf{V}_i)_{i\in I}$ of a graph canonically induces an edge partition $(\mathsf{E}_{ij})_{i,j\in I}$, where the cells E_{ij} consist of all edges with initial endpoint in V_i and terminal endpoint in V_j. As above we write

$$|\mathsf{E}_{ij}|_\rho := \sum_{\mathsf{e}\in\mathsf{E}_{ij}} \rho(\mathsf{e}), \qquad i, j \in I. \tag{8.20}$$

If $(\mathsf{V}_i)_{i\in I}$ is an almost equitable partition with respect to ν, then summing all identities of the form (8.17) over $\mathsf{v} \in \mathsf{V}_i$ and over $\mathsf{w} \in \mathsf{V}_j$ yields

$$|\mathsf{E}_{ij}|_\rho = c_{ij}|\mathsf{V}_i|_\nu = c_{ji}|\mathsf{V}_j|_\nu \qquad \text{for all } i, j \in I \text{ such that } i \neq j. \tag{8.21}$$

(Let us remark that if $(\mathsf{V}_i)_{i\in I}$ is equitable, then additionally

$$|\mathsf{E}_{ii}|_\rho = \frac{1}{2}c_{ii}|\mathsf{V}_i|_\nu \qquad \text{for all } i \in I.$$

If in particular $|\mathsf{V}_i|_\nu < \infty$, then also $|\mathsf{E}_{ij}|_\rho < \infty$ for each $j \neq i$, and also for $i = j$ if the partition is equitable.)

We can finally show a commutation result for the C_0-semigroups $(e^{-t\mathscr{L}^N})_{t\geq 0}$ and $(e^{-t\mathscr{L}^D})_{t\geq 0}$ that act on $\ell^2_{\deg_\rho}(\mathsf{V})$: These semigroups exist by Lemma 4.4.

Theorem 8.28. *Let* $\mathsf{G} = (\mathsf{V}, \mathsf{E}, \rho)$ *have an almost equitable partition (with respect to the node weight* $\nu \equiv 1$*). Assume its cells* $(\mathsf{V}_i)_{i\in I}$ *to be finite for all* $i \in I$ *and consider the associated cellwise averaging operator* P *defined in* (8.19)*. Then*

$$e^{-t\mathscr{L}^N} P = P e^{-t\mathscr{L}^N} \quad \text{and} \quad e^{-t\mathscr{L}^D} P = P e^{-t\mathscr{L}^D} \qquad \text{for all } t \geq 0.$$

A graph with an equitable partition should be thought of as consisting of different layers (the cells) that are connected by edges that display a certain regularity in their distribution. Because of the invariance of $\mathscr{L}$ under edge re-orienting of G, without loss of generality we may and do assume all edges to be oriented in such a way that

$$\boxed{c_{ij}^{\text{in}} = 0, \quad c_{ij}^{\text{out}} = c_{ij}, \quad c_{ji}^{\text{out}} = 0, \quad c_{ji}^{\text{in}} = c_{ji} \qquad \text{for all } i, j \in I \text{ with } i < j.}$$

Proof. We have to check condition (e) in Corollary 8.10 for the symmetric form a introduced in (6.23), i.e.,

$$\|\mathscr{I}^T Pf\|^2_{\ell^2_\rho(\mathsf{E})} \leq \|\mathscr{I}^T f\|^2_{\ell^2_\rho(\mathsf{E})} \qquad \text{for all } f \in w^{1,2}_\rho(\mathsf{V}). \tag{8.22}$$

First of all, define a cellwise averaging operator by

$$\tilde{P}u(\mathsf{e}) := \frac{1}{|\mathsf{E}_{ij}|_\rho} \sum_{\mathsf{f}\in\mathsf{E}_{ij}} u(\mathsf{f})\rho(\mathsf{f}), \qquad u \in \ell^2_\rho(\mathsf{E}),\ \mathsf{e} \in \mathsf{E}_{ij}, \tag{8.23}$$

which is an orthogonal projector of the Hilbert space $\ell^2_\rho(\mathsf{E})$ onto the closed subspace of all functions that are constant along cells E_{ij}. For all $f \in \ell^2(\mathsf{V})$ and all $i, j \in I$

- if $\mathsf{e} \in \mathsf{E}_{ii}$, then $\mathscr{I}^T Pf(\mathsf{e}) = 0$ since $Pf(\mathsf{v}) = Pf(\mathsf{w})$ for all $\mathsf{v}, \mathsf{w} \in \mathsf{V}_i$,
- if $\mathsf{e} \in \mathsf{E}_{ij}$ with $i \neq j$, and hence $\mathsf{e}_{\mathrm{term}} \in \mathsf{V}_j$ and $\mathsf{e}_{\mathrm{init}} \in \mathsf{V}_i$, then

$$\begin{aligned}
\mathscr{I}^T Pf(\mathsf{e}) &= Pf(\mathsf{e}_{\mathrm{term}}) - Pf(\mathsf{e}_{\mathrm{init}}) \\
&= \frac{1}{|\mathsf{V}_j|} \sum_{\mathsf{v}\in\mathsf{V}_j} f(\mathsf{v}) - \frac{1}{|\mathsf{V}_i|} \sum_{\mathsf{w}\in\mathsf{V}_i} f(\mathsf{w}) \\
&\overset{(*)}{=} \frac{c_{ji}}{|\mathsf{E}_{ij}|_\rho} \sum_{\mathsf{v}\in\mathsf{V}_j} f(\mathsf{v}) - \frac{c_{ij}}{|\mathsf{E}_{ij}|_\rho} \sum_{\mathsf{w}\in\mathsf{V}_i} f(\mathsf{w}) \\
&= \frac{c_{ji}^{\mathrm{in}}}{|\mathsf{E}_{ij}|_\rho} \sum_{\mathsf{v}\in\mathsf{V}_j} f(\mathsf{v}) - \frac{c_{ij}^{\mathrm{out}}}{|\mathsf{E}_{ij}|_\rho} \sum_{\mathsf{w}\in\mathsf{V}_i} f(\mathsf{w}) \\
&\overset{(**)}{=} \frac{1}{|\mathsf{E}_{ij}|_\rho} \sum_{\mathsf{e}\in\mathsf{E}_{ij}} f(\mathsf{e}_{\mathrm{term}})\rho(\mathsf{e}) - \frac{1}{|\mathsf{E}_{ij}|_\rho} \sum_{\mathsf{e}\in\mathsf{E}_{ij}} f(\mathsf{e}_{\mathrm{init}})\rho(\mathsf{e}) \\
&= \frac{1}{|\mathsf{E}_{ij}|_\rho} \sum_{\mathsf{e}\in\mathsf{E}_{ij}} \left(f(\mathsf{e}_{\mathrm{term}}) - f(\mathsf{e}_{\mathrm{init}})\right) \rho(\mathsf{e}) \\
&= \tilde{P} \mathscr{I}^T f(\mathsf{e}).
\end{aligned}$$

Here $(*)$ follows from (8.21) and $(**)$ is a direct consequence of Lemma 2.1 applied to the subgraph of G that consists solely of the edges in E_{ij} and their endpoints. (The rearranging of the sums in the last step is justified by the fact the edge cells E_{ij} are finite, as so are the node cells $\mathsf{V}_i, \mathsf{V}_j$ by assumption.)

In other words, for all $i, j \in I$ and all $\mathsf{e} \in \mathsf{E}$

$$\mathscr{I}^T Pf(\mathsf{e}) = \begin{cases} \tilde{P}\mathscr{I}^T f(\mathsf{e}) & \mathsf{e} \in \mathsf{E}_{ij},\ i \neq j, \\ 0 & \mathsf{e} \in \mathsf{E}_{ii}. \end{cases}$$

In view of the above observations, one has for all $f \in w_\rho^{1,2}(\mathsf{V})$

$$\begin{aligned}
\sum_{\mathsf{e}\in\mathsf{E}} \rho(\mathsf{e})|\mathscr{I}^T Pf(\mathsf{e})|^2 &= \sum_{\substack{i,j\in I\\ i\neq j}} \sum_{\mathsf{e}\in\mathsf{E}_{ij}} \rho(\mathsf{e})|\mathscr{I}^T Pf(\mathsf{e})|^2 + \sum_{i\in I} \sum_{\mathsf{e}\in\mathsf{E}_{ii}} \rho(\mathsf{e})|\mathscr{I}^T Pf(\mathsf{e})|^2 \\
&= \sum_{\substack{i,j\in I\\ i\neq j}} \sum_{\mathsf{e}\in\mathsf{E}_{ij}} \rho(\mathsf{e})|\tilde{P}\mathscr{I}^T f(\mathsf{e})|^2 \\
&= \sum_{\substack{i,j\in I\\ i\neq j}} \sum_{\mathsf{e}\in\mathsf{E}_{ij}} \rho(\mathsf{e}) \left| \frac{1}{|\mathsf{E}_{ij}|_\rho} \sum_{\mathsf{f}\in\mathsf{E}_{ij}} \rho(\mathsf{f})\mathscr{I}^T f(\mathsf{f}) \right|^2
\end{aligned}$$

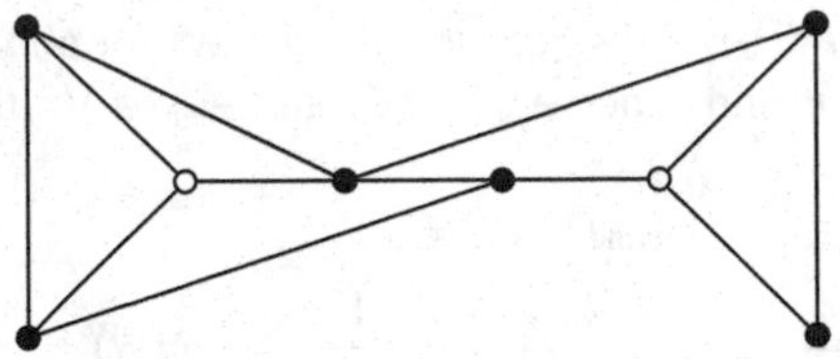

Fig. 8.9 If f attains the same value in the black nodes, and if it attains the same value in the white nodes, so does $e^{-t\mathscr{L}} f$ for all $t > 0$

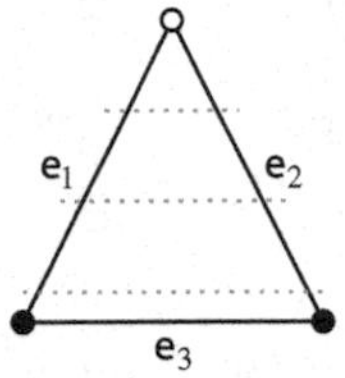

Fig. 8.10 Shorting the points of a graph in a radial manner does in general not yield a subspace that remains invariant under time evolution. In this case, "radial" means that we are shorting—i.e., identifying—all points of the graph which we have drawn as having the same ordinate, including all those on the edge e_3

$$\begin{aligned}
&\overset{(***)}{\leq} \sum_{\substack{i,j\in I\\ i\neq j}} \sum_{\mathsf{e}\in\mathsf{E}_{ij}} \frac{\rho(\mathsf{e})}{|\mathsf{E}_{ij}|_\rho} \sum_{\mathsf{f}\in\mathsf{E}_{ij}} \rho(\mathsf{f}) \left|\mathscr{I}^T f(\mathsf{f})\right|^2 \\
&= \sum_{\substack{i,j\in I\\ i\neq j}} \sum_{\mathsf{f}\in\mathsf{E}_{ij}} \rho(\mathsf{f}) \left|\mathscr{I}^T f(\mathsf{f})\right|^2 \\
&\leq \sum_{\mathsf{e}\in\mathsf{E}} \rho(\mathsf{e}) |\mathscr{I}^T f(\mathsf{e})|^2,
\end{aligned}$$

where $(***)$ follows from Bessel's inequality. This shows that (8.22) holds and completes the proof. □

Example 8.29. The almost equitable partition for the graph in Fig. 8.9 shows how rough can be graphs that satisfy the assumptions of Theorem 8.28.

Let us turn to metric graphs $\mathfrak{G} = (\mathsf{V}, \mathfrak{E})$ built upon discrete graphs $\mathsf{G} = (\mathsf{V}, \mathsf{E})$ with an equitable partition. If one looks for an analog of Theorem 8.28, one may conjecture that the correct pendant of the P in (8.28) is the operator that projects onto "radial" functions or rather, more generally, the one that shorts all points with the same coordinate belonging over all (metric) edges in the same cell E_{ij} for $i \neq j$; and that shorts *all* points in the same cell E_{ii} for each i. Unfortunately, by Corollary 8.10 this projector does not commute with the semigroup even in very simple cases, as the first condition in (b) therein is not satisfied.

Indeed, shorting the graph in Fig. 8.10 in this radial manner one obtains a corresponding orthogonal projector defined by

$$u \mapsto \begin{pmatrix} \frac{u_1+u_2}{2} \\ \frac{u_1+u_2}{2} \\ \int_0^1 u_3(y)\, dy \cdot 1 \end{pmatrix},$$

but the range of this operator also contains functions that are not continuous in the nodes: Take e.g. any function $u \geq 0$ whose nonempty support is contained in the interior of e_3.

However, the following analog of Theorem 8.28 does hold.

Theorem 8.30. *Let* $\mathsf{G} = (\mathsf{V}, \mathsf{E}, \rho)$ *have an almost equitable partition (with respect to the node weight* $\nu \equiv 1$*). Assume its cells* $(\mathsf{V}_i)_{i \in I}$ *to be finite for all* $i \in I$ *and additionally that*

$$\boxed{c_{ij}^{in} = 0, \quad c_{ij}^{out} = c_{ij}, \quad c_{ji}^{out} = 0, \quad c_{ji}^{in} = c_{ji} \qquad \text{for all } i, j \in I \text{ with } i < j} \tag{8.24}$$

and

$$\boxed{c_{ii}^{in} = c_{ii}^{out} \qquad \text{for all } i \in I,} \tag{8.25}$$

with the notation of Definition 8.24.

Consider the averaging operator

$$\Pi u_{\mathsf{e}}(x) := \frac{1}{|\mathsf{E}_{ij}|_\rho} \sum_{\mathsf{f} \in \mathsf{E}_{ij}} u_{\mathsf{f}}(x) \rho(\mathsf{f}), \qquad u \in L^2\big((0,1); \ell^2_\rho(\mathsf{E})\big),\ x \in (0,1),\ \mathsf{e} \in \mathsf{E}_{ij}. \tag{8.26}$$

If c *is cell-wise constant, i.e., for all* $i, j \in I$ *there exist functions* $c_{\mathsf{E}_{ij}}$*, such that*

$$c_{\mathsf{e}}(x) = c_{\mathsf{E}_{ij}}(x) \quad \text{for all } \mathsf{e} \in \mathsf{E}_{ij} \text{ and a.e. } x \in (0,1),$$

then

$$e^{t\nabla(c^2\nabla)}\Pi = \Pi e^{t\nabla(c^2\nabla)} \qquad \text{for all } t \geq 0.$$

Here $\nabla(c^2\nabla)$ *is the elliptic operator with node conditions (Cc) – (Kc).*

Under the assumptions (8.24)–(8.25), and with the notation in (8.18), one has

$$\mathsf{E}_{ij} = \emptyset \quad \text{if } j \neq i \text{ or } j \neq i+1. \tag{8.27}$$

Furthermore, for all $i \in I$ the subgraph of G induced by V_i is orientedly Eulerian, cf. Definition A.3. Observe that

$$\Pi u(x) = \tilde{P}(u(x)), \qquad \text{for a.e. } x \in (0,1), \tag{8.28}$$

where $\tilde{P}$ is the orthogonal projector defined in (8.26). In other words, Π is an orthogonal projector that acts by shorting all points which—with respect to the chosen parametrization of the set of metric edges—share the same coordinate.

Proof. We focus for the sake of simplicity on the case of $\mathscr{W} = 0$, i.e., of Kirchhoff node conditions without any absorption term: In this way $\nabla(c^2\nabla)$ is certainly a self-adjoint operator, hence the associated form is symmetric. The general case will be briefly discussed afterwards.

It suffices to check condition (e) in Corollary 8.10 for the quadratic form $a_{\mathscr{W}}$ (with $\mathscr{W} = 0$) associated with $\nabla(c^2\nabla)$, cf. Sect. 6.5, i.e.,

$$a(u,v) := \sum_{\mathsf{e}\in\mathsf{E}_{ij}} \int_0^1 c_{\mathsf{e}}^2(x) u_{\mathsf{e}}'(x)\overline{v_{\mathsf{e}}'(x)}\rho(\mathsf{e})\, dx,$$

defined on the space of $W^{1,2}$-functions on the metric graph $\mathfrak{G}$ that are continuous in the nodes of G, i.e., on

$$W_Y^{1,2}\big((0,1);\ell^2_\rho(\mathsf{E})\big) := \left\{u \in W^{1,2}\big((0,1);\ell^2_\rho(\mathsf{E})\big) : \begin{pmatrix} u(1) \\ u(0) \end{pmatrix} \in \operatorname{Rg}\begin{pmatrix} (\mathscr{I}^+)^T \\ (\mathscr{I}^-)^T \end{pmatrix}\right\}.$$

By definition $\tilde{P}$ and hence Π act as linear combinations, hence they commute with the first derivative: Therefore, $\Pi u \in W^{1,2}\big((0,1);\ell^2_\rho(\mathsf{E})\big)$ whenever $u \in W^{1,2}\big((0,1);\ell^2_\rho(\mathsf{E})\big)$. We still have to check that Πu is continuous in the nodes, provided so is u. Let $\mathsf{v} \in \mathsf{V}_j$. If $\mathsf{e} \in \mathsf{E}_{ij}$ with $i = j-1$, then

$$\begin{aligned}(\Pi u)_{\mathsf{e}}(\mathsf{v}) &= \frac{1}{|\mathsf{E}_{ij}|_\rho} \sum_{\mathsf{f}\in\mathsf{E}_{ij}} u_{\mathsf{f}}(\mathsf{f}_{\mathrm{term}})\rho(\mathsf{f}) \\ &\overset{(*)}{=} \frac{1}{|\mathsf{E}_{ij}|_\rho} \sum_{\mathsf{v}\in\mathsf{V}_j} c_{ji} u(\mathsf{v}) \\ &\overset{(**)}{=} \frac{1}{|\mathsf{V}_j|} \sum_{\mathsf{v}\in\mathsf{V}_j} u(\mathsf{v}),\end{aligned}$$

where $(*)$ and $(**)$ hold by Lemma 2.1 and by (8.21), respectively. Similar computations show that $(\Pi u)_{\mathsf{e}}(\mathsf{v}) = \frac{1}{|\mathsf{V}_j|} \sum_{\mathsf{v}\in\mathsf{V}_j} u(\mathsf{v})$ if $\mathsf{e} \in \mathsf{E}_{jj}$ or $\mathsf{e} \in \mathsf{E}_{jk}$ for $k = j+1$. Thus, the value of Πu in the nodes is independent of the incident edge, hence the continuity condition (Cc) is satisfied.

In view of Corollary 8.10, and observing that a is symmetric, it now suffices to show that $a(\Pi u, \Pi u) \le a(u,u)$ holds for all $u \in W^{1,2}\big((0,1);\ell^2_\rho(\mathsf{E})\big)$. Indeed, one has by (8.27)

$$a(\Pi u, \Pi u) = \sum_{\mathsf{e}\in\mathsf{E}} \int_0^1 c_{\mathsf{e}}^2(x) |(\Pi u)_{\mathsf{e}}'(x)|^2 \rho(\mathsf{e})\, dx$$

$$= \sum_{\substack{i,j \in I \\ i \leq j}} c^2_{\mathsf{E}_{ij}}(x) \sum_{\mathsf{e} \in \mathsf{E}_{ij}} \int_0^1 \left| \frac{1}{|\mathsf{E}_{ij}|_\rho} \sum_{\mathsf{f} \in \mathsf{E}_{ij}} u'_\mathsf{f}(x)\rho(\mathsf{f}) \right|^2 \rho(\mathsf{e})\, dx$$

$$\stackrel{(***)}{\leq} \sum_{\substack{i,j \in I \\ i \leq j}} c^2_{\mathsf{E}_{ij}}(x) \sum_{\mathsf{e} \in \mathsf{E}_{ij}} \frac{\rho(\mathsf{e})}{|\mathsf{E}_{ij}|_\rho} \sum_{\mathsf{f} \in \mathsf{E}_{ij}} \int_0^1 |u'_\mathsf{f}(x)|^2 \rho(\mathsf{f})\, dx$$

$$= \sum_{\substack{i,j \in I \\ i \leq j}} \sum_{\mathsf{f} \in \mathsf{E}_{ij}} \int_0^1 c^2_\mathsf{e}(x) |u'_\mathsf{f}(x)|^2 \rho(\mathsf{f})\, dx$$

$$\leq a(u,u),$$

where $(***)$ is justified again by Bessel's inequality. □

8.3.1 General Symmetries of Equations on Metric Graphs

Shorting the points of a network, in the sense of Definition 8.20, means imposing the same value (of temperature, density, etc.—depending on the model) on the points of all edges in the metric graph that share the same parametrization, provided they belong to a certain relevant subset of the metric edge set $\mathfrak{E}$. In this section we are going to make a case for the introduction of a generalized notion of shorting.

Let $\tilde{P}$ be an orthogonal projector of $\ell^2_\rho(\mathsf{E})$ and consider

$$\left\{ u \in L^2\big((0,1); \ell^2_\rho(\mathsf{E})\big) : u(x) \in \operatorname{Rg} \tilde{P} \text{ for a.e. } x \in (0,1) \right\}. \tag{8.29}$$

Invariance of this kind of subspaces under the C_0-semigroup $(e^{t\nabla(c^2\nabla)})_{t\geq 0}$ (cf. Sect. 6.5) will be the topic of this section.

The orthogonal projector Π of $L^2\big((0,1); \ell^2_\rho(\mathsf{E})\big)$ onto the closed subspace defined in (8.29) is formally given in (8.28), even if $\tilde{P}$ is now a more general operator. The shorting operators introduced in Sect. 8.3 are special instances of operators of this kind: For this reason we refer to such Π as *generalized shorting operators*.

Example 8.31. The range of Π—i.e., the space in (8.29)—is finite dimensional only if $\tilde{P} \equiv 0$. Hence, no rank-1 operator (see Definition 4.36)—and in particular not the functional that maps each function into its average—can be a generalized shorting operator.

With the notations of (6.32) and (6.29), by Corollary 8.10 the semigroup generated by the elliptic operator $\nabla(c^2\nabla)$ with standard node conditions (Cc) –

(KRc) (which is possibly non self-adjoint, depending on the operator $\mathcal{W}$ that appears in the node conditions) leaves invariant Rg Π if and only if

- $\Pi\left(W_Y^{1,2}((0,1);\ell^2_\rho(\mathsf{E}))\right) \subset W_Y^{1,2}((0,1);\ell^2_\rho(\mathsf{E}))$ and
- $a(\Pi u, u - \Pi u) = 0$ for all $u \in W_Y^{1,2}((0,1);\ell^2_\rho(\mathsf{E}))$.

The latter condition is equivalent to

$$\int_0^1 (c^2 \Pi u' \mid u' - \Pi u')_{\ell^2_\rho(\mathsf{E})}\, dx + (\mathcal{W}(\Pi u)_{|\mathsf{V}} \mid (u - \Pi u)_{|\mathsf{V}})_{\ell^2_{\deg_\rho}(\mathsf{V})} \stackrel{!}{=} 0$$

for all $u \in W_Y^{1,2}((0,1);\ell^2_\rho(\mathsf{E}))$. (Observe that in order to write $(\Pi u)_{|\mathsf{V}}$ one already has to know that Πu is continuous in the nodes, i.e., the first condition already has to hold.) But the first addend always vanishes if (an analog of) Theorem 8.30 applies, hence the condition boils down to checking that

$$(\mathcal{W}(\Pi u)_{|\mathsf{V}} \mid (u - \Pi u)_{|\mathsf{V}})_{\ell^2_{\deg_\rho}(\mathsf{V})} \stackrel{!}{=} 0 \qquad \text{for all } u \in W_Y^{1,2}((0,1);\ell^2_\rho(\mathsf{E})).$$

In the remainder of this section we will mostly focus on the analysis of the first condition in dependence on the orthogonal projector $\tilde{P}$ on $\ell^2_\rho(\mathsf{E})$ associated with Π—an issue that turns out to be intimately related to structural property of graphs.

Definition 8.32. We call the generalized shorting operator $\tilde{P}$ *admissible* if Π in (8.28) maps $W_Y^{1,2}((0,1);\ell^2_\rho(\mathsf{E}))$ into itself.

We stress that admissibility does depend on the orientation of the graph. Now, the proof of Theorem 8.30 prevails for generalized shorting operators whenever $\tilde{P}$ is admissible. Thus the main question is to understand how admissibility can be characterized.

Lemma 8.33. *Let* G *be finite and let* Π *be a generalized shorting operator associated (via* (8.28)*) with an orthogonal projector* $\tilde{P}$ *on* $\ell^2_\rho(\mathsf{E})$*). If* $\tilde{P}$ *is admissible, then* $\mathbf{1}$ *is an eigenvector of* $\tilde{P}$.

Proof. Consider the constant function $\mathbf{1} : [0,1] \ni x \mapsto (1,\dots,1)^\top \in \mathbb{C}^{\mathsf{E}}$ and observe that $\mathbf{1} \in W_Y^{1,2}((0,1);\ell^2_\rho(\mathsf{E}))$ and $\Pi\mathbf{1}(x) = \tilde{P}\mathbf{1}$ for a.e. $x \in (0,1)$. This shows that on each edge $\Pi\mathbf{1}$ is a constant function.

Since by hypothesis $\Pi\left(W_Y^{1,2}((0,1);\ell^2_\rho(\mathsf{E}))\right) \subset W_Y^{1,2}((0,1);\ell^2_\rho(\mathsf{E}))$ and hence $\Pi\mathbf{1} \in W_Y^{1,2}((0,1);\ell^2_\rho(\mathsf{E}))$, all these (constant) edge values coincide, hence $\Pi\mathbf{1} = \alpha\mathbf{1}$ for a scalar α. □

Remark 8.34. Observe that $\tilde{P}\mathbf{1} \in \{0, \mathbf{1}\}$, since the only eigenvalues of an orthogonal projector are 0 and 1, and that $\mathbf{1} \in \operatorname{Ker}(\mathrm{Id} - \tilde{P})$ if $\mathbf{1} \in \operatorname{Rg}\tilde{P}$. Moreover, $\tilde{P}$ is admissible if and only if $\mathrm{Id} - \tilde{P}$ is admissible. Therefore, we can assume without loss of generality that $\mathbf{1} \in \operatorname{Rg}\tilde{P}$.

This observation can be used especially to deliver an alternative proof of irreducibility of the semigroup generated by the elliptic operator $\nabla(c^2\nabla)$ with node conditions (Cc) – (KRc) (cf. Proposition 6.77), provided G is finite and connected: To this aim we have to show that there exists no proper subgraph G' of G (say, with edge set E') such that the closed subspace

$$\{f \in L^2\big((0,1);\ell^2_\rho(\mathsf{E})\big) : f_\mathsf{e} \equiv 0 \text{ for all } \mathsf{e} \in \mathsf{E}'\}$$

is invariant under $(e^{t\nabla(c^2\nabla)})_{t\geq 0}$. The orthogonal projector onto it is given by Π as in (8.28), where $\tilde{P}$ is the $\mathsf{E}' \times (\mathsf{E} \setminus \mathsf{E}')$ diagonal block matrix

$$\tilde{P} = \begin{pmatrix} \mathrm{Id} & 0 \\ 0 & 0 \end{pmatrix}.$$

Since $\mathbf{1}$ is not an eigenvector of $\tilde{P}$, by Lemma 8.33 and Corollary 8.10 $\{f \in L^2(\mathfrak{G}) : f_{|\mathfrak{G}'} = 0\}$ is not invariant under $(e^{\nabla(c^2\nabla)})_{t\geq 0}$, independently of the elliptic coefficient c^2 and on the term $\mathscr{W}$ that appears in the node conditions.

Some classes of oriented graphs can even be characterized in terms of the admissibility of the all-edge-shorting

$$\tilde{P} u_\mathsf{e} = \frac{1}{|\mathsf{E}|_\rho} \sum_{\mathsf{f} \in \mathsf{E}} u_\mathsf{f} \rho(\mathsf{f}), \qquad \mathsf{e} \in \mathsf{e},\ u \in \mathbb{C}^\mathsf{E}. \tag{8.30}$$

Theorem 8.35. *Let G have finite volume. Consider the orthogonal projector $\tilde{P}$ defined in* (8.30) *and let Π be the associated generalized shorting operator on $L^2\big((0,1);\ell^2_\rho(\mathsf{E})\big)$ as in* (8.28). *Then $\tilde{P}$ is admissible if and only if G is orientedly bipartite or orientedly Eulerian, cf. Definition A.3.*

Proof. By definition, $\tilde{P}$ is admissible if and only if Πu is continuous in the nodes for all $u \in W^{1,2}_Y\big((0,1);\ell^2_\rho(\mathsf{E})\big)$. Denote by V_0 (resp., by V_1) the subset of V consisting of all nodes with nonzero outdegree (resp., nonzero indegree).

Let first $\tilde{P}$ be admissible. If $\mathsf{V}_0 \cap \mathsf{V}_1 = \emptyset$, then G is by definition an orientedly bipartite graph. Let on the other hand $\mathsf{V}_0 \cap \mathsf{V}_1 \neq \emptyset$. By admissibility of $\tilde{P}$, a vector $(\Pi u)_{|\mathsf{V}} \in \mathbb{C}^\mathsf{V}$ of joint node values exists for any $u \in W^{1,2}_Y\big((0,1);\ell^2_\rho(\mathsf{E})\big)$ if and only if

$$\frac{1}{|\mathsf{E}|_\rho} \sum_{\mathsf{e} \in \mathsf{E}} u_\mathsf{e}(0)\rho(\mathsf{e}) = \frac{1}{|\mathsf{E}|_\rho} \sum_{\mathsf{e} \in \mathsf{E}} u_\mathsf{e}(1)\rho(\mathsf{e}). \tag{8.31}$$

For an arbitrary node $\mathsf{v} \in \mathsf{V}$ choose $u \in W^{1,2}_Y\big((0,1);\ell^2_\rho(\mathsf{E})\big)$ in such a way that $u(\mathsf{v}) = 1$ and $u(\mathsf{w}) = 0$ for all $\mathsf{w} \neq \mathsf{v}$. Then

$$\sum_{\mathsf{e} \in \mathsf{E}} u_\mathsf{e}(0)\rho(\mathsf{e}) = \sum_{\mathsf{e} \in \mathsf{E}} \iota^-_{\mathsf{ve}} \rho(\mathsf{e}) = \deg^{\text{in}}_\rho(\mathsf{v}),$$

as well as

$$\sum_{\mathsf{e}\in\mathsf{E}} u_{\mathsf{e}}(1)\rho(\mathsf{e}) = \sum_{\mathsf{e}\in\mathsf{E}} \iota^+_{\mathsf{ve}}\rho(\mathsf{e}) = \deg^{\text{out}}_\rho(\mathsf{v}),$$

where (ι^+_{ve}), (ι^-_{ve}) denotes the incoming and outgoing incidence matrix, respectively. Therefore, G is by definition orientedly Eulerian.

If conversely G is orientedly bipartite, then for an arbitrary $u \in W^{1,2}_Y\big((0,1);\ell^2_\rho(\mathsf{E})\big)$ there holds

$$(\Pi u)(\mathsf{v}) = \frac{1}{|\mathsf{E}|_\rho} \sum_{\mathsf{e}\in\mathsf{E}} u_{\mathsf{e}}(0)\rho(\mathsf{e}) \qquad \text{for all } \mathsf{v} \in \mathsf{V}_0$$

and

$$(\Pi u)(\mathsf{v}) = \frac{1}{|\mathsf{E}|_\rho} \sum_{\mathsf{e}\in\mathsf{E}} u_{\mathsf{e}}(1)\rho(\mathsf{e}) \qquad \text{for all } \mathsf{v} \in \mathsf{V}_1.$$

This shows continuity of Πu in the nodes and hence admissibility of $\tilde{P}$. It can be proved likewise that (8.31) holds, and hence that $\tilde{P}$ is admissible, if instead G is orientedly Eulerian. □

We conclude mentioning a further structural result.

Theorem 8.36. *If* G *is finite, then the following assertions hold.*

(1) All nodes of G *have degree 1 if and only if all orthogonal projectors* $\tilde{P}$ *of* $\ell^2_\rho(\mathsf{E})$ *are admissible.*

(2) G *is a star if and only if all orthogonal projectors* $\tilde{P}$ *of* $\ell^2_\rho(\mathsf{E})$ *with eigenvector* **1** *are admissible.*

8.4 Variational Symmetries via Shortings

Let us conclude this chapter by suggesting a further motivation for investigating admissible (generalized) shorting operators. To this aim we will come back to the ideas sketched in Sect. 8.2 and consider a new notion of symmetry group that is more easily checked than that of point symmetry group, but only fits differential equations with a Lagrangian structure.

Definition 8.37. For a bounded open domain $U \subset \mathbb{R}^d$ and a Lagrangian $L \in C^2(J^1)$ we call a point transformation group $\mathscr{T} := (T_\epsilon(\mathbf{x}))_{\epsilon\in I_{\mathbf{x}},\,\mathbf{x}\in J^0}$ ($I_{\mathbf{x}} \subset \mathbb{R}$ for each $\mathbf{x} \in J^0$) a *variational symmetry group* of the Euler–Lagrange equation associated with L if for the 0-th jet $(x_\epsilon, u_\epsilon) := (j_0 T_\epsilon)(x, u)$ and $\omega_\epsilon := \{x_\epsilon \in \mathbb{R}^d : x \in \omega\}$ the identity

$$\int_{\omega} L(j_1 T_\epsilon(x, j_1 u))dx = \int_{\omega_\epsilon} L(x, j_1 u)dx \tag{8.32}$$

holds for all open domains $\omega \subset U$, all $|\epsilon|$ small enough that (8.32) makes sense, and all $u \in C^1(U)$.

The following fundamental result was proved by E. Noether in [19]. It shows a tight relation between invariance of linear subspaces and the notion of symmetry—as made precise in Definition 8.37—that has been much exploited over the last century in the classical and quantum field theories. In a modern language based on jet formalism Noether's theorem can be expressed as follows.

Theorem 8.38. *Let U be an open bounded domain of $\mathbb{R}^d$ with smooth boundary and consider a Lagrangian $L \in C^2(J^1)$. Then for each variational symmetry group $(T_\epsilon(\mathbf{x}))_{\epsilon \in I_{\mathbf{x}}, \mathbf{x} \in J^0}$ with generator*

$$A := (\xi_1, \dots, \xi_n, \phi)$$

of the Euler–Lagrange equation associated with L the identity

$$\sum_{k=1}^{n} \frac{\partial}{\partial x_k} \left(\tilde{\phi}(x, j_1 u) \frac{\partial L}{\partial u_{x_k}}(x, j_1 u(x)) - L(x, j_1 u(x)) \xi_k(x, j_1 u) \right) = 0$$

holds for all $(x, u) \in J^1$ in the solution manifold of the same equation, provided each ξ_i depends on the independent variables only, i.e. $\xi_i \neq \xi_i(u)$, $i = 1, \dots, n$. Here $\tilde{\phi}$ is defined by

$$(\tilde{\xi}_1, \dots, \tilde{\xi}_n, \tilde{\phi})(x, f) := \left(\lim_{\epsilon \to 0} \frac{x_\epsilon - x}{\epsilon}, \lim_{\epsilon \to 0} \frac{f_\epsilon(x_\epsilon) - f(x)}{\epsilon} \right).$$

Roughly speaking, Noether's theorem states that if a Lie group acts on the phase space of a differential equation in terms of variational symmetries, then there exist a number of independent conserved quantities of the differential equation equal to the dimension of the Lie algebra.

Unlike the heat equation, which cannot be seen as a Euler–Lagrange in a natural way, cf. [20, Chapter 4], the Schrödinger equation has a natural Lagrangian structure: This observation goes back to R. Feynman [6]. Hence, we can apply to it Noether's theorem.

Example 8.39. With a computation similar to that performed to determine the point symmetry groups of the heat equation as in Example 8.4, the point symmetry groups of the one-dimensional Schrödinger equation can be shown to be

$$i \frac{\partial \psi}{\partial t} = \frac{\partial^2 \psi}{\partial x^2}, \qquad t \in \mathbb{R},\ x \in \mathbb{R},$$

are those given by

$$
\begin{aligned}
T_\epsilon^{(1)} &: (t,x,\psi) \mapsto (t+\epsilon, x, \psi), && \epsilon \in \mathbb{R},\\
T_\epsilon^{(2)} &: (t,x,\psi) \mapsto (e^{2\epsilon}t, e^{\epsilon}x, \psi), && \epsilon \in \mathbb{R},\\
T_\epsilon^{(3)} &: (t,x,\psi) \mapsto \left(\tfrac{t}{1-4\epsilon t}, \tfrac{x}{1-4\epsilon t}, \sqrt{1-4\epsilon t}\, e^{\frac{-\epsilon x^2}{1-4\epsilon t}}\psi\right), && \epsilon < \tfrac{1}{4t},\\
T_\epsilon^{(4)} &: (t,x,\psi) \mapsto (t, x+\epsilon, \psi), && \epsilon \in \mathbb{R},\\
T_\epsilon^{(5)} &: (t,x,\psi) \mapsto (t, x+2\epsilon t, e^{-i\epsilon x - i\epsilon t^2}\psi), && \epsilon \in \mathbb{R},\\
T_\epsilon^{(6)} &: (t,x,\psi) \mapsto (t, x, e^{i\epsilon}\psi), && \epsilon \in \mathbb{R}.
\end{aligned}
$$

We will be particularly interested in $(T_\epsilon^{(6)})_{\epsilon\in\mathbb{R}}$: On one hand, it is still a point symmetry group if we consider the Schrödinger equation in a rougher environment, like a metric graph; on the other hand, the conserved quantity whose existence is guaranteed by Noether's Theorem is rather important: In quantum mechanics, this conserved quantity is interpreted as mass.

As long as a wavefunction is scalar-valued—as it is in Example 8.39—there is but one possibility of rotating its phase. But if the wavefunction is vector-valued, like the solution of a Schrödinger equation in the Hilbert space $L^2\big((0,1);\ell^2_\rho(\mathsf{E})\big)$, cf. Proposition 7.26, then "partial" phase rotations may actually define further variational symmetry groups.

Theorem 8.40. *Let* $\mathsf{G} = (\mathsf{V}, \mathsf{E}, \rho)$ *have* N *almost equitable partitions (with respect to the node weight* $\nu \equiv 1$*). For each of them, assume its cells* $(\mathsf{V}_i)_{i\in I}$ *to be finite for all* $i \in I$ *and additionally that*

$$c_{ij}^{in} = 0, \quad c_{ij}^{out} = c_{ij}, \quad c_{ji}^{out} = 0, \quad c_{ji}^{in} = c_{ji} \qquad \text{for all } i, j \in I \text{ with } i < j$$

and

$$c_{ii}^{in} = c_{ii}^{out} \qquad \text{for all } i \in I,$$

with the notation of Definition 8.24. Then there exists N *conserved quantities for the Schrödinger equation on the metric graph* $\mathfrak{G}$ *over* G.

One can interpret them as conservation of mass on reduced solution manifolds. An analogous assertion holds for the wave equation, for which the Lagrangian structure is well-known.

Proof. For each of the almost equitable partitions, consider the associated shorting operator Π on the Hilbert space $L^2\big((0,1);\ell^2_\rho(\mathsf{E})\big)$. Now, Theorem 8.30 applies and Π commutes with the semigroup $(e^{t\Delta})_{t\geq 0}$ generated by Δ with standard node conditions (Cc) – (Kc), and by Corollary 8.11 this is equivalent to the fact that Π commutes with the unitary group $(e^{is\Delta})_{s\in\mathbb{R}}$. Since Π is self-adjoint, by Stone's Theorem 7.7 one can consider the unitary group $(e^{is\Pi})_{s\in R}$. By Example 4.1, Π commutes with $(e^{is\Delta})_{s\in\mathbb{R}}$ if and only if so does $(e^{is\Pi})_{s\in\mathbb{R}}$. Thus, we have found a variational symmetry group $(e^{is\Pi})_{s\in\mathbb{R}}$ for the Schrödinger equation. □

8.5 Notes and References

Section 8.1. According to J. Neuberger [18, § 5.2], formula (8.3) is due to J. von Neumann.

Section 8.2. There is an extensive theory of applications of Lie groups to ordinary and partial differential equations. A comprehensive introduction can be found in [20]. The interplay between automorphisms of (discrete) graphs and the various semigroups that govern diffusion-type equations on the graphs is studied in [15, 16].

Section 8.3. The results in this section have been extensively developed in [4, 16].

The analogy between projecting a function to its cell-wise average and shorting all the nodes in the same cell is quite old and has been variously exploited. Shorting techniques have become popular when they were proposed in [17] by C.St.J.A. Nash-Williams in order to prove recurrence of the random walk on the lattice $\mathbb{Z}^2$, cf. the exposition in [7, § 2.2]. A manifold of further potential theoretic problems can nowadays be treated by shorting methods, see e.g. [24, §§ 2–3 and references therein]. Related notions of symmetries on metric graphs have been discussed by several authors, see e.g. [3, 9, 13, 23]. Proposition 8.22 is [4, Prop. 3.13].

In the case of unweighted graphs, the concept of almost equitable partition has been introduced in [5] as a generalization of notions proposed in [21], [1, Chapter 20], and [22], cf. also [14] for an extension to infinite, uniformly locally finite graphs.

An equivalent but perhaps more eidetic explanation of the result discussed in Remark 8.25 relies upon the notion of *quotient graph*: If G is a graph with an equitable partition, then its quotient graph is the directed, weighted directed graph with node set of cardinality $|I|$ (the i-th node corresponding to the cell V_i) such that $(\mathsf{V}_i, \mathsf{V}_j)$ is an edge (and if so, with weight c_{ij}) if and only if $c_{ij} \neq 0$. Again, the spectrum of the adjacency matrix of G contains the spectrum of the adjacency matrix of its weighted quotient graph, see e.g. [12, Thm. 9.3.3]. As we have seen in Example 8.26.(3), the (unweighted) Petersen graph is an example of graph with an equitable partition. Its quotient graph with respect to its equitable partition which we have denoted by $\mathscr{V}^{(3)}$ is shown in Fig. 8.11. The corresponding adjacency matrix is

$$\begin{pmatrix} 0 & 3 & 0 \\ 1 & 0 & 2 \\ 0 & 1 & 2 \end{pmatrix}.$$

Section 8.4. The defining condition of variational symmetry groups looks like a weak formulation of that of point symmetry group. Each point symmetry groups of a Euler–Lagrange differential equation is also a variational symmetry group, cf. [20, §4.2]. A modern treatment of Noether's theory can be found e.g. [20, § 4.4] or [8, § 8.6].

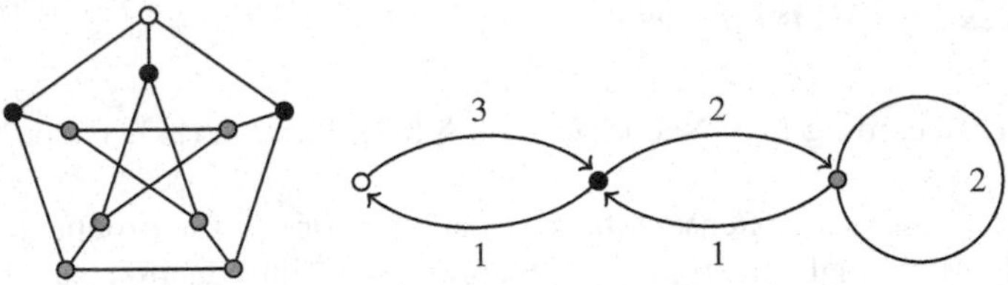

Fig. 8.11 An equitable partition of the Petersen graph and the associated quotient graph

References

1. N.L. Biggs, *Algebraic Graph Theory*. Cambridge Tracts Mathematics, vol. 67 (Cambridge University Press, Cambridge, 1974)
2. B. Bollobás, *Modern Graph Theory* (Springer, Berlin, 1998)
3. J. Boman, P. Kurasov, Symmetries of quantum graphs and the inverse scattering problems. Adv. Appl. Maths. **35**, 58–70 (2005)
4. S. Cardanobile, D. Mugnolo, R. Nittka, Well-posedness and symmetries of strongly coupled network equations. J. Phys. A **41**, 055102 (2008)
5. D.M. Cardoso, C. Delorme, P. Rama, Laplacian eigenvectors and eigenvalues and almost equitable partitions. Europ. J. Combin. **28**, 665–673 (2007)
6. D. Derbes, Feynman's derivation of the schrödinger equation. Am. J. Phys. **64**, 881–884 (1996)
7. P.G. Doyle, J.L. Snell, *Random Walks and Electric Networks*. Carus Mathematical Monographs, vol. 22 (Mathematical Association of America, Washington, DC, 1984)
8. L.C. Evans, *Partial Differential Equations*, 2nd edn. Graduate Studies in Mathematics, vol. 19 (American Mathematical Society, Providence, 2010)
9. P. Exner, P. Šeba, Free quantum motion on a branching graph. Rep. Math. Phys. **28**, 7–26 (1989)
10. C.D. Godsil, *Algebraic Combinatorics*. Chapman and Hall Mathematics (Chapman & Hall, New York, 1993)
11. C.D. Godsil, B.D. McKay, Feasibility conditions for the existence of walk-regular graphs. Lin. Algebra Appl. **30**, 51–61 (1980)
12. C. Godsil, G. Royle, *Algebraic Graph Theory*, Graduate Texts in Mathematics, vol. 207 (Springer, Berlin, 2001)
13. M. Keller, D. Lenz, R.K. Wojciechowski, Volume growth, spectrum and stochastic completeness of infinite graphs. Math. Z. **274**, 905–932 (2013)
14. B. Mohar, M. Omladič, Divisors and the spectrum of infinite graphs. Lin. Algebra Appl. **91**, 99–106 (1987)
15. D. Mugnolo, A Frucht's theorem for quantum graphs, in *Spectral Theory, Mathematical System Theory, Evolution Equations, Differential and Difference Equations*, ed. by W. Arendt, J.A. Ball, J. Behrndt, K.-H. Förster, V. Mehrmann, C. Trunk. Operator Theory, Advances and Applications, vol. 221 (Birkhäuser, Basel, 2012), pp. 481–490
16. D. Mugnolo, Parabolic theory of the discrete p-Laplace operator. Nonlinear Anal. Theory Methods Appl. **87**, 33–60 (2013)
17. C.St.J.A. Nash-Williams, Random walk and electric currents in networks. Math. Proc. Cambridge Phil. Soc. **55**, 181–194 (1959)
18. J.W. Neuberger, *Sobolev Gradients and Differential Equations*. Lecture Notes in Mathematics, vol. 1670 (Springer, Berlin, 1997)
19. E. Noether, Invariante Variationsprobleme. Gött. Nachr. 235–257 (1918)
20. P.J. Olver, *Applications of Lie Groups to Differential Equations*. Graduate Texts in Mathematics, vol. 107 (Springer, New York, 1993)

21. H. Sachs, Simultane überlagerungen gegebener graphen. Publ. Math. Inst. Hung. Acad. Sci. (Budapest) **9**, 415–427 (1960)
22. A. Schwenk, Computing the characteristic polynomial of a graph, in *Graphs and Combinatorics (Proc. Washington D.C. 1973)*. Lecture Notes in Mathematics, vol. 406 (Springer, New York, 1974), pp. 153–172
23. S. Severini, G. Tanner, Regular quantum graphs. J. Phys. A **37**, 6675–6686 (2004)
24. W. Woess, *Random Walks on Infinite Graphs and Groups*. Cambridge Tracts Mathematics, vol. 138 (Cambridge University Press, Cambridge, 2000)

Appendix A
Basics on Graph Theory

We review in this chapter the most elementary notions of graph theory and present in passing a few graph-theoretical theorems of combinatorial nature that have actually played a role in other chapters. Throughout the book we have tacitly adopted all notations presented in this Appendix.

Definition A.1. A *directed graph*, or *digraph*, is a pair $\mathsf{G} = (\mathsf{V}, \mathsf{E})$, where V is a (finite or countable) set and E is a subset of $\mathsf{V} \times \mathsf{V}$. We refer to the elements of V and E as *nodes* and *edges*, respectively. A digraph is said to be *simple* if for any two elements $\mathsf{v}, \mathsf{w} \in \mathsf{V}$

(S1) at most one of the pairs $\mathsf{e} := (\mathsf{v}, \mathsf{w}), \bar{\mathsf{e}} := (\mathsf{w}, \mathsf{v})$ is an element of E and
(S2) the pair (v, v) is not an element of E.

Simple digraph are called *oriented graphs*. We call G *finite* if so are V and hence E. The *initial* and *terminal endpoint* (or sometimes: *tail* and *head*) of $\mathsf{e} \equiv (\mathsf{v}, \mathsf{w})$ are v and w, respectively. We denote them by

$$\mathsf{e}_{\mathrm{init}} := \mathsf{v} \quad \text{and} \quad \mathsf{e}_{\mathrm{term}} := \mathsf{w}, \tag{A.1}$$

and say that they are *adjacent* (shortly: $\mathsf{v} \sim \mathsf{w}$) and more precisely that v *precedes* w, or that w *follows* v; or that they are *neighbors*. One also says that e is *incident* in v (as well as in w); and that two edges are adjacent if they share and endpoint.

(Observe that this notion of edge adjacency is independent of orientation.)

Conditions (S1) and (S2) stipulate that there are no multiple edges between two nodes, and no loops connecting one node to itself.

Definition A.2. A *simple graph* is an oriented graph the orientations of whose edges are ignored; i.e., any edge is a set of the form $\{\mathsf{v}, \mathsf{w}\}$ for $\mathsf{v}, \mathsf{w} \in \mathsf{V}$ with $\mathsf{v} \neq \mathsf{w}$.

Conversely, an oriented graph *over* a given simple graph is determined fixing the orientation of each of its edges—clearly, there are $2^{|\mathsf{E}|}$ different oriented graphs over the same simple graph. One can show by a simple double counting argument that for all simple graphs the *Handshaking Lemma*

D. Mugnolo, *Semigroup Methods for Evolution Equations on Networks*, Understanding Complex Systems, DOI 10.1007/978-3-319-04621-1,

$$|\mathsf{E}| = \frac{1}{2}\sum_{\mathsf{v}\in\mathsf{V}} \deg(\mathsf{v}) \tag{A.2}$$

holds.

Definition A.3. Let $\mathsf{G} = (\mathsf{V}, \mathsf{E})$ be an oriented graph.

1. A *subgraph* of G is an oriented graph $\tilde{\mathsf{G}} = (\tilde{\mathsf{V}}, \tilde{\mathsf{E}})$ such that $\tilde{\mathsf{V}} \subset \mathsf{V}$ and $\tilde{\mathsf{E}} \subset \mathsf{E}$. The subgraph is called *induced* if additionally

$$\mathsf{v}, \mathsf{w} \in \tilde{\mathsf{V}} \text{ and } (\mathsf{v}, \mathsf{w}) \in \mathsf{E} \qquad \text{implies} \qquad (\mathsf{v}, \mathsf{w}) \in \tilde{\mathsf{E}}.$$

2. If any two nodes are adjacent, then G is called *complete*.
3. Nodes with no neighbors are said to be *isolated*. Nodes with only one neighbor are called *leaves*. A node is *inessential* if it has exactly two neighbors. *Ramification nodes* are those with at least three neighbors.
4. If $\mathsf{v}, \mathsf{w} \in \mathsf{V}$ and $n \in \mathbb{N}$, then an *n-path* from v to w is a pair of sequences

$$(\mathsf{v}_1, \ldots, \mathsf{v}_{n+1}) \in \mathsf{V}^{n+1} \qquad \text{and} \qquad (\mathsf{e}_1, \ldots, \mathsf{e}_n) \in \mathsf{E}^n,$$

where $\mathsf{v}_1 = \mathsf{v}$, $\mathsf{v}_{n+1} = \mathsf{w}$, and for all $i = 1, \ldots, n$ either $\mathsf{e}_i = (\mathsf{v}_i, \mathsf{v}_{i+1})$ or $\mathsf{e}_i = (\mathsf{v}_{i+1}, \mathsf{v}_i)$. A path from v to w is called *closed* if $\mathsf{v} = \mathsf{w}$; it is called *oriented* if $\mathsf{e}_i = (\mathsf{v}_i, \mathsf{v}_{i+1})$ for all $i = 1, \ldots, n$. A path all of whose edges are distinct is called a *trail*.
5. One defines an equivalence relation identifying all closed paths that consist of sequences of nodes and edges that are equal up to some shift. The representative of a corresponding equivalence class is called a *circuit*, or sometimes a *periodic orbit*. A circuit is uniquely determined by the sequence of its edges.
6. Because a circuit is allowed to contain the same nodes and even the same edges, one can produce a new circuit $\mathsf{C} = n \cdot \mathsf{C}_0$ out of an old one C_0 by repeating n times all its nodes and edges in the same sequence. Given a circuit C, its *gene* (or sometimes: *primitive*) $\mathrm{Gen}(\mathsf{C})$ is the circuit such that $\mathsf{C} = n \cdot \mathrm{Gen}(\mathsf{C})$ for the largest possible $n \in \mathbb{N}$.
7. A *cycle* is a closed path whose nodes (with the exception of the initial and terminal one) and edges are pairwise different. An *oriented cycle* is an oriented path that is a cycle.
8. If for any two nodes $\mathsf{v}, \mathsf{w} \in \mathsf{V}$ there is a path from v to w, then G is called *connected*, and *strongly connected* if the path can be chosen to be oriented. A *connected/strongly connected component* of G is a largest connected/strongly connected subgraph.
9. A subgraph of G is called a *forest* if none of its subgraphs is a cycle as subgraph, and a *tree* if additionally it is connected. It is called a *spanning tree* of G if its node set agrees with V. One checks that G is a forest if and only if

$$|\mathsf{E}| - |\mathsf{V}| + \kappa = 0, \tag{A.3}$$

where κ is the number of connected components. Indeed, $|\mathsf{E}| - |\mathsf{V}| + \kappa$ is the number of independent cycles contained in G, cf. [15, § 5], and is thus called *cyclomatic number* (or sometimes *first Betti number*) of G.

10. If there is one node $\mathsf{v} \in \mathsf{V}$—which is then called *center*—such that any further $\mathsf{w} \in \mathsf{V}$ is adjacent to v, and only to it, then G is a *star*. If the center is initial/terminal endpoint of each edge, then G is called *outbound/inbound star*, respectively; in either case it is an *oriented star*.
11. If V can be partitioned in two subsets $\mathsf{V}_1, \mathsf{V}_2$ such that $(\mathsf{v}, \mathsf{w}) \notin \mathsf{E}$ for any two nodes $\mathsf{v}, \mathsf{w} \in \mathsf{V}_i$, $i = 1, 2$, then G is called *bipartite*. If furthermore each node v in V_1 and each node in V_2 is initial and terminal endpoint of any incident edge, respectively, then G is called *orientedly bipartite*.
12. An *Eulerian tour* of G is a closed trail C in G whose edge set $\tilde{\mathsf{E}}$ agrees with E. If G contains an Eulerian tour C, then G is called *Eulerian*. If its orientation makes C an *oriented* cycle, then we call G *orientedly Eulerian*.

Example A.4. $(\mathsf{v}, \mathsf{w}, \mathsf{z}, \mathsf{v})$, $((\mathsf{v}, \mathsf{w}), (\mathsf{w}, \mathsf{z}), (\mathsf{z}, \mathsf{v}))$ on the one hand, and $(\mathsf{w}, \mathsf{z}, \mathsf{v}, \mathsf{w})$ and $((\mathsf{w}, \mathsf{z}), (\mathsf{z}, \mathsf{v}), (\mathsf{z}, \mathsf{w}))$ on the other hand, are two representatives of the same circuit.

Remarks A.5. 1) We do not regard paths as subgraphs of G, since they may in general contain the same nodes and even the same edges more than once; but suitably identifying all such paths yields one representative that is indeed an induced subgraph of G. (Loosely speaking, this is the "union" of all the nodes and edges that belong to the given path.)

2) A cycle can be equivalently defined as a connected graph each of whose nodes has exactly two neighbors.

3) A graph is a closed trail if and only if it is a circuit that agrees with its gene. Each closed trail C in G, and in particular each cycle, can thus be identified with one vector $z \in \mathbb{C}^{\mathsf{E}}$ defined by

$$z(\mathsf{e}) := \begin{cases} 1 & \text{if } \mathsf{e} \text{ belongs to the edge set of } \mathsf{C}, \\ 0 & \text{otherwise.} \end{cases}$$

4) Non-oriented Eulerian graphs can be given an orientation in a natural way. With respect to this orientation, they are also strongly connected.

If we consider two *triangle graphs*, i.e., two simple graphs consisting of three nodes and three edges each, then as soon as one draws them one sees that they can be identified. We can formalize this intuition as follows.

Definition A.6. Let $\mathsf{G} = (\mathsf{V}, \mathsf{E}), \tilde{\mathsf{G}} = (\tilde{\mathsf{V}}, \tilde{\mathsf{E}})$ be oriented graphs.

A bijective mapping $O : \mathsf{V} \to \tilde{\mathsf{V}}$ is called an *isomorphism* whenever for all $\mathsf{v}, \mathsf{w} \in \mathsf{V}$ $(O\mathsf{v}, O\mathsf{w}) \in \tilde{\mathsf{E}}$ or $(O\mathsf{w}, O\mathsf{v}) \in \tilde{\mathsf{E}}$ if and only if $(\mathsf{v}, \mathsf{w}) \in \mathsf{E}$ or $(\mathsf{w}, \mathsf{v}) \in \mathsf{E}$; and a *automorphism* if $\mathsf{G} = \tilde{\mathsf{G}}$. The set of all automorphisms of G forms the *automorphism group* of G, which we denote by $\mathrm{Aut}(\mathsf{G})$.

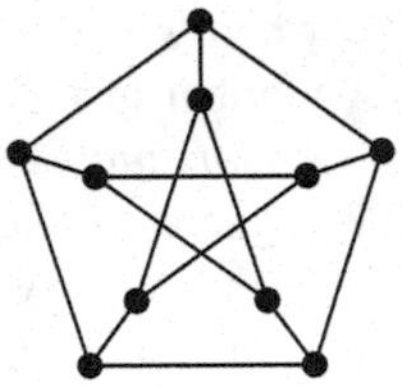

Fig. A.1 The Petersen graph

Fig. A.2 A graph for which Aut(G) and Aut′(G) are not isomorphic

A bijective mapping $U : \mathsf{E} \to \tilde{\mathsf{E}}$ is called an *edge isomorphism* whenever for all $\mathsf{e}, \mathsf{f} \in \mathsf{E}$ $U(\mathsf{e}), U(\mathsf{f})$ are adjacent (in the sense of Definition A.1) if and only if e, f are adjacent; and an *edge automorphism* if $\mathsf{G} = \tilde{\mathsf{G}}$. The set of all edge automorphisms of G forms the *edge automorphism group* of G, which we denote by $\mathrm{Aut}^*(\mathsf{G})$.

Observe that all nodes that belong to the same orbit induced by some subgroup of the automorphism group of a graph have necessarily the same number of neighbors.

Example A.7. The Petersen graph in Fig. A.1 is probably the single most famous finite graph. The automorphism group of this graph is isomorphic to the symmetric group S_5. It acts on the graph's nodes by arbitrary permutations of the outer nodes: this determines uniquely the action on the inner nodes, too.

In other words, automorphisms (resp., edge automorphisms) are node (resp., edge) permutations that preserve node (resp., edge) adjacency. Corresponding notions hold if $\mathsf{G}, \tilde{\mathsf{G}}$ are simple graphs, as the above notion does not depend on orientation.

Now, observe that each symmetry $O \in \mathrm{Aut}(\mathsf{G})$ naturally induces an edge symmetry $U := \tilde{O} \in \mathrm{Aut}^*(G)$: simply define

$$\tilde{O}(e) := (O\mathsf{v}, O\mathsf{w}) \qquad \text{whenever } \mathsf{e} = (\mathsf{v}, \mathsf{w}).$$

While clearly

$$\mathrm{Aut}'(\mathsf{G}) := \{\tilde{O} : O \in \mathrm{Aut}(\mathsf{G})\}$$

(whose elements we call *induced edge automorphisms*) is a group, it can be strictly smaller than Aut(G)—simply think of the graph G defined in Fig. A.2. There, $\mathrm{Aut}(\mathsf{G}) = C_2 \times C_2$ (independent switching of the adjacent nodes and/or of the isolated nodes) but Aut′(G) is trivial, so Aut(G) and Aut′(G) are not isomorphic.

However, this is an exceptional case. The following has been proved by G. Sabidussi and H. Whitney, cf. [13, Thm. 1] and [2, Cor. 9.5b].

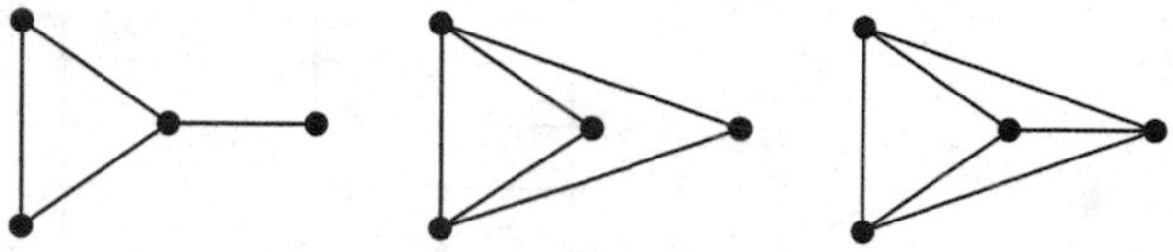

Fig. A.3 The only graphs on four or more nodes for which Aut(G), Aut*(G) are not isomorphic

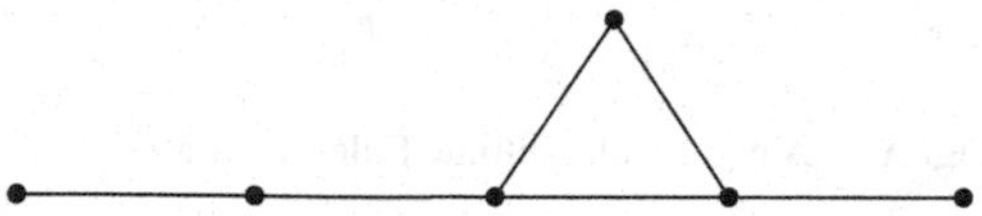

Fig. A.4 A graph without any non-trivial automorphisms

Lemma A.8. *Let* G *be an (either oriented or simple) finite graph. Then the groups* Aut(G) *and* Aut′(G) *are isomorphic provided that* G *contains at most one isolated node and no isolated edge.*

If additionally G *is connected and has at least three nodes, then the three groups* Aut(G), Aut′(G), Aut*(G) *are pairwise isomorphic if and only if G is different from each of the graphs in Fig. A.3.*

The following *Frucht's theorem* is one of the most interesting result in the theory of graph automorphisms. It was proved in [9] and strengthened in [14, 21, 22].

Theorem A.9. *For any group Γ there are uncountably many connected* cubic *graphs* G*—i.e., graphs each of whose nodes has exactly three neighbors—such that* Aut(G) *is isomorphic to Γ.*

If the automorphism group of G is transitive, then G is said to be *node transitive*. A necessary condition for a graph to be node transitive is that each node has the same number of neighbors.

It was observed in [7] that finite graphs can only exceptionally have non-trivial automorphisms: This assertion can be given a precise meaning in the theory of random graphs. An elementary example of a graph with only trivial automorphisms is depicted in Fig. A.4.

L. Euler found in [8] a sufficient and necessary condition for a finite graph to be Eulerian. His result was extended to the infinite case in [6].

Theorem A.10. *Let* $\mathsf{G} = (\mathsf{V}, \mathsf{E})$ *be a connected (non-oriented) graph.*

(1) If G *is finite, then it is Eulerian if and only if each of its nodes has an even number of incident edges; it is orientedly Eulerian if and only if each of its nodes has an equal number of incoming and outgoing edges.*
(2) If G *is infinite, then it is Eulerian if and only if*

- E *is at most countable,*
- *each node has an even or infinite number of neighbors,*
- *if* $\mathsf{E}' \subset \mathsf{E}$ *is finite, then the connected components of* $\mathsf{G}' := (\mathsf{V}, \mathsf{E} \setminus \mathsf{E}')$ *are at most two, and in fact exactly one if each node has an even number of neighbors in* G′.

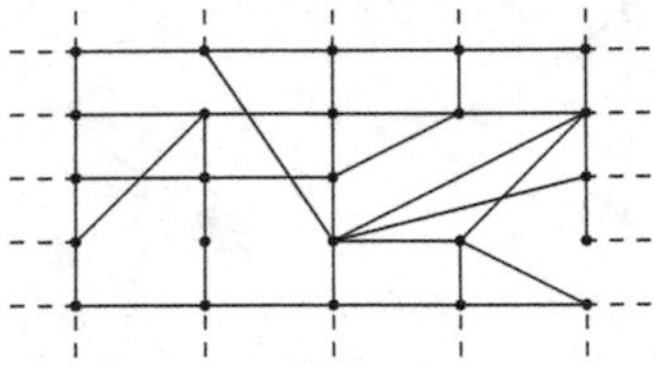

Fig. A.5 A non-regular infinite Eulerian graph

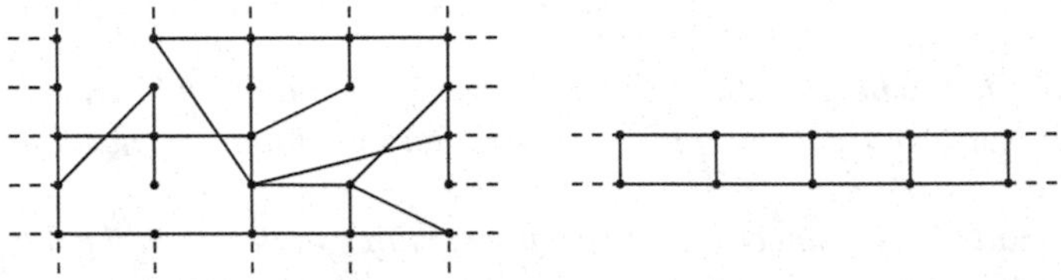

Fig. A.6 Two infinite non-Eulerian graphs

Example A.11. An example of infinite Eulerian graph is given as follows: The *oriented d-dimensional lattice* is the oriented graph (V, E) whose set is $\mathsf{V} = \mathbb{Z}^d$ and whose edge set E is defined as follows: For any two vectors $x, y \in \mathbb{Z}^d$, $(x, y) \in \mathsf{E}$ if and only if there is exactly *one* $k_0 \in \{1, \ldots, d\}$ such that

$$y_{k_0} - x_{k_0} = 1 \qquad \text{and} \qquad y_k = x_k \quad \text{for all } k \neq k_0.$$

I.e., all edges of $\mathbb{Z}^d$ are of the form $(x, x + e_k)$ for some $x \in \mathbb{Z}^d$ and $k = 1, \ldots, d$, where e_k is the k-th vector of the canonical basis of $\mathbb{Z}^d$. Accordingly, each edge is oriented in the direction of the first orthant.

If we discard this orientation, we obtain the *plain d-dimensional lattice*, which by Theorem A.10.(2) is Eulerian if and only if $d > 1$. One can easily modify this setting to obtain non-regular Eulerian graphs, e.g. through a local perturbation like in Fig. A.5.

Definition A.12. The *doubling* of a digraph $\mathsf{G} = (\mathsf{V}, \mathsf{E})$ is the digraph $\mathsf{G}^{\|} := (\mathsf{V}, \overline{\mathsf{E}})$, where $\overline{\mathsf{E}} := \{\mathsf{e}, \bar{\mathsf{e}} : \mathsf{e} \in \mathsf{E}\}$.

Hence, the doubling of a digraph is the digraph obtained adding to each oriented edge $\mathsf{e} \in \mathsf{E}$ its reverse $\bar{\mathsf{e}}$. Clearly, $\mathsf{G}^{\|}$ is not simple even if G is so (more precisely, (S2) is not satisfied by G).

Remark A.13. The doubling $\mathsf{G}^{\|}$ is at most countable and each node has an even or infinite number of neighbors for any digraph G. However, the third condition in Theorem A.10.(2) may well fail for $\mathsf{G}^{\|}$, too: think e.g. of the doubling of the infinite path graph $\mathbb{Z}$, and more generally of any infinite tree. Also the right graph in Fig. A.6 has a non-Eulerian doubling. However, this condition is satisfied by the doubling of the planar lattice $\mathbb{Z}^2$, and more generally by any finite perturbation of a graph associated with a tiling of the plane, provided all tiles are bounded: The left graph in Fig. A.6 is not Eulerian, but its doubling is.

Definition A.14. A *weighted digraph* is a triple $\mathsf{G} = (\mathsf{V}, \mathsf{E}, \rho)$, where (V, E) is a digraph and $\rho : \mathsf{E} \to (0, \infty)$ is some given function such that $\rho(\mathsf{e}) = \rho(\bar{\mathsf{e}})$ whenever both $\mathsf{e}, \bar{\mathsf{e}} \in \mathsf{E}$. A *weighted oriented graph* is a weighted, simple digraph.

Weighted oriented graphs with both sinks and sources have been often called *networks* since the 1940s, in particular by the Cambridge school during their investigations on the axiomatic theory of electric circuits, cf. [4] and subsequent papers.

Conventions A.15. *(1) We always regard unweighted graphs as weighted ones upon imposing $\rho \equiv 1$. All notions defined for weighted graphs thus extend to unweighted ones. Conversely, all notions that only rely upon a graph's connectivity (e.g., existence of an Eulerian graph) remain unchanged if a weight function is introduced.*

(2) If G is weighted, we regard its subgraphs as weighted with respect to the function $\tilde{\rho} := \rho_{|\tilde{\mathsf{E}}}$.

Example A.16. Weights can be used to effectively deform a given graph without affecting its connectivity. E.g., the lattice graph $\mathbb{Z}^d$, cf. Example A.11, can be turned into a weighted oriented graph $(\mathbb{Z}^d, \rho)$ in which each pair (v, w) of adjacent nodes is distant $\rho\big((\mathsf{v}, \mathsf{w})\big)$ length units. If in particular $\rho \equiv \rho_0 > 0$, then ρ_0 is referred to as *lattice constant*. Clearly, modifying ρ_0 leads to a rescaling of the whole lattice. Heuristically, one expects that discretized equations on $\mathbb{Z}^d$ tend to their continuous counterparts in $\mathbb{R}^d$ whenever the lattice constant tends to 0. This is a favorite empirical principle in the treatment of a number of problems in physics, and in particular in the so-called *lattice field theory*.

Definition A.17. Let $\tilde{\mathsf{G}}$ be a subgraph of a weighted oriented graph $\mathsf{G} = (\mathsf{V}, \mathsf{E}, \rho)$. Then *capacity* and *volume* of $\tilde{\mathsf{G}}$ are defined by

$$\mathrm{cap}_\rho(\tilde{\mathsf{G}}) := \prod_{\mathsf{e} \in \tilde{\mathsf{E}}} \rho(\mathsf{e}) \qquad \text{and} \qquad \mathrm{vol}_\rho(\tilde{\mathsf{G}}) := |\tilde{\mathsf{E}}|_\rho \sum_{\mathsf{e} \in \tilde{\mathsf{E}}} \rho(\mathsf{e}),$$

respectively. The *surface of $\tilde{\mathsf{G}}$ with respect to a weight function* $\nu : \mathsf{V} \to (0, \infty)$, or simply *surface* if $\nu \equiv 1$, is defined as

$$\mathrm{surf}_\nu(\tilde{\mathsf{G}}) := |\tilde{\mathsf{V}}|_\nu := \sum_{\mathsf{v} \in \tilde{\mathsf{V}}} \nu(\mathsf{v}).$$

Using the notation of Chap. 3, it is clear that a graph has finite surface with respect to ν (resp., finite volume with respect to ρ) if and only if $\nu \in \ell^1(\mathsf{V})$ (resp., if and only if $\rho \in \ell^1(\mathsf{E})$).

Clearly, the surface of $\tilde{\mathsf{G}}$ is simply the cardinality of $\tilde{\mathsf{V}}$ if $\nu \equiv 1$; and by the Handshaking Lemma (cf. (A.2)) $\mathsf{G} = (\mathsf{V}, \mathsf{E}, \rho)$ has finite volume if and only if G has finite surface with respect to $\nu = \deg_\rho$.

Remark A.18. In the specific case of paths or cycles, the word *length* is universally used in the literature instead of *volume*. We also adopt this convention and for a path $\tilde{\mathsf{G}} := \mathsf{P}$ we denote its length by $\mathrm{len}_\rho(\mathsf{P})$. The same notation extends in a natural way to circuits.

Clearly, in the unweighted case surface and volume of a subgraph are simply the cardinalities of its node and edge set, respectively, whereas its capacity is always 1 unless G is empty.

One of the reasons why graphs are so popular in the applied sciences is that it is easy to identify real-life objects with graph-like structures. Conversely, also the following holds.

Lemma A.19. *Each (oriented or simple) graph* $\mathsf{G} = (\mathsf{V}, \mathsf{E})$ *can be embedded in* $\mathbb{R}^3$*, i.e., it is possible*

(1) to associate to each node $\mathsf{v} \in \mathsf{V}$ *a point* $x_\mathsf{v} \in \mathbb{R}^3$ *in an injective way, and*
(2) to connect two points $x_\mathsf{v}, x_\mathsf{w}$ *by an three-dimensional simple arc* s_{vw} *if and only if* v, w *are adjacent, in such a way that different arcs do not share any internal points.*

The embedding can be e.g., performed by associating each $\mathsf{v} \in \mathsf{V}$ with a different point of the x-axis and then connecting each pair (v, w) of adjacent nodes by a simple arc that lies in one of the uncountably many different planes that contain the x-axis, choosing a different plane for each pair.

Accordingly, we may identify edges with simple arcs, although this identification clearly depends on the chosen embedding. This is the first step towards the development of the theory of metric graphs presented in Sect. 3.2.

It is sometimes useful to switch from a description of a system based on agents (persons, particles, nations...) to one based on their interactions: E.g., Feynman diagrams in quantum field theory [24], p-graphs in anthropology [25], or highway networks from our Model 2 in Chap. 1 are based on this idea. The following formalism goes back to [12].

Definition A.20. Let $\mathsf{G} = (\mathsf{V}, \mathsf{E})$ be a simple graph. Its *line graph* is the simple graph $\mathsf{G}_L := (\mathsf{V}_L, \mathsf{E}_L)$ with node set $\mathsf{V}_L := \mathsf{E}$ and such that for any $\mathsf{e}, \mathsf{f} \in \mathsf{V}_L$ $(\mathsf{e}, \mathsf{f}) \in \mathsf{E}_L$ if and only if e, f have a common endpoint.

Definition A.21. Let $\mathsf{G} = (\mathsf{V}, \mathsf{E})$ be an oriented graph. Its *line graph* is the oriented graph $\mathsf{G}_L := (\mathsf{V}_L, \mathsf{E}_L)$ with node set $\mathsf{V}_L := \mathsf{E}$ and such that for any $\mathsf{e}, \mathsf{f} \in \mathsf{V}_L$ $(\mathsf{e}, \mathsf{f}) \in \mathsf{E}_L$ if and only if $\mathsf{e}_{\mathrm{term}} = \mathsf{f}_{\mathrm{init}}$.

If now $\mathsf{G} = (\mathsf{V}, \mathsf{E}, \rho)$ is a *weighted* oriented graph, then its line graph is $\mathsf{G}_L := (\mathsf{V}_L, \mathsf{E}_L, \rho_L)$, where $\mathsf{V}_L, \mathsf{E}_L$ are constructed as before and the weight function ρ_L is defined by $\rho_L((\mathsf{e}, \mathsf{f})) := \rho(\mathsf{f})$.

Comparing Definitions A.20 and A.21 one sees that for two nodes $\mathsf{e}, \mathsf{f} \in \mathsf{V}_L$ the edge (e, f) may belong to the edge set of the *non-oriented* version of the line graph even if neither (e, f) nor (f, e) belong to the edge set of the *oriented* line graph: this is the case precisely when they share either the initial endpoint or the terminal endpoint—i.e., e, f form an oriented star, rather than an oriented path.

Appendix B
Basics on Sobolev Spaces

One of the most fruitful mathematical ideas of the last century is the weak formulation of differential equations. One weakens the notion of solution of a boundary value differential problem, looks for a solution in a suitably larger class (which typically allows one to use standard Hilbert space methods, like the Representation Theorem of Riesz–Fréchet) and eventually proves that the obtained solution is in fact also a solution in a classical sense. The essential idea behind this approach is that of *weak derivative*, one which is based on replacing the usual property of differentiability by a prominent quality of differentiable functions—the possibility to integrate by parts.

Definition B.1. Let $I \subset \mathbb{R}$ be an open interval, whose boundary we denote by ∂I, and $p \in [1,\infty]$. A function $f \in L^p(I)$ is said to be *weakly differentiable* if there exists $g \in L^p(I)$ such that

$$\int_I f h' = -\int_I g h \qquad \text{for all } h \in C_c^\infty(I). \tag{B.1}$$

The function g is unique and is called the *weak derivative* of f, shortly: $g := f'$. The set of weakly differentiable functions $f \in L^p(I)$ such that $f' \in L^p(I)$ is denoted by $W^{1,p}(I)$ and called *Sobolev space of order 1*. We define the Sobolev space $W^{k,p}(I)$ of order $k \geq 2$ recursively as the set of those functions $f \in W^{k-1,p}(I)$ such that $f^{(k-1)}$, the weak derivative of order $k-1$, belongs to $W^{1,p}(I)$.

If $c \in L^\infty(I)$, then $\tilde{g} \in \mathbb{K}^{\partial I}$ is said to be *conormal derivative of f (with respect to c)* if

$$\int_I c f' h' + \int_I (c f')' h = \int_{\partial I} \tilde{g} h_{|\partial I} \qquad \text{for all } h \in W^{1,p}(I). \tag{B.2}$$

In this case, $\tilde{g}$ is unique and is denoted by $\frac{\partial_c f}{\partial \nu}$.

If I is bounded, then by Hölder's inequality $W^{1,p}(I)$ is continuously embedded in $W^{1,q}(I)$ for all $p,q \in [1,\infty]$ such that $p \geq q$.

D. Mugnolo, *Semigroup Methods for Evolution Equations on Networks*, Understanding Complex Systems, DOI 10.1007/978-3-319-04621-1,

Lemma B.2. *Let I be an open interval, $k \in \mathbb{N}$ and $p \in [1, \infty]$. Then $W^{k,p}(I)$ is a Banach space with respect to the norm defined by*

$$\|f\|^p_{W^{k,p}(I)} := \sum_{j=1}^{k} \|f^{(j)}\|^p_{L^p(I)}.$$

It is separable if $p \in [1, \infty)$ and reflexive if $p \in (1, \infty)$.

Furthermore, $W^{k,2}(I)$ is a Hilbert space with respect to the inner product

$$(f|g)_{W^{k,2}(I)} := \sum_{j=1}^{k} (f^{(j)}|g^{(j)})_{L^2(I)} = \sum_{j=1}^{k} \int_0^1 f^{(j)}(x)\overline{g^{(j)}(x)}dx.$$

Furthermore, the space $C^1(\overline{I})$ is densely and continuously embedded in $W^{1,p}(I)$, since any $f \in C^1(\overline{I})$ satisfies (B.1) (which for continuously differentiable functions is nothing but the usual formula of integration by parts) with $g := -f'$.

The following results are special cases of [1, Thm. 4.12], [3, Thm. 8.8, Rem. 8.10, and § 8.1.(iii)] and [10, Satz 1]. We denote by $C_b(\overline{I})$ the space of continuous functions on $\overline{I}$ and by $C_0(\overline{I})$ its closed subspace consisting of those bounded continuous functions on $\overline{I}$ such that for all $\epsilon > 0$ $\{x \in I : |f(x)| \geq \epsilon\}$ is compact. (Clearly, $C_b(\overline{I}) = C_0(I)$ if I is finite; otherwise, it consists of those continuous functions that vanish at ∞.) Lemma B.3.(2) is usually referred to as *Rellich–Kondrachov Theorem*.

Lemma B.3. *Let $I \subset \mathbb{R}$ be an open interval. Then the following assertions hold.*

(1) $W^{1,p}(I) \overset{d}{\hookrightarrow} C_0(I)$, i.e., $W^{1,p}(I)$ is densely and continuously embedded in $C_0(I)$ for all $p \in [1, \infty]$.
(2) If I is bounded, then both the embeddings of $W^{1,1}(I)$ in $L^q(I)$ and of $W^{1,p}(I)$ in $C(\overline{I})$ are compact, for all $p, q < \infty$.
(3) The embedding of $W^{1,2}(I)$ in $L^2(I)$ is a Hilbert–Schmidt operator, provided I is bounded.
(4) $C^\infty(I)$ is dense in $W^{1,p}(I)$.
(5) $C_c^\infty(\mathbb{R})$ is dense in $W^{1,p}(\mathbb{R})$ but $C_c^\infty(I)$ is not dense in $W^{1,p}(I)$ whenever I is bounded.

Remark B.4. In order to explain some notions used in Lemma B.3, let us recall that a linear operator T from H_1 to H_2, where H_1, H_2 are Hilbert spaces, is said to be of *p-th Schatten class* (short: $T \in \mathcal{L}_p(H_1, H_2)$) for $p \in [1, \infty)$ if it is compact and

$$\|T\|_{\mathcal{L}_p} := \|(s_n)_{n\in\mathbb{N}}\|_{\ell^p} < \infty\},$$

where s_n is the nth eigenvalue of $\sqrt{T^*T}$. An operator is called *of trace class* or *Hilbert–Schmidt* if it is of p-th Schatten class for $p = 1$ or $p = 2$, respectively. One can show that $\mathcal{L}_p(H_1, H_2)$ is for each p a Banach space (a Hilbert space for $p = 2$) and an ideal, in the sense that the composition of two operators is of p-Schatten class already if either of them is of p-Schatten class, as soon as the other operator is merely bounded. Because of Hölder's inequality for sequence spaces, $\mathcal{L}_p(H_1, H_2) \subset \mathcal{L}_q(H_1, H_2)$ for all $p \le q$, and for instance the composition of two Hilbert–Schmidt operators is of trace class.

Roughly speaking, in the following a Banach space is referred to as *lattice* if it carries some order structure that is compatible with its norm. We refer to [17] for a general introduction to this theory, but in the present book we will not deal with any Banach lattices but their most elementary cases, viz spaces of continuous functions over locally compact spaces and L^p-spaces with respect to some σ-finite measure.[1]

Following [19] and [20, Chapter 2] we adopt throughout this book a rather general notion of lattice ideal (more general, e.g., than the one in [18, 23]).

Definition B.5. Given U, V subspaces of $L^p(\Omega, \mathsf{u})$, U is said to be a *lattice ideal* of V if

- $u \in U$ implies $|u| \in V$ and
- $v \operatorname{sgn} u \in U$ provided $u \in U$ and $v \in V$ such that $|v| \le |u|$.

Then one checks the following, see e.g. [20, Thm. 4.21]. We use the fact that any function $f \in W^{1,p}(I)$ has well-defined boundary values provided I has a nonempty boundary ∂I—in fact, the trace operator is bounded.

Lemma B.6. *Let $I \subset \mathbb{R}$ be an open interval. Let*

$$\mathring{W}^{1,p}(I) := \{f \in W^{1,p}(I) : f_{|\partial I} = 0\}, \qquad 1 \le p < \infty,$$

and let V be any closed subspace of $W^{1,p}(I)$ that contains $\mathring{W}^{1,p}(I)$. Then for all $p \in [1, \infty)$ $\mathring{W}^{1,p}(I)$ is a lattice ideal of V.

Sobolev spaces can also be defined if one replaces the open interval I by an open domain in $\mathbb{R}^n$, but continuity of $W^{1,p}$-functions is peculiar to the case of $n = 1$.

The proof of Lemma B.3.(2) is based on the Ascoli–Arzelà Theorem. In other cases it may be necessary to prove, more simply, precompactness of subspaces of L^p-spaces—i.e., compactness of the embedding operator from $W^{1,p}(I)$ to $L^q(I)$ for some $p, q \in [1, \infty]$. The classical way of doing so is to apply the *Fréchet–Kolmogorov Theorem*, which has been recently generalized as a consequence of the following *Hanche-Olsen–Holden Lemma*, cf. [11, Lemma 1].

Lemma B.7. *Let (M, d_M) be a metric space. Then M is totally bounded if for all $\varepsilon > 0$ there is $\delta > 0$, a metric space (W, d_W), and a mapping $\Phi : M \to W$ such that*

[1] We stress that this concept of lattice has nothing to do with the *physical* notion of lattice at the core of Example A.11, but we keep this contradictory terminology as we believe that there is very little danger of confusion.

- *$\Phi(M)$ is totally bounded and*
- *for all $x, y \in M$ $d_W(\Phi(x), \Phi(y)) < \delta$ implies $d_M(x, y) < \epsilon$.*

Lemma B.8. *Let $I \subset \mathbb{R}$ be an open interval. Then the following assertions hold.*

(1) If $\partial I \neq \emptyset$, then $C_c^\infty(I)$ is dense in $\mathring{W}^{1,p}(I)$ for all $p \in [1, \infty)$.
(2) There exists $C > 0$ such that

$$\|u\|_{L^2}^3 \le C \|u'\|_{L^2} \|u\|_{L^1}^2, \qquad u \in W^{1,2}(\mathbb{R}) \cap L^1(\mathbb{R}). \tag{B.3}$$

(3) If $I = (\alpha, \beta)$ for some $\alpha, \beta \in \mathbb{R}$, then there exists $C > 0$ such that

$$\|u\|_{L^\infty} \le C \|u'\|_{L^1} + |u(\alpha)|, \qquad u \in W^{1,1}(I). \tag{B.4}$$

(4) If I is bounded, then there exists $C > 0$ such that

$$\|u\|_{L^\infty} \le C \|u'\|_{L^1}, \qquad u \in \mathring{W}^{1,1}(I). \tag{B.5}$$

(5) If I is bounded, then for all $p, q, r \in [1, \infty]$ such that $q \le p$ there exists $C > 0$ such that

$$\|u\|_{L^p} \le C \|u\|_{W^{1,r}}^a \|u\|_{L^q}^{1-a}, \qquad u \in W^{1,r}(I), \tag{B.6}$$

where

$$a := \frac{q^{-1} - p^{-1}}{(1 + q^{-1} - r^{-1})}.$$

In particular,

$$\|u\|_{L^\infty} \le C \|u\|_{W^{1,2}}^{\frac{1}{2}} \|u\|_{L^2}^{\frac{1}{2}}, \qquad u \in W^{1,2}(I). \tag{B.7}$$

Both estimates (B.4)–(B.5) are referred to as *Poincaré inequality*, whereas (B.3) and (B.6) are usually called *Nash inequality* and *Gagliardo-Nirenberg inequality*, respectively. They have important consequences for the long-time behavior of diffusion equations. Observe that the Poincaré inequality shows that

$$(f, g) \mapsto (f'|g')_{L^2(I)} + f(\alpha)\overline{g(\alpha)}$$

defines an equivalent inner product on $\mathring{W}^{1,2}(\alpha, \beta)$.

Example B.9. Consider the operator $S : f \mapsto i\hbar f'$, which in mathematical physics is called the *momentum operator*—in fact, it is the quantum mechanical observable associated to the classical momentum. Then the linear operator S is not bounded on

$L^2(\mathbb{R})$, but is indeed bounded from $W^{1,p}(\mathbb{R})$ to $L^2(\mathbb{R})$. Likewise, $\Delta : f \mapsto f''$ is a bounded linear operator from $H^2(\mathbb{R})$ to $L^2(\mathbb{R})$.

One of the reasons for introducing the Sobolev spaces $W^{k,p}(I)$, in particular for $p = 2$, is that they allow for a convenient operator theoretical setting for studying those operators that are not bounded on usual L^2-spaces, as is typically the case for differential operators; and for solving elliptic problems by means of the following *Lax–Milgram Lemma*, proved in [16].

Lemma B.10. *Let V be a Hilbert space and a a sesquilinear mapping from $V \times V$ to $\mathbb{K}$. Let a be bounded and coercive. Then, for any $\phi \in V'$ there is a unique solution $u =: T\phi \in V$ to $a(u, v) = \langle \phi, v\rangle$—which also satisfies $\|u\| \le \frac{1}{c}\|\phi\|_{V'}$. Moreover, T is an isomorphism from V' to V.*

We conclude this appendix briefly explaining how to treat weakly differentiable vector-valued functions. For the sake of simplicity we focus on the Hilbert case, which is the most interesting for our purposes.

Definition B.11. Let W be a separable Hilbert space and I be an open interval.

(1) The space $L^2(I; W)$ is the set of all weakly measurable functions $f : I \to W$ such that $x \mapsto \|f(x)\|_W$ is of class $L^2(I;\mathbb{R})$, which is a normed space with respect to the norm

$$\|f\|_{L^2(I;W)} := \int_I |f(x)|_W^2 dx.$$

(2) The space $W^{1,2}(I; W)$ is the set

$$W^{1,2}(I;W) := \Big\{f \in L^2(I;W) : \exists f' := g \in L^2(I;W) \text{ s.t.} \\ \textstyle\int_I f(x)h'(x)dx = -\int_I g(x)h(x)dx \text{ for all } h \in C_c^\infty(I;\mathbb{R})\Big\}, \tag{B.8}$$

which is a normed space with respect to the norm

$$\|f\|_{W^{1,2}(I;W)} := \int_I \left(|f(x)|_W^2 + |f'(x)|_W^2\right) dx.$$

The following has been proved in [5].

Lemma B.12. *Let W be a separable Hilbert space, $p \in (1, \infty)$, and I be an open interval. The following assertions hold.*

(1) If $G : W \to W$ is a Lipschitz continuous mapping and if

- *$G(0) = 0$, or*
- *I is finite,*

then $G \circ f \in W^{1,p}(I; W)$ whenever $f \in W^{1,p}(I; W)$.

(2) *Let W be a complex Hilbert lattice. If $u \in W^{1,p}(I;W)$, then also its real and complex parts* $\mathrm{Re}\, u, \mathrm{Im}\, u$ *belong to $W^{1,p}(I;W)$, and so do the positive and negative parts* $(\mathrm{Re}\, u)^+, (\mathrm{Re}\, u)^-$ *of its real part, with*

$$(\mathrm{Re}\, u)' = \mathrm{Re}(u'), \qquad ((\mathrm{Re}\, u)^+)' = \mathrm{Re}(u')\mathbf{1}_{\{u \geq 0\}}.$$

Proof. We start observing that the proof of [3, Prop. 8.5] holds also in the vector-valued case with minor changes. In other words, if $f \in L^p(I;W)$, then $f \in W^{1,p}(I;W)$ is equivalent to the existence of a positive constant C with the property that for all open bounded $\omega \subset I$ and all $h \in \mathbb{R}$ with $|h| \leq \mathrm{dist}(!, \partial \mathrm{I})$ one has

$$\int_\omega \|f(x+h) - f(x)\|_W^p dx \leq C|h|^p. \tag{B.9}$$

Assume G to be Lipschitz with constant L. First we note that the estimate

$$\int_\omega \|G(f(x+h)) - G(f(x))\|_W^p dx \leq \int_\omega L^p \|f(x+h) - f(x)\|_W^p dx \leq CL^p|h|^p$$

holds for every $f \in W^{1,p}(\Omega;W)$, since by assumption f satisfies (B.9). It remains to show that $G \circ f \in L^p(\Omega;W)$.

If $G(0) = 0$, it suffices to observe that

$$\begin{aligned}\int_I \|G(f(x))\|_W^p dx &= \int_I \|G(f(x)) - 0\|_W^p dx \\ &= \int_I \|G(f(x)) - G(0)\|_W^p dx \leq L^p \int_I \|f(x)\|_W^p dx < \infty.\end{aligned}$$

Thus $G \circ f \in L^p(I;W)$ and the above criterion applies.

Let now I be bounded. Fix an arbitrary vector $w \in W$ and estimate

$$\begin{aligned}\|G(f(x))\|_W &= \|G(f(x)) - G(w) + G(w)\|_W \\ &\leq L\|f(x) - w\|_W + \|G(w)\|_W \\ &\leq L\|f(x)\|_W + L\|w\|_W + \|G(w)\|_W.\end{aligned}$$

Taking the pth power and integrating on I with respect to x we obtain a finite number, since I has finite measure. □

Remark B.13. The analog of Lemma B.12.(2) does not hold for Sobolev spaces of higher order: Consider $u : (-2,2) \ni x \mapsto x^2 - 1 \in \mathbb{R}$, which belongs to $W^{k,p}(-2,2)$ for any $k \in \mathbb{N}$, $p \in [1,\infty]$, but whose positive part has discontinuous derivative, so that $(u^+)' \notin W^{1,p}(-2,2)$ by Lemma B.3.(1).

References

1. R.A. Adams, J.J.F. Fournier, *Sobolev Spaces* (Elsevier, Amsterdam, 2003)
2. M. Behzad, G. Chartrand, L. Lesniak-Foster, *Graphs & Digraphs* (Prindle, Weber & Schmidt, Boston, 1979)
3. H. Brezis, *Functional Analysis, Sobolev Spaces and Partial Differential Equations* (Universitext. Springer, Berlin, 2010)
4. R.L. Brooks, C.A.B. Smith, A.H. Stone, W.T. Tutte, The dissection of rectangles into squares. Duke Math. J. **7**, 312–340 (1940)
5. S. Cardanobile, D. Mugnolo, Parabolic systems with coupled boundary conditions. J. Differ. Equ. **247**, 1229–1248 (2009)
6. P. Erdős, T. Grünwald, E. Weiszfeld, On Eulerian lines in infinite graphs. Mat. Fiz. Lapok **43**, 129–140 (1936)
7. P. Erdős, A. Rényi, Asymmetric graphs. Acta Math. Acad. Sci. Hung. **14**, 295–315 (1963)
8. L. Euler, Solutio problematis ad geometriam situs pertinentis. Comm. Acad. Scient. Petropol. **8**, 128–140 (1741)
9. R. Frucht, Graphs of degree three with a given abstract group. Can. J. Math **1**, 365–378 (1949)
10. B. Gramsch, Zum Einbettungssatz von Rellich bei Sobolevräumen. Math. Z. **106**, 81–87 (1968)
11. H. Hanche-Olsen, H. Holden, The Kolmogorov–Riesz compactness theorem. Expos. Mathematicae **28**, 385–394 (2010)
12. F. Harary, R.Z. Norman, Some properties of line digraphs. Rend. Circ. Mat. Palermo **9**, 161–168 (1960)
13. F. Harary, E.M. Palmer, On the point-group and line-group of a graph. Acta Math. Acad. Sci. Hung. **19**, 263–269 (1968)
14. H. Izbicki, Unendliche Graphen endlichen Grades mit vorgegebenen Eigenschaften. Monats. Math. **63**, 298–301 (1959)
15. G. Kirchhoff, Ueber die Auflösung der Gleichungen, auf welche man bei der Untersuchung der linearen Vertheilung galvanischer Ströme geführt wird. Ann. Physik **148**, 497–508 (1847)
16. P.D. Lax, A.N. Milgram, Parabolic equations, in *Contributions to the Theory of Partial Differential Equations*, vol. 33 (Princeton University Press, Princeton, NJ, 1954), pp. 167–190
17. P. Meyer-Nieberg, *Banach Lattices* (Universitext. Springer, Berlin, 1991)
18. R. Nagel (ed.), *One-Parameter Semigroups of Positive Operators*. Lecture Notes in Mathematics, vol. 1184 (Springer, Berlin, 1986)
19. E.M. Ouhabaz, Invariance of closed convex sets and domination criteria for semigroups. Potential Anal. **5**, 611–625 (1996)
20. E.M. Ouhabaz, *Analysis of Heat Equations on Domains*. London Mathematical Society Monograph Series, vol. 30 (Princeton University Press, Princeton, 2005)
21. G. Sabidussi, Graphs with given group and given graph-theoretical properties. Can. J. Math. **9**, 515–525 (1957)
22. G. Sabidussi, Graphs with given infinite group. Monats. Math. **64**, 64–67 (1960)
23. A. Schwenk, Computing the characteristic polynomial of a graph, in *Graphs and Combinatorics (Proc. Washington D.C. 1973)*. Lecture Notes in Mathematics, vol. 406 (Springer, New York, 1974), pp. 153–172
24. M. Veltman, *Diagrammatica: The Path to Feynman Diagrams*. Lecture Notes in Physics, vol. 4 (Cambridge University Press, Cambridge, 1994)
25. D.R. White, P. Jorion, Representing and computing kinship: a new approach. Curr. Anthropol. **33**, 454–463 (1992)

Index

D. Mugnolo, *Semigroup Methods for Evolution Equations on Networks*,
Understanding Complex Systems, DOI 10.1007/978-3-319-04621-1,

Printed in the United States
By Bookmasters